Bedford Cultural Editions

STEPHEN CRANE

Maggie: A Girl of the Streets (A Story of New York)

Bedford Cultural Editions

STEPHEN CRANE

Maggie: A Girl of the Streets (A Story of New York)

EDITED BY

Kevin J. Hayes

University of Central Oklahoma

Palgrave Macmillan

For Bedford/St. Martin's
Developmental Editors: Katherine A. Retan and John E. Sullivan
Editorial Assistant: Katherine Gilbert
Production Supervisor: Joe Ford
Project Management: Publisher's Studio, a division of Stratford Publishing Services, Inc.
Marketing Manager: Charles Cavaliere
Cover Design: Donna Dennison
Cover Photo: Harper's New Monthly Magazine, May 1883. Courtesy of the Trustees of the Boston Public Library.
Composition: Stratford Publishing Services, Inc.
Printing and Binding: Haddon Craftsmen, an R. R. Donnelley & Sons Company

President: Charles H. Christensen
Editorial Director: Joan E. Feinberg
Director of Editing, Design, and Production: Marcia Cohen
Manager, Publishing Services: Emily Berleth

Library of Congress Catalog Card Number: 98-86156

Softcover reprint of the hardcover 1st edition 1999 978-0-312-21824-9

Manufactured in the United States of America.

4 3 2 1 0 9
f e d c b a

For information, write: Bedford/St. Martin's, 75 Arlington Street, Boston, MA 02116 (617-426-7440)

ISBN: 978-0-312-15266-6 (paperback)
ISBN 978-1-349-62050-0 ISBN 978-1-137-10011-5 (eBook)
DOI 10.1007/978-1-137-10011-5

About the Series

The need to "historicize" literary texts — and even more to analyze the historical and cultural issues all texts embody — is now embraced by almost all teachers, scholars, critics, and theoreticians. But the question of how to teach such issues in the undergraduate classroom is still a difficult one. Teachers do not always have the historical information they need for a given text, and contextual documents and sources are not always readily available in the library — even if the teacher has the expertise (and students have the energy) to ferret them out. The Bedford Cultural Editions represent an effort to make available for the classroom the kinds of facts and documents that will enable teachers to use the latest historical approaches to textual analysis and cultural criticism. The best scholarly and theoretical work has for many years gone well beyond the "new critical" practices of formalist analysis and close reading, and we offer here a practical classroom model of the ways that many different kinds of issues can be engaged when texts are not thought of as islands unto themselves.

The impetus for the recent cultural and historical emphasis has come from many directions: the so-called new historicism of the late 1980s, the dominant historical versions of both feminism and Marxism, the cultural studies movement, and a sharply changed focus in older movements such as reader response, structuralism, deconstruction, and psychoanalytic theory. Emphases differ, of course, among schools and individuals, but what these movements and approaches have in common is a commitment to explore — and to have students in the classroom study interactively — texts in their full historical and cultural dimensions. The aim is to discover how older texts (and those from other traditions) differ from our own assumptions and expectations, and thus the focus in teaching falls on cultural and historical difference rather than on similarity or continuity.

The most striking feature of the Bedford Cultural Editions — and the one most likely to promote creative classroom discussion — is the inclusion of a generous selection of historical documents that contextualize the main text in a variety of ways. Each volume contains works (or passages from works) that are contemporary with the main accounts, histories, sections from conduct books, travel books, poems, novels, and other historical sources. These materials have several uses. Often they provide information beyond what the main text offers. They provide, too, different perspectives on a particular theme, issue, or event central to the text, suggesting the range of opinions contemporary readers would have brought to their reading and allowing students to experience for themselves the details of cultural disagreement and debate. The documents are organized in thematic units — each with an introduction by the volume editor that historicizes a particular issue and suggests the ways in which individual selections work to contextualize the main text.

Each volume also contains a general introduction that provides students with information concerning the political, social, and intellectual contexts for the work as well as information concerning the material aspects of the text's creation, production, and distribution. There are also relevant illustrations, a chronology of important events, and, when helpful, an account of the reception history of the text. Finally, both the main work and its accompanying documents are carefully annotated in order to enable students to grasp the significance of historical references, literary allusions, and unfamiliar terms. Everywhere we have tried to keep the special needs of the modern student — especially the culturally conscious student of the turn of the millennium — in mind.

For each title, the volume editor has chosen the best teaching text of the main work and explained his or her choice. Old spellings and capitalizations have been preserved (except that the long "s" has been regularized to the modern "s") — the overwhelming preference of the two hundred teacher-scholars we surveyed in preparing the series. Original habits of punctuation have also been kept, except for occasional places where the unusual usage would obscure the syntax for modern readers. Whenever possible, the supplementary texts and documents are reprinted from the first edition or the one most televant to the issue at hand. We have thus meant to preserve — rather than counter — for modern students the sense of "strangeness" in older texts, expecting that the oddness will help students to see where older texts are *not* like modern ones, and expecting too that today's historically informed teachers will find their own creative ways to make something of such historical and cultural differences.

In developing this series, our goal has been to foreground the kinds of issues that typically engage teachers and students of literature and history now. We have not tried to move readers toward a particular ideological, political, or social position or to be exhaustive in our choice of contextual materials. Rather, our aim has been to be provocative — to enable teachers and students of literature to raise the most pressing political, economic, social, religious, intellectual, and artistic issues on a larger field than any single text can offer.

J. Paul Hunter, University of Chicago
William E. Cain, Wellesley College
Series Editors

About This Volume

Maggie Johnson remains the most memorable prostitute in American literature, yet the changes in social and sexual behavior that have occurred during the century since Stephen Crane's *Maggie: A Girl of the Streets* first appeared sometimes make it difficult for modern readers to appreciate her story fully. Attitudes toward a women's place in the home, at work, and on the streets have changed significantly. Besides the complete text of the novel, this volume contains many supporting documents that have been chosen to help reconstruct the historical, cultural, and social milieu of late-nineteenth-century America and thus to help modern readers understand *Maggie* with a view closer to that of Crane's contemporary readers. Some of the included documents depict living conditions in the impoverished tenement districts of New York where the novel is set. Others describe the amusements of the day, those places where the people who lived and worked in the tenement districts escaped to during their all-too-brief moments of spare time: beer gardens, concert halls, dime museums, saloons, and other shops and stores along the Bowery. Further articles describe women's life and work, surveying the various employment opportunities available to the single woman in the late nineteenth century and examining American society's attitudes toward the working woman. Another group of documents has been included to help modern readers understand historical attitudes toward prostitution. The volume closes with a group of essays and selections from fictional works that situate *Maggie* in its aesthetic context, looking at what the literary theorists of the late nineteenth century said about the art of fiction and the place of realism, naturalism, and impressionism in literature, and closing with extracts from several works portraying life in the tenement districts and fallen women.

ACKNOWLEDGMENTS

Special thanks go to Stanley Wertheim and Paul Sorrentino, whose biographical research has challenged many of the longstanding assumptions concerning Stephen Crane. Provocatively, they have shown that Thomas Beer's 1923 biography, *Stephen Crane: A Study in American Letters*, which Crane scholars had relied on for decades, was assembled from largely fabricated evidence. It is daunting to realize that much Crane scholarship has been based on a fictionalized biography, but Wertheim and Sorrentino deserve credit for inaugurating a new era of Crane studies, for everything that has been written about Stephen Crane and his work since Beer's biography appeared in 1923 must be reevaluated. Wertheim's and Sorrentino's *Crane Log* provides a good basis for further research, and it has been immensely valuable for assembling the present work, especially the "Chronology of Crane's Life and Times." Both have read portions of this volume, and I am grateful for their advice.

I would also like to thank several others for reading and commenting on the present work during various stages of composition: Nancy Bentley, Leonard Cassuto, Robert S. Levine, and Dana D. Nelson. Additional thanks go to Kenneth T. Jackson, whose monumental *Encyclopedia of New York City* greatly simplified the task of annotating *Maggie* and the supplementary documents. I. N. P. Stokes's *Iconography of Manhattan Island,* a venerable work, also proved extremely useful, especially for compiling the "Chronology of Crane's Life and Times." I do not cite Jackson's *Encyclopedia* or Stokes's *Iconography* separately, so this note of thanks must serve in lieu of further documentation. I also thank those who have helped me at various libraries along the way: New-York Historical Society, New York Public Library, and Canaday Center at the University of Toledo. Further, I would like to thank the newspaper and microfilm librarians at the University of Central Oklahoma and the University of Oklahoma. I am especially grateful to Doylene Manning and the interlibrary loan department at the Max Chambers Library, University of Central Oklahoma.

I would also like to express my thanks to the staff at Bedford/St. Martin's, especially Kathy Retan, who first encouraged me to undertake this project, and John E. Sullivan, Katherine Gilbert, Emily Berleth, and the staff at Publisher's Studio and Stratford Publishing Services, who helped see it through every stage of its composition.

Kevin J. Hayes
University of Central Oklahoma

Contents

Illustrations

Bedford Cultural Editions

STEPHEN CRANE

Maggie: A Girl of the Streets (A Story of New York)

Part One

Maggie: A Girl of the Streets (A Story of New York) The Complete Text

Corwin Knapp Linson, *Portrait of Stephen Crane* (1894). Courtesy of the Barrett Collection, University of Virginia.

Introduction: Cultural and Historical Background

"It is inevitable that you be greatly shocked by this book but continue, please with all possible courage to the end. For it tries to show that environment is a tremendous thing in the world and frequently shapes lives regardless. If one proves that theory, one makes room in Heaven for all sorts of souls (notably an occasional street girl) who are not confidently expected to be there by many excellent people. It is probable that the reader of this small thing may consider The Author to be a bad man, but obviously, this is a matter of small consequence to The Author." So Stephen Crane inscribed copies of the privately printed first edition of *Maggie: A Girl of the Streets (A Story of New York)* which he gave to Hamlin Garland, the Reverend Thomas Dixon, and other friends, reformers, and literary critics.

In the 1890s, when an author published a book with an established firm, he or she usually could rely on the publisher to advertise it in the trade magazines and literary periodicals and on the spare pages at the back of the firm's other publications. Such advertisements briefly described the work and sometimes provided excerpts from its best reviews. Encouraging people to purchase the book, these "puffs," as they were called, helped establish readers' expectations. Before trying to publish *Maggie* in book form, Crane approached Richard Watson Gilder, editor of *Century Magazine,* who rejected the *Maggie* manuscript for serial publication. Crane either could not find another publisher or realized that no reputable firm would publish the narrative as it stood with its profanity and graphic detail. Rather than expurgating his work, Crane exhausted a small inheritance and financed the book's publication himself. While its large print, generous white spaces, and relatively high cover price of fifty cents ranked it above the inexpensive literature of the day, *Maggie* was a poorly produced volume with mustard yellow paper covers and multiple printing errors. By self-publishing *Maggie* Crane eliminated the usual opportunities for marketing, advertising, and distribution and assumed the responsibility for promoting the book himself. This task was more difficult because Crane published it pseudonymously as "Johnston Smith." He sent inscribed copies of *Maggie* to those who might review the novel in the newspapers or magazines or, at least, promote it through word of mouth. With no publisher's puffs, Crane's inscription alone shaped contemporary readers' expectations of the work.

The inscription says much about Crane's attitude toward his book and its potential audience. For one thing, it acknowledges his awareness that the book would shock its readers. *Maggie*'s subtitle, *A Girl of the Streets,* blatantly informed them that this was a story about the lowest form of prostitute, the streetwalker. Republished in England some years later, the subtitle was changed to *A Child of the Streets* to soften the initial response of British readers. The change may not have had much effect, however; "Maggie" was a common slang term used to identify a prostitute, usually a street walker. In addition to the title character, the story's other degraded, impoverished characters would also have shocked contemporary readers. The inscription, however, encouraged them to persist regardless of any personal objections they might have to the subject matter.

Should Crane's encouragement prove insufficient to motivate hesitant readers to continue, he also supplied an underlying sociological justification for the book: the environment in which people lived influenced their lives powerfully enough to render them helpless to resist external forces. Those who acknowledged the effects of environment, however, may not have accepted the conclusion Crane makes in the inscription's third sentence, that a fallen woman could deserve a place in heaven. Far from encouraging readers, this sentence seems to discourage them as it plainly admits that the book embodies a belief contrary to that of "many excellent people," specifically, the bourgeois Christian readers who formed a large majority of the late-nineteenth-century readership of polite letters. The sentence challenges the reader to think differently from the "excellent people."

The inscription's last sentence conveys the likelihood that readers would not accept the views *Maggie* embodied and would, therefore, reject it as a bad book and its author as a bad man. The sentence is reminiscent of a remark Herman Melville made to Nathaniel Hawthorne shortly after Melville had finished writing *Moby-Dick* (1851): "I have written an evil book and feel as spotless as a lamb" (*Correspondence* 212). Crane, like Melville, had said what he wanted to say and refused to compromise. He concluded the inscription by asserting that it little mattered what the reader thought of the author. There may be some truth to the assertion, but, in retrospect, it seems a pose, for the twenty-one-year-old Crane wanted very much to make writing his career, an endeavor that necessarily required public acceptance. Since Crane's inscription helped shape the expectations of his contemporary readers, it provides a good way of understanding *Maggie* over a century later.

THE URBAN ENVIRONMENT

By sending copies of *Maggie* to literary people and reformers, Crane indicated his belief that the book was both a work of literature and a social document. The approach was not uncommon. American literature and American reform had crossed paths before. Harriet Beecher Stowe's *Uncle Tom's Cabin* (1852), for example, the most popular book in the history of American literature, was also the single most well-known and influential book of the antislavery movement. The history of the reform movement in America predates the Civil War, a time when the basic human right to freedom provided a prominent cause for public-minded middle-class Americans in the northern states to rally around. Slavery, however, was not the only wrong the antebellum reformers sought to right: many reform movements — feminism, the temperance movement, and labor reform — had their starts before the Civil War.

The war disrupted the reform movements, which largely had to remake themselves during the late nineteenth century. During these decades, those who sought various reforms often linked their efforts to antebellum causes. This new generation of reformers sought a cause that approached the seriousness and importance of slavery. They found it in urban America. After the Civil War, the demographics of the United States changed significantly. More and more Americans began migrating from rural to urban areas in search of employment, and immigration from Eastern Europe increased greatly. The large influx of people into major urban areas created serious overcrowding problems. As the urban problem worsened through the waning decades of the nineteenth century, the new reformers realized that they had a cause of wide-ranging social significance. Some even associated the drive to improve the poor, overcrowded urban areas with the abolition movement. Prominent among them was Alice Wellington Rollins, who described the plight of the urban poor in "The New Uncle Tom's Cabin" (pp. 106–12), an article which harked back to Stowe's novel. Rollins subsequently wrote a novel exemplifying the squalor of urban living conditions called *Uncle Tom's Tenement* (1888).

The idea that environment significantly shaped human behavior was increasingly accepted during Crane's lifetime. Many discussions of New York tenement life published during the half-dozen years before *Maggie* emphasized how growing up in the tenements adversely shaped a person's thoughts, beliefs, and behavior. The prevailing attitudes toward the influence of environment were mostly derived from the British philosopher Herbert Spencer, who was considered by many contemporaries to be the nineteenth century's most important thinker after Charles Darwin. In *The Data of Ethics* (1879), a work later incorporated in *The Principles of Ethics* (1892–93), Spencer argued that people adapted themselves to their environment by submitting to the will of the social organism or community to which they belonged. Others applied Spencer's basic theory to moral issues. In "Tenement-House Morality," for example, the Reverend James O. S. Huntington argued that the tenement house environment contributed to the moral decay of its inhabitants (see pp. 120–28). Discussing the variety of ethnic groups represented and languages spoken in the tenements, Huntington characterized the tenement dwellers collectively as "a restless, seething mass of human beings, unable to talk together, unable to think together, able only, under some overmastering passion, to act together" (see p. 123).

Tenements originated in the 1830s. The influx of working-class Irish and German immigrants to U.S. cities in the 1830s and 1840s created a significant demand for low-income housing. In New York, to meet this demand, older buildings — warehouses, mansions, churches — were converted to multifamily residences. Their spacious rooms were subdivided into many smaller rooms with little regard for light, ventilation, or sanitary conditions. These large old buildings became known as rookeries. Stephen Crane would describe one in his slum novel *George's Mother* (1896):

> Bleecker lived in an old three-storied house on a side-street. A Jewish tailor lived and worked in the front parlor, and old Bleecker lived in the back parlor. A German, whose family took care of the house, occupied the basement. Another German, with a wife and eight children, rented the dining-room. The two upper floors were inhabited by tailors, dress-makers, a pedler, and mysterious people who were seldom seen. (*Works* 1: 141)

The rookeries soon inspired tenement buildings, inexpensive, multi-family dwellings specifically constructed for the purpose of housing a maximum number of people in a minimum amount of space. Like the separate living spaces within the rookeries, the individual tenements were small, poorly ventilated, and poorly lit. Their tiny bedrooms often had no windows. Furthermore, their plumbing facilities were woefully inadequate. One sink with a hand-operated pump located in the hallway provided the only water source for the entire floor. Privies were located in the small yards in back. Other tenement buildings known as rear houses, with even poorer facilities, were built in the yards behind the street-facing tenements, eliminating nearly all of the open space, and leaving an intricate network of crooked passageways. Legislation enacted in the 1860s and 1870s sought to improve the conditions of the tenements, and buildings erected after these laws came into effect were improvements over the old structures. Beyond mandating several thousand holes knocked in the walls for additional ventilation, the laws had little effect on the old buildings or on avaricious landlords eager to maximize their profits. Landlords often let every available space, from the dank, pestilent basements to the stiflingly hot, low-ceilinged attics. Economic necessity sometimes compelled tenants to take in boarders, further exacerbating overcrowded conditions.

Tenement legislation could not keep pace with immigration. The 1880s saw a huge rise in the number of immigrants from Italy, Russia, and Hungary as well as many other Eastern European countries. New York's population swelled by one quarter during the decade, and most of the newcomers ended up in the already crowded tenement districts. The large influx of immigrants further contributed to public fears about tenement house morality. The prevailing stereotypes that ranked Irish, German, Italian, and Eastern European immigrants well below the native-born Anglo-Saxons in the racial hierarchy, combined with the Spencerian idea that the tenement district would act as an organism, led many to believe that tenement life would devolve into a kind of savage existence.

No one was more responsible for bringing the miserable circumstances of New York tenement life to the public's attention than Jacob A. Riis. As a journalist, Riis had long crusaded for New York's poor, but it was not until he took advantage of the new invention of flash photography that he successfully brought the condition of the city's poor to light. Never before had those who lived outside the tenements seen what life was like inside them. Riis photographed tenement interiors and displayed the living and working conditions of the city's poor to middle-class audiences in person and in print. The public lecture had been an important form of bourgeois entertainment through much of the nineteenth century, and the development of the magic lantern, an early version of the slide projector, allowed speakers late in the century to enhance their lectures with visual imagery. Travelogues displaying exotic scenes of beauty from faraway places were popular. Riis showed his bourgeois audiences nearby scenes of ugliness. Advances in print technology gave Riis the opportunity to disseminate further his images of poverty. The newly perfected photogravure process allowed him to include many of his photographs in his first book, *How the Other Half Lives: Studies among the Tenements of New York* (1890).

Stephen Crane likely read *How the Other Half Lives* not long after its initial publication, but he also knew Riis's photographic studies from his public lectures. Riis spoke in Asbury Park, New Jersey, in July 1892, and Crane reported the lecture in an article in the *New-York Tribune* (*Works* 8: 514). No doubt it contained new material Riis used in his next book, *The Children of the Poor* (1892), which Crane may have read shortly before he finished writing *Maggie*. In his illustrated lectures and books, Riis perpetuated the idea of the negative influence of environment, asserting that "in the tenements all the influences make for evil." He let his bourgeois audiences know that "the gap that separates the man with the patched coat from his wealthy neighbor is, after all, perhaps but a tenement" (Riis *Other Half* 60, 87). Riis's work significantly influenced *Maggie*: Crane recognized that a good writer could do in prose what Riis had done in visual images. The *Academy* (January 16, 1897):76, which called *Maggie* "one of the most downright earnestly-written books ever published," characterized it as a series of "instantaneous literary photographs of slum life."

ALCOHOL AND TENEMENT LIFE

In *How the Other Half Lives,* Jacob Riis emphasized the damaging effects of alcohol on the tenement population and depicted the interiors of saloons and stale-beer dives. Stephen Crane hardly needed to read Riis to understand the degrading effects of alcohol, for he was already well coached in that area. Born in Newark, New Jersey, in 1871, Stephen Crane was the fourteenth and last child (nine surviving infancy) of Mary Helen Crane and the Reverend Dr. Jonathan Townley Crane, a Methodist minister. Both parents were vehement temperance advocates. The year before his son was born, the Reverend Dr. Crane published *The Arts of Intoxication: The Aim and the Results,* a staunch condemnation of alcohol. Crane's mother, a leader of the Women's Christian Temperance Union, frequently lectured against the evils of drink. A reporter who attended one of her lectures commented, "Mrs. Crane is a practical, common-sense talker, and her efforts to demonstrate to the intemperate the folly of their course, proves her to have made intemperance and its terrible consequences a close study." She became such a vocal advocate of temperance during her youngest son's childhood that one local observer remarked that "she ought to stay at home and take care of her large family, instead of making so many speeches" (*Log* 14, 22).

Given his parents' views on temperance, it is no wonder that Stephen Crane recognized the demoralizing effects of alcohol when he began to study the conditions of slum life. Living quarters in the slums were so small and confining that tenement dwellers took any opportunity to escape. Every tenement building had a saloon on the ground floor, and, at night, fathers often fled the tenement's stifling atmosphere for the saloon downstairs. During the daytime, mothers who remained at home while their husbands worked had little difficulty getting alcohol (see pp. 174–82). Female alcoholism was a prominent social problem of the day, and, in characterizing Maggie Johnson's mother as an alcoholic, Crane made it an important aspect of his novel. Children had access to alcohol as well. Often parents sent them to the saloon downstairs to bring back beer. Reformers frequently complained about the ease with which children could obtain beer, but their complaints did little to rectify the problem. Adolescent boys often formed "growler gangs," the growler being a tin pail that they could fill at a local saloon for a few cents, pass around, and continue to refill as long as their pennies lasted. Then, as now, alcohol provided a temporary respite from a life of poverty, yet it often led to numerous other problems, not the least of which was family violence.

In *George's Mother,* Crane would return to the problem of alcoholism in the slums and again emphasize the violence associated with drink. One reviewer found that the book "painfully suggests the awful warnings of hundreds of temperance tracts" (*Log* 190). Set before the end of *Maggie* and in the same neighborhood, *George's Mother* includes Maggie Johnson in a brief appearance. George Kelcey has a crush on Maggie and imagines rescuing her from the squalor of her home life: "He was relatively happy sometimes when Maggie's mother would get drunk and make terrific uproars. He used then to sit in the dark and make scenes in which he rescued the girl from her hideous environment" (*Works* 1: 138). Crane's use of the word *environment* here recalls his inscription to *Maggie*. Mrs. Johnson's frequent drinking bouts and the resultant violence and neglect create a hostile environment that Maggie longs to escape.

WOMEN'S LIFE AND WORK

The working lives of American women changed drastically over the course of the nineteenth century. At its start, few women worked for wages. Those who did usually worked as domestics. With advances in textile production during the 1820s and 1830s, American women began working outside the home. The development of the public school system and the medical field offered women professional career opportunities as teachers and nurses. The retail department stores that emerged after the Civil War further expanded the female work force. With the invention of the typewriter, the secretarial field became women's work. In the last decade of the nineteenth century, many manufacturing industries attracted female workers, and they produced a wide variety of goods — cigars, cigarettes, coats, gloves, paper boxes, shoes, soap, and toys. The largest industry employing female workers was the shirt industry. Women not employed in the factories often had the opportunity to do needlework piecemeal in their homes.

As women became an important part of the labor force, they recognized their collective power and began joining together for their mutual benefit. From the mid–nineteenth century, working women formed unions to safeguard their rights and to improve working conditions. The most important of these, the Working Women's Protective Union, was established in 1863 under the auspices of *New York Sun* editor Moses Beach. It advocated shorter hours and higher wages, yet devoted most of its efforts to providing legal services for women victimized by unscrupulous employers. The Knights of Labor, a union formed in 1869 among skilled garment workers in Philadelphia, expanded greatly over the next two decades and extended membership to most workers, including women. The Working Woman's Society was established in 1888 to protect the interests of the working woman.

An 1889 U.S. Department of Labor report surveyed working conditions among women in large cities across the country. The statistics from New York shirt factories provide a useful perspective on Maggie's job in a "collars and cuffs" factory. The average woman in a New York shirt factory began working when she was sixteen and spent approximately six and a half years there. In other words, she started at the factory as soon as she was old enough to work and continued until she married. The statistics indicate that the average age of women in New York shirt factories was twenty-five, but most were younger, with a handful of widows and spinsters bringing up the average. Eighty-five per cent of the women working in the New York shirt factories were unmarried. Nearly half were native born while 15 percent had been born in Ireland and 5 percent in Germany. Less than 10 percent of their mothers were native born, however. Sixty percent had been born in Ireland and 15 percent in Germany. Over 80 percent lived at home, and most assisted with the housework there. Over three-quarters of those who lived at home contributed their earnings to the family income. On the average, they had five people in their family, three of whom worked. The average yearly income of a female shirt-factory worker in New York was just over three hundred dollars. The wages were not much, far less than men working the same job, but they were marginally adequate for a young single woman whose income represented one-third of her family's total income. Maggie is not dissimilar to the many thousand young women who lived and worked in New York during the early 1890s. Presumably, she is in her mid-teens when she begins work at the factory, her first occupation. While working, she still lives at home and, in light of her mother's destructive alcoholic behavior, probably does most of the housework.

Women who did not live with their parents or other family had difficulty finding work sufficient to support themselves. Given the opportunity, employers much preferred hiring young women who lived at home because they could be sure of their respectability. According to the mores of the time, women were expected to live with their parents until they married and established a family of their own. The prevailing stereotype held that only women with loose morals lived away from their families. One contemporary employer interviewed by Clara Lanza welcomed female employees, yet admitted that in order to avoid any accusations of impropriety, he only hired women who lived with their parents. The woman trying to make it on her own need not apply (Lanza 488).

As Joanne L. Meyerowitz has shown, wage-earning women who lived alone in boarding houses or shared apartments with others became known as "women adrift" (xvii). They faced difficulty obtaining work, and when they found employment, their wages were seldom sufficient to pay room and board. One explanation is that employers set wage rates based on the expectation that female workers lived at home, but a contemporary report of the Working Women's Society, which both Jacob Riis and Katherine Pearson Woods reiterated, made a harsher accusation: "It is a known fact that men's wages cannot fall below a limit upon which they can exist, but woman's wages have no limit, since the paths of shame are always open to her" (see p. 202 and p. 214).

Though women in the late nineteenth century had the freedom to earn wages outside the home, they did not have the freedom to live their own lives. The single woman who lived apart from her family was viewed with suspicion and scorn. People often assumed that there was only one reason a young single woman did not live with her family: she had, to use the language of the day, sacrificed her "most precious jewel," or succumbed to a "fate worse than death." For today's audience, the most difficult part about reading *Maggie* is understanding what the fuss is all about, yet in Stephen Crane's America, a single woman could commit no greater sin than losing her virginity. She would be better off, many believed, to sacrifice her life before her maidenhood. At the time, people made virtually no distinction between a seduced woman and a prostitute.

THE SOCIAL EVIL

Before Maggie gets a job in the shirt factory, her brother Jimmie brashly presents her options: "Yeh've edder got teh go teh hell or go teh work!" While the phrase "go to hell" is used today as a general, all-purpose curse, in Crane's day it often had a more specific connotation: to go to hell meant to become a prostitute. Despite its vulgar insensitivity, Jimmie's statement represents an important attitude toward prostitution that few writers before Crane had articulated or would accept: the idea that women became prostitutes of their own volition. In the nineteenth century, people generally believed that women who became prostitutes did so from moral weakness. A woman succumbed to one man's seduction, and then, realizing no man would want to marry a "fallen" woman, turned to a life of prostitution. This led to great personal shame, which, in the short term, it was supposed, compelled her to seek oblivion in a bottle, and, in the long run, death at her own hands.

William W. Sanger's *History of Prostitution* (1859) tells a much different story (see pp. 266–72). Sanger, resident physician at the correctional facility on Blackwell's Island, New York, undertook a survey of 2,000 prostitutes incarcerated there. His results show that the reasons women turned to prostitution often had little to do with moral turpitude. Over a quarter of the women said that poverty had led them to prostitution. Before Sanger's study, few recognized or acknowledged that hunger or homelessness might lead a woman to prostitute herself. Another quarter of the women Sanger interviewed admitted that they became prostitutes out of "inclination" (Sanger's word). The idea that women had strong sexual desires was another difficult concept for Victorian society to understand or to acknowledge publicly. Though people generally assumed that young women turned to prostitution because they had been seduced, only 13 percent of the women in Sanger's survey gave "seduced and abandoned" as their reason for becoming prostitutes.

Not only was mid-nineteenth-century America as a whole unwilling to accept these findings, but Sanger himself was reluctant to accept them. In his analysis of the data he downplayed the more statistically significant categories and enhanced the importance of the "seduced and abandoned" category. Considering the women who attributed their illicit sexual behavior to "inclination," Sanger asserted that there were "other and controlling influences," for "the full force of sexual desire is seldom known to a virtuous woman." The number of women in the "seduced and abandoned" category, he further asserted, provided "but a faint idea of the actual total that should be recorded under the designation, as many who are included in other classes should doubtless have been returned in this" (see p. 270).

In terms of literary history, the seduction plot was a commonplace theme of the novel during the eighteenth century, and it continued so through the nineteenth. Two American novels that appeared before

Maggie make particularly good examples. In *We and Our Neighbors: or, Records of an Unfashionable Street* (1875), Harriet Beecher Stowe tells the story of a young woman also named Maggie, a secondary character in the book, who leaves her position as a domestic and later finds employment in a department store, where she meets a man who seduces her and then places her in a house of prostitution. After some time, Stowe's Maggie seeks shelter in a Magdalen home, a fictional house of refuge patterned after the homes run by local Magdalen societies that sheltered prostitutes and helped them leave their trade and regain their dignity. (The name is a reference to biblical prostitute Mary Magdalen, who encounters Christ and reforms.) The heroine of Stowe's novel, Eva Henderson, takes Maggie into her home as a servant, an effort her family strongly disapproves. Eva tolerates the verbal abuse of her mother and aunt, but Maggie leaves their home to save her benefactor from such cruelty and returns to the vice district to ply her old trade. Eva eventually locates Maggie and brings her to a Methodist mission home where she becomes a devout Christian. Maggie then returns to Eva and works for her as a seamstress.

In his inscription to *Maggie,* Crane suggested that "many excellent people" would not assign streetwalkers a place in heaven. Stowe, however, seems quite willing to allow this. The outcome of *We and Our Neighbors* reflects an attitude shared by those upright, well-to-do women who contributed to the Magdalen societies. They, too, were willing to allow repentant prostitutes a place in heaven. Indeed, they were more likely to grant them eternal happiness than any kind of earthly felicity. In *We and Our Neighbors,* Stowe's Maggie becomes a good seamstress as well as a good Christian, but she can never be a good wife. Her first transgression eliminated that possibility. American literature would have to wait until Eugene O'Neill's *Anna Christie* (1922) before a prostitute would be forgiven *and* allowed to marry.

In *The Evil That Men Do* (1889; see pp. 353–61), a work that has been called the best fictional treatment of the working girl and the prostitute to appear before *Maggie* (Wyman 174), Edgar Fawcett also uses a typical seduction plot, but otherwise the novel has little in common with Stowe's. Fawcett takes the tenement district as his setting. *The Evil That Men Do* tells the story of Cora Strang, a young woman adrift who works a series of low-paying jobs. Ultimately finding herself unemployed and without a place to live, she calls on her friend Em Cratchett who, doing needlework at home, battles to support her family and live an upright life but eventually dies from starvation. After her encounter with Em, Cora obtains a position as personal maid in a fine household where she is discovered by the wealthy young man-about-town Caspar Drummond. Predictably, Drummond seduces Cora with the promise of marriage and abandons her to marry a society woman. Cora goes to a disorderly house, as houses of prostitution were called, turns prostitute, takes to drink, and ultimately descends to streetwalking. A friend kills Cora to put her out of her misery. In terms of realism, *The Evil That Men Do* marks an advance over earlier prostitute novels, for Fawcett emphasizes how poverty contributes to prostitution. Still, he stops short of making poverty the main cause of prostitution. There is no ambiguity in *The Evil That Men Do*; Fawcett makes absolutely clear from his title where he locates the responsibility for the social evil.

Maggie, too, incorporates a traditional seduction plot, but Crane changed it sufficiently to lift his story above earlier seduction novels. Maggie Johnson's seducer differs greatly from the stereotype. In *The Evil That Men Do,* Caspar Drummond is rich and promises Cora marriage and a life of wealth. In Crane's novel, Pete and Maggie basically come from the same class. The promise of a better life exists in Maggie's romantic imagination, not in Pete's words. The figure of Maggie's brother Jimmie creates moral ambiguity and makes the novel more complex than earlier tales of seduction. A staunch defender of his sister's purity, Jimmie has more than one woman seeking child support from him. To enhance the story's moral ambiguity, Crane avoided explicit moralizing. As noted in *The Critic* (June 13, 1896): 421, Crane "indulges in no rhetoric and is never denunciatory."

LITERARY MOVEMENTS

Stephen Crane was a precocious child. According to one account, he read the novels of James Fenimore Cooper when he was only four years old (*Log* 12). Though born at the time literary realism was emerging, Crane grew up reading the literature of an earlier generation, the historical romance. The reputation of Sir Walter Scott, the finest writer of historical romances in the English language, had grown through the nineteenth century. The year of Crane's birth marked the centennial of Scott's, and that year the cornerstone of a statue to commemorate Scott was laid in Central Park. Scott's works likely formed a significant part of Crane's boyhood reading. A later inventory of the books in Crane's library lists *Ivanhoe* (1819), *Kenilworth* (1821), and *The Talisman* (1825) (Kibler 239).

Like other boys his age, young Crane would have avidly read popular nickel and dime novels — though he would have had to read them behind his father's back. Two years before Stephen Crane's birth, the Reverend Dr. Crane had published *Popular Amusements* (1869), a book that critiqued many pastimes, including novel-reading. He continued to blast popular fiction during his youngest son's boyhood. Ten years after *Popular Amusements,* as Stephen was reaching the age when boys took an interest in adventure stories, Dr. Crane lectured "at length on the growing tendency of the people for reading trashy literature," spoke "strongly against the reading of story papers of the more ordinary class," and sought "to impress upon the minds of the mothers present that it was their duty to their children to supply them with suitable reading, to the end that as they progressed and became older their minds would be so trained that the low, sensational trash would be odious to them" (*Log* 19–20). It seems likely that his father's disapproval of cheap literature only encouraged Stephen's interest, and the cheap books could be had in any nearby drugstore or newsstand. Even if his father refused him money to buy trashy literature, Stephen could have purchased a nickel novel by hoarding his candy money for less than a week.

At the time the Reverend Dr. Crane presented his lecture blasting the story papers, *The Saturday Journal,* the day's most popular story paper, was serializing many different works of fiction. A sample of its contents indicates the popular fiction available during Crane's youth. Corinne Cushman's "A Wild Girl," partially set in Brooklyn, described a young woman tricked into marrying a bogus count whose death allows her to marry a real count. Anthony P. Morris's "Man of Steel" was a historical romance set in revolutionary Paris. Philip S. Warne's "Mississippi River Life in '56" presented a series of western character sketches. Prentiss Ingraham's "Merle, the Mutineer" told an adventure story of early-nineteenth-century Caribbean pirates. Albert W. Aiken's "Fresh of Frisco" was a story of a California mining camp in 1850. And Rett Winwood's "Bride at Sixteen" described a young wife wronged by her scheming sister (Johannsen 1: 437–38).

As these sample titles suggest, popular fiction offered tales with a variety of settings, from distant historical periods to the present day, from distant parts of the globe to nearby neighborhoods. Regardless of their differences, these popular stories shared many common characteristics. They used plenty of dialogue. All were plot driven and action packed. They contained much violence, deception, and suspense. Many were love stories. Despite their often sensational contents, the nickel and dime novels remained fairly innocuous. The Reverend Dr. Crane's attack on cheap literature was largely an overreaction, for, like most products of mass culture, it embodied a conventional, even conservative, morality. In the cheap novels of Crane's day, the heroines remained chaste and the heroes were honorable.

As literature, the nickel and dime novels occupied a middle ground between romanticism and realism. Their plots and sentimentality often echoed the historical romances of Scott's day, yet their use of dialect, dialogue, and occasional urban settings anticipated literary realism. Albert W. Aiken's *The Two Detectives; or, The Fortunes of a Bowery Girl* (1884), for example, tells the story of a young woman raised in the slums of New York. It captures the violence endemic to her urban world and replicates the dialect of the street life (see pp. 338–45). Far from succumbing to her urban environment, however, Aiken's Bowery girl rises above it. Like other heroines of sentimental fiction, she retains her honor and purity, and her story ends happily.

As he matured, Crane's reading became more sophisticated. A school chum remembered him as "a voracious reader of all the nineteenth century English writers" (*Log* 45). Of course it is impossible to identify which writers Crane's friend meant, but, along with Sir Walter Scott, Charles Dickens and William Makepeace Thackeray were considered the century's most important British writers. Crane likely knew *Vanity Fair* (1848), *Oliver Twist* (1838), and *David Copperfield* (1850), among other works. At the University of Syracuse, where Crane had transferred after spending his fall semester at Lafayette College in Easton, Pennsylvania, he discovered the great Russian author Leo Tolstoy. During the spring of 1891, he read *Anna Karenina* (1877) and *War and Peace* (1869) (*Log* 59). [Crane's interest in Tolstoy would continue: some years later he acquired Jonas Jonsson Stadling's *In the Land of Tolstoi* (1897).]

By the time he came to Tolstoy, Crane had already begun an early version of *Maggie*. While his reading gave him much inspiration, Crane realized that books alone provided insufficient source material. Like many college students, then as now, he questioned the value of book learning over that of personal experience. To write realistically Crane recognized that he needed to observe what the reformer Charles Loring Brace had labeled the "dangerous classes" (see p. 273). When some fraternity brothers later remembered Crane, several reminisced about how he would leave the Delta Upsilon house to visit the Syracuse police court and the city's slums. While in Syracuse, he devoted more effort to playing baseball and writing than to his course work. He only took one course, English literature, during his first term and took none his second. He later wrote a correspondent, "I did little work at school, but confined my abilities, such as they were, to the diamond. Not that I disliked books, but the cut-and-dried curriculum of the college did not appeal to me. Humanity was a much more interesting study" (*Correspondence* 99).

A sketch that appeared anonymously in the *New York Herald* (July 5, 1891) called "Where 'De Gang' Hears the Band Play" gives a good indication of the quality of Crane's early version of *Maggie* (see pp. 165–69). Like the published novel, the sketch is set in the tenement district of New York's Lower East Side; it contains poignant details that emphasize the squalor of tenement life; and it tells the story of a brother and sister named Jimmy and Maggie. Jimmy in the sketch is much the same as Jimmie in the published novel. He is a tough youth who enjoys drinking growlers with his friends. His relationship to his sister is ambivalent. He does not hesitate to hock her watch to buy beer, yet he remains her protector. When they both plan to attend an evening band concert, Jimmy threatens, "An', say, if I catch you doin' the walk to-night with that dude mash I'll spoil his face, see?" (see p. 166). Jimmy also expects loyalty from his sister. When Maggie plans to attend an excursion, apparently organized by a labor association to which Jimmy no longer belongs, he says he will steal her shoes to prevent her from going. Unlike Jimmy, Maggie in the sketch differs significantly from Maggie in the novel. While both work in the factory, Maggie in the sketch does not possess the romantic longings of the later Maggie. Instead, she is a hard-minded, gum-chewing tough girl.

The multiple references to the association picnic suggest that the published sketch was part of a larger work, but Crane would make major changes as he reconceived his story, for neither the sketch nor picnic scene would become part of the book. Most important, the character of Maggie changed significantly. In the sketch, Maggie is too savvy and streetwise to let romantic delusions cloud her judgment. Instead of imagining Pete the way she wanted him to be, she would have seen him for what he was. The original version had plenty of realistic detail — one of Crane's fraternity brothers recalled that it was "saturated with obscenity and profanity" (*Log* 62) — but Crane continued to enhance the story's realism as he revised it. Simultaneously, however, he interjected elements of romanticism to contrast the gritty reality of the slums with Maggie's romantic and unrealistic hope for a better life.

After the school year at Syracuse ended, Crane relocated to Asbury Park, New Jersey, where he worked as a reporter for the *New-York Tribune*. The opportunity put him in contact with an ambitious young writer from the Upper Midwest, Hamlin Garland, who came to New Jersey during the summer of 1891 to lecture on literature and art. That same year, Garland made his most important contribution to American literary realism, *Main-Travelled Roads,* a series of dark, gripping stories set in his native region. On August 17, he discussed the well-respected man of letters, William Dean Howells, and Crane reported the lecture for the *New-York Tribune.* According to Crane's account, Garland lauded Howells's *Hazard of New Fortunes* (1890) and *A Modern Instance* (1882) as two of the finest novels ever written. Furthermore, Garland reiterated Howells's advice that "the novelist be true to himself and to things as he sees them" (*Works* 8: 507–08).

It is unknown how well-acquainted Crane and Garland became that August, but the next summer Garland returned to Asbury Park, and he and Crane spent time playing baseball and talking about literature (*Log* 80). Garland's essay "The Future of Fiction" and his literary manifesto, *Crumbling Idols: Twelve Essays on Art and Literature* (1894), would not be published until after *Maggie* appeared, but Garland's aesthetic theories were already well formed by the time he and Crane became friends. In 1886, Garland had read Eugène Véron's *Aesthetics* (1879), and that work significantly affected his critical outlook (see pp. 305–11). In *Crumbling Idols,* Garland expresses skepticism toward a university education and emphasizes the importance of a writer's contact with real life, ideas that Crane already recognized, gauging by his behavior at Syracuse. For his own literary approach, Garland coined the term *veritism,* the essence of which, he explained, was to "Write of those things of which you know most, and for which you care most. By so doing you will be true to yourself, true to your locality, and true to your time" (*Log* 82; *Crumbling Idols* 35).

In the fall of 1892, Crane went to New York City and moved into a rooming house inhabited by a group of medical students that included his friend Frederick M. Lawrence. They called the place the Pendennis Club, a joking reference to Thackeray's novel *The History of Pendennis* (1850). With Lawrence, Crane shared a second-story room that overlooked the East River. Lawrence later described a typical lounging activity: "Feet on the windowsills, pipes in our mouths, we spent long

Party at Pendennis Club **to celebrate the publication of *Maggie* (March 18, 1893). Crane and his roommates jokingly called their apartment the Pendennis Club, a reference to a Thackeray novel. Crane is the one with the banjo. Courtesy of Syracuse University.**

hours watching the busy traffic of the river and the long, slow-crawling caterpillars over on Blackwell's Island, which we knew to be lines of lock-stepping convicts" (6). The image of Blackwell's Island would serve as an ominous backdrop to the action in *Maggie*. Together Crane and Lawrence visited many of the locations that would figure prominently in the book: the tenements on the Lower East Side, the beer gardens, and the Bowery. Lawrence recalled Crane's enthusiasm for the people-filled streets of the tenement district: "One day he came in, his usually somber face alight, and queried abruptly. 'Did you ever see a stone-fight?' When I replied in the negative, he launched into a glowing description of one that he had just seen" (6). The stone fight would become *Maggie*'s opening scene.

While real-life incidents enrich the book, Crane was also influenced by his reading. Among the volumes in their room, Lawrence remembered "one of Zola's books, I think it was *Pot-bouille*" (6). The French novelist Émile Zola first came to prominence in the 1860s, and he was sometimes hailed (and often reviled) as the founder of a new, naturalistic approach to writing fiction. As his notoriety grew, his works were translated into English. *Pot-bouille* (1882), a story of adultery among the bourgeoisie, was part of Zola's multivolume Rougon-Macquart series, which traced the fortunes of one family's descendants. Crane likely read other volumes in the series, including the two that preceded *Pot-bouille, L'Assommoir* (1879), a story describing the ravages of alcohol among the working class, and *Nana* (1880), a story of a prostitute. Crane probably also knew *Thérèse Raquin* (1867), Zola's most important book before the Rougon-Macquart series.

Crane's contemporary reviewers saw he was doing something that went beyond the literary realism William Dean Howells represented, that is, writing that combined realistic description with a conservative, even predictable moral stance (good wins over evil). They were unsure how to label Crane's writing, however. The *New-York Tribune* (May 31, 1896):26, always critical of Crane's work, characterized his style as "aggressive realism," and the London *Bookman* (October 1896): 19–20, came up with the idea of "stern realism," finding that Crane surpassed "nearly all his models of the sternly realistic school, who fail so often in their finer, their more beautiful portraits." Henry D. Traill, a prominent British political journalist, characterized Crane's technique as "new realism," but, specifically discussing *Maggie,* he derisively queried, "Is it art? If so, is the making of mud-pies an artistic occupation, and are the neglected brats who are to be found rolling in the gutters of every great city unconscious artists?" (64). Many who advocated literary realism nevertheless believed that literature should function to refine, improve, educate, and acculturate mankind. A depiction of men and women living out their baser instincts served no elevating purpose. Coining a new term, the reviewer A. G. Sedgwick called Crane a writer of the "animalistic school," defining animalism as "a species of realism which deals with man considered as an animal, capable of hunger, thirst, lust, cruelty, vanity, fear, sloth, predacity, greed, and other passions and appetites that make him kin to the brutes, but which neglects, so far as possible, any higher qualities which distinguish him from his four-footed relatives, such as humor, thought, reason, aspiration, affection, morality, and religion" (15).

While Sedgwick's term *animalism* never really stuck, his definition of it is not inapplicable to literary naturalism. Yet that term, too, inadequately describes Crane's work. The most important American literary naturalist writers, Frank Norris and Theodore Dreiser, combined determinism with a preponderance of detail to overwhelm their readers. The best naturalistic novels require their readers to see how every event and encounter contributes to the protagonist's fate. While *Maggie* contains some heavy-handed, naturalistic passages — the British litterateur Arnold Bennett said, "To read *Maggie* is to put one's ears into the bell of a cornet blown by giant lungs"(7) — Crane's technique differs considerably from naturalist writers. He is much more selective about the detail he includes.

Hamlin Garland's term *veritism* well describes Crane's technique, yet, like Sedgwick's animalism, the term never entered the critical language. As Garland defined it, veritism combined local color with literary impressionism. One early reviewer aligned *Maggie* with the local color movement. The Port Jervis, New York, *Union* remarked, "The dialect of the New York slums, which is reproduced in this volume with absolute accuracy, is, we take it, something new in literature. It is certainly as legitimate a subject of literary and artistic treatment as the dialect of the Georgia negro or Tennessee mountaineer and even more interesting to the average New Yorker" (Gullason 302).

There is little doubt that Crane captured the argot of New York's tenement district as well as other writers captured the vernacular of their localities, but the local color movement, as this reviewer's comment suggests, was generally associated with rural areas. The other component to Garland's veritism, literary impressionism, as James Nagel has recognized, better characterizes Crane's style. A key difference between realism and impressionism is that realism seeks to render its subject objectively while an impressionistic writer creates a subjective picture, depicting a scene the way he or she sees it. Impressionistic descriptions are realistic, yet they are also highly personal and very selective. The impressionistic writer picks and chooses his detail, compressing much information into a brief amount of space.

With its impressionism, its gritty detail, and its underlying moral stance, *Maggie* was unlike anything American readers had seen. As the editor Richard Watson Gilder well knew when he rejected the manuscript, the nineteenth-century reading public would be reluctant to accept such a story. Stephen Crane was foolhardy to imagine that he could make his literary reputation on the privately printed, personally distributed book. Beyond those he gave away, he sold few copies of *Maggie*. As he began his next book, *The Red Badge of Courage* (1895), boxes of unsold copies of *Maggie* surrounded him and served as a reminder of his failure as a professional writer. Nevertheless, *Mag-*

gie earned him the support of Hamlin Garland and, more important, William Dean Howells.

With the publication of *Red Badge* in 1895 and its immediate and resounding success, that work's publisher, D. Appleton and Company, agreed to publish *Maggie* — but not without changes. Appleton had Crane make some changes, and their editors emended the text further. In the 1896 Appleton *Maggie,* for example, Jimmie's "Yeh've edder got teh go teh hell or go teh work!" becomes "Yeh've edder got t' go on d' toif er go t' work!" The excision of the fat man paragraph, the second to last paragraph of Chapter XVII, perhaps the most crucial passage in the book, made Maggie's last night difficult to comprehend (Parker and Higgins 146–48). Rupert Hughes, who had become a diehard Crane enthusiast after reading the 1893 edition, reviewed the 1896 Appleton edition, noticed the changes, and found "almost all of them ill-advised" (318).

While the changes made by the editors at Appleton are regrettable, there can be little doubt that their edition of *Maggie* did more to advance Crane's reputation than his privately printed edition had done. Still, Crane's decision to have *Maggie* privately printed is admirable. Though it may have been naive, it was also idealistic, for Crane sacrificed commercial support and publicity for the sake of artistic integrity. By avoiding the often rigorous editorial intrusion to which commercial firms subject their books, perhaps Crane's first edition text of *Maggie* better reflects his unique personal vision than anything else he published.

Chronology of Crane's Life and Times

1871

November 1: Stephen Crane is born in Newark, New Jersey, the fourteenth and last child of the Reverend Dr. Jonathan Townley Crane (1819–1880), presiding elder of Methodist churches in the Newark district, and Mary Helen (Peck) Crane (d. 1891).

Autumn: Henry M. Stanley meets David Livingstone at Ujiji on Lake Tanganyika.

Charles Darwin (1809–1882), *The Descent of Man.* Edward Eggleston (1837–1902), *The Hoosier Schoolmaster.* George Eliot (1819–1880), *Middlemarch.* William Dean Howells (1837–1920), *Their Wedding Journey.*

1872

November: Ulysses S. Grant is elected to second term as president.

Mark Twain (1835–1910), *Roughing It.*

1873

November: William M. "Boss" Tweed is convicted on 204 charges of fraud and sentenced to twelve years in prison.

The first practical typewriter is invented and manufactured.

Herbert Spencer (1820–1903), *The Study of Sociology.* Jules Verne (1828–1905), *Twenty Thousand Leagues under the Sea.*

1874

Women's Christian Temperance Union (WCTU) is established at Cleveland, Ohio.

Society for the Prevention of Cruelty to Children is formed by Elbridge T. Gerry (1837–1927) and John D. Wright.

Thomas Hardy (1840–1928), *Far from the Madding Crowd.*

1875

The Art Students League is organized in New York City.

Henry James (1843–1916), *A Passionate Pilgrim and Other Tales.*

1876

March: Words are first transmitted by electric telephone, at Boston, between Alexander Graham Bell and his assistant, Thomas A. Watson.

April: The Crane family moves to Paterson, New Jersey. Stephen receives primary education from his sister Agnes Elizabeth.

May: The Centennial Exhibition at Philadelphia opens.

June: At the Battle of Little Big Horn in Montana, General George Custer and all of the 265 men of the Seventh Cavalry are killed by Sitting Bull's Sioux Indians.

November: Rutherford B. Hayes is elected president.

Henry James, *Roderick Hudson.* Herman Melville (1819–1891), *Clarel.* Mark Twain, *The Adventures of Tom Sawyer.*

1877

January: Queen Victoria is officially proclaimed "Empress of India." Thomas A. Edison receives the first patent for a phonograph.

June: Society of American Artists is organized.

Henry James, *The American.* Sarah Orne Jewett (1849–1909), *Deephaven.* Émile Zola (1840–1902), *L'Assommoir.*

1878

January: The Knights of Labor is established.

April: Dr. Crane becomes pastor of Drew Methodist Church in Port Jervis, New York.

September: Stephen Crane enrolls in public school.

October: Edison Electric, the first electric light company, is formed in New York City.

December: The Third Avenue Elevated Railroad is completed. The line runs from South Ferry to 129th Street.

The manufacture of bicycles in the United States is begun by A. A. Pope in Hartford, Connecticut.

The Society for the Prevention of Crime is formed.

Henry James, *The Europeans*.

1879

January: The Gilbert and Sullivan operetta *H. M. S. Pinafore* is played in New York for the first time.

May: St. Patrick's Cathedral is dedicated.

October: Thomas Edison perfects the first incandescent lamp.

Mary Baker Eddy organizes the Church of Christ, Scientist.

Henry George (1839–1897), *Progress and Poverty*. William Dean Howells, *A Lady of Aroostook*. Henry James, *Daisy Miller*.

1880

February: The Reverend Jonathan T. Crane dies.

March: The Metropolitan Museum of Art opens its building in Central Park.

May: The New York Free Circulating Library opens its first library.

November: James A. Garfield is elected president.

December: Broadway, from Fourteenth to Twenty-sixth Streets, is lighted with the Brush electric arc light.

Mark Twain, *A Tramp Abroad*. Lew Wallace (1827–1905), *Ben-Hur*. Émile Zola, *Nana*.

1881

July: President Garfield is shot in the Baltimore & Potomac Railroad Station by C. J. Guiteau.

September: President Garfield dies, and Chester A. Arthur is inaugurated president.

Alexander III ascends the throne of Russia and persecution of Jews leads tens of thousands to emigrate to the United States.

William Dean Howells, *Dr. Breen's Practice* and *A Modern Instance*. Henry James, *Portrait of a Lady* and *Washington Square*. Joel Chandler Harris (1848–1908), *Uncle Remus, His Songs and Sayings*. Émile Zola, *Pot-bouille*.

1882

January: Oscar Wilde (1854–1900) delivers his first lecture in America.

May: The Charity Organization Society of the City of New York is incorporated.

Mark Twain, *The Prince and the Pauper*.

1883

May: The Brooklyn Bridge is formally opened to the public.

June: Mrs. Crane moves her family to the resort town of Asbury Park, New Jersey.

Summer: Stephen Crane's brother Townley begins operating a summer news agency in Asbury Park for the *New-York Tribune* and the Associated Press.

October: Matthew Arnold arrives in New York on a visit to America.

Mark Twain, *Life on the Mississippi*.

1884

November: Grover Cleveland is elected president.

Mary N. Murfree (1850–1922), *In the Tennessee Mountains*. Herbert Spencer, *The Man versus the State*. Mark Twain, *Adventures of Huckleberry Finn*.

1885

September: Crane enrolls in the Pennington Seminary in Pennington, New Jersey.

William Dean Howells, *The Rise of Silas Lapham*. Émile Zola, *Germinal*.

1886

October: The Statue of Liberty is dedicated.

Henry James, *The Princess Casamassima*. Robert Louis Stevenson (1850–1894), *The Strange Case of Dr. Jekyll and Mr. Hyde*.

1887

November or December: Crane abruptly leaves Pennington Seminary.

H. Rider Haggard (1856–1925), *She*. Mary E. Wilkins Freeman (1852–1930), *A Humble Romance*. Émile Zola, *La Terre*.

1888

January: Crane registers at Claverack College and Hudson River Institute, a military school, as a student in the classical department but later transfers to the academic department.

May: DeWolf Hopper, a popular actor, gives Ernest Thayer's "Casey at the Bat" its first public recitation at Wallack Theater, New York.

June: The New York legislature passes a law providing that all criminals sentenced to death shall be executed using electricity.

July: Crane begins assisting his brother Townley with his news-reporting agency in Asbury Park. He would return each summer through 1892 in a similar capacity.

November: Benjamin Harrison is elected president.

The Kodak camera comes on the market, instantly making photography an affordable and practical amateur hobby.

Edward Bellamy (1850–1898), *Looking Backward.* Henry James, *The Aspern Papers.* Herman Melville, *John Marr and Other Sailors.*

1889

March: The first performance in America of Richard Wagner's *Der Ring des Nibbelungen* takes place in New York at the Metropolitan Opera House.

May: The Johnstown, Pennsylvania, flood occurs, resulting in over two thousand deaths and millions of dollars in property damage.

Jane Addams establishes Hull-House in the slums of Chicago.

Mark Twain, *A Connecticut Yankee in King Arthur's Court.*

1890

July: The Sherman Anti-Trust Act becomes law.

September: Crane enrolls as a freshman in the mining-engineering program at Lafayette College, Easton, Pennsylvania.

December: Mary Helen Crane dies.

Emily Dickinson (1830–1886), *Poems.* William Dean Howells, *A Hazard of New Fortunes.* William James (1842–1910), *The Principles of Psychology.* Henry M. Stanley (1841–1904), *In Darkest Africa.* Oscar Wilde, *The Picture of Dorian Gray.*

1891

January: Crane transfers to the University of Syracuse where he spends most of his time writing and playing baseball. He publishes sketches in the *University Herald,* and works as a stringer for the *New-York Tribune.*

March: Congress passes the International Copyright Act.

Spring?: Crane begins writing an early version of *Maggie* in the Delta Upsilon House at the University of Syracuse.

May: Carnegie Hall is officially opened.

June: Crane joins friends on a camping trip in Sullivan County, an excursion which forms the basis for Crane's Sullivan County sketches published in the *New-York Tribune* the next year.

August: Hamlin Garland (1860–1940) lectures on American literature and art at Avon-by-the-Sea.

Autumn: Crane explores the tenement districts of New York's Lower East Side.

Ambrose Bierce (1842–1914?), *Tales of Soldiers and Civilians*. Emily Dickinson, *Poems . . . Second Series*. Hamlin Garland, *Main-Travelled Roads*. Thomas Hardy, *Tess of the D'Urbervilles*. Herman Melville, *Timoleon*.

1892

January: The federal immigration center on Ellis Island is officially opened.

July: Jacob A. Riis (1849–1914) lectures at Avon-by-the-Sea, which Crane reports in the *New-York Tribune*.

August: Crane reports the annual "American Day" parade of the Junior Order of United American Mechanics (JOUAM) of New Jersey in Asbury Park in the *New-York Tribune*. The derogatory report prompts rejoinders by members of JOUAM and apologies from the *Tribune*, which publishes nothing of Crane's after 1892 and derides his works through the remainder of his life.

August: Hamlin Garland returns to Asbury Park, and he and Crane play baseball and discuss literature. Crane shows Garland a version of *Maggie*, and he recommends Crane send it to Richard Watson Gilder, editor of *Century Magazine*.

September: The first successful gasoline-powered automobile is built by Charles and Frank Duryea.

October: Crane moves to New York City, sharing a rooming house with several medical students.

October: The World's Fair formally opens at Chicago.

November: Grover Cleveland is elected president.

Arthur Conan Doyle (1859–1930), *The Adventures of Sherlock Holmes*. Rudyard Kipling (1865–1936), *Barrack-Room Ballads*. Jacob Riis, *Children of the Poor*. Émile Zola, *La Débâcle*.

1893

January: The Library of Congress receives Crane's copyright application for *Maggie*, the typewritten title page reading, "A Girl of the Streets, / A Story of New York. / — By — / Stephen Crane."

Late February–Early March: *Maggie: A Girl of the Streets (A Story of New York)* appears.

March: Crane's roommates at the Pendennis Club throw a party to celebrate the publication of *Maggie*.

July: Frederick Jackson Turner delivers "The Significance of the Frontier in American History" before the American Historical Association.

October?: Crane moves into the old Art Students' League building at 143–147 East Twenty-third Street.

December: *The Red Badge of Courage* appears in an abridged version, syndicated in the *Philadelphia Press*, the *New York Press*, and several others papers across the country.

The Anti-Saloon League is organized at Oberlin, Ohio.

Ambrose Bierce, *Can Such Things Be?*

1894

June: Congress passes a law making the first Monday in September, "Labor Day," a legal holiday.

Lafcadio Hearn (1850–1904), *Glimpses of Japan*. Anthony Hope (1863–1933), *The Prisoner of Zenda*. Benjamin Kidd (1858–1916), *Social Evolution*. Rudyard Kipling, *The Jungle Book*. Mark Twain, *The Tragedy of Pudd'nhead Wilson and the Comedy, Those Extraordinary Twins*.

1895

January: Crane leaves New York for an extensive tour of the West and Mexico to write feature articles for the Bacheller syndicate.

February: The revolt of Cuba against Spain breaks out.

February: Crane arrives in Nebraska where he experiences a snowstorm which inspires his short story "The Blue Hotel." While in Nebraska, he meets Willa Cather.

March: Crane arrives in Texas, visiting Galveston and San Antonio.

April: Crane visits Mexico.

May: Copeland and Day publish Crane's first collection of verse, *The Black Riders, and Other Lines*. Crane returns to New York City.

September: D. Appleton and Company publish *The Red Badge of Courage*.

December: Crane attends the Philistine dinner in his honor in Buffalo, New York.

Wilhelm Röntgen discovers X rays. Marconi invents radio telegraphy.

H. G. Wells (1866–1946), *The Time Machine.*

1896

April: The first modern Olympic Games are held in Athens, Greece.

May: *George's Mother.*

June: Appleton publishes expurgated edition of *Maggie.*

September: Heinemann publishes the British edition of *Maggie* entitled *Maggie: A Child of the Streets,* which is prefaced with "An Appreciation" by William Dean Howells.

November: *The Little Regiment.*

November: William McKinley is elected president.

Late November: Crane leaves New York for Jacksonville, Florida, where he meets Cora Taylor, the estranged wife of an English aristocrat, Captain Donald William Stewart. In Jacksonville Cora operates a house of assignation known as the Hotel de Dream.

Emily Dickinson, *Poems . . . Third Series.* Harold Frederic (1856–1898), *The Damnation of Theron Ware.* Sarah Orne Jewett, *The Country of the Pointed Firs.*

1897

January: The *Commodore,* sailing from Jacksonville toward Cuba, sinks; Crane escapes in an open boat, an experience that leads to his fine short story "The Open Boat."

March: Crane returns to New York and signs a contract with William Randolph Hearst's *New York Journal* to report the impending Greco-Turkish War. He and Cora travel to Europe, arriving in London in late March and Athens in April.

April: Turkey declares war on Greece.

April: Grant's Tomb is dedicated.

May: *The Third Violet.*

June: Stephen and Cora, who calls herself Mrs. Stephen Crane, settle at Ravensbrook, a small villa in Surrey. Their neighbors include several prominent men of letters: Robert Barr, Joseph Conrad, Harold Frederic, Edward Garnett, Ford Madox Hueffer (later known as Ford Madox Ford), and Henry James.

October: Crane meets Joseph Conrad.

November: The Astoria Hotel and the adjoining Waldorf are opened as the Waldorf-Astoria on Fifth Avenue between 33rd and 34th Streets (later the site of the Empire State Building).

Gold is discovered in the Klondike.

Richard Harding Davis (1864–1916), *Soldiers of Fortune.* Henry James, *The Spoils of Poynton* and *What Maisie Knew.* Moses Coit Tyler, *The Literary History of the American Revolution.*

1898

February: The USS *Maine* is exploded in Havana Harbor, killing two officers and 258 crew members.

April: Congress authorizes the use of armed forces to end the civil war in Cuba. President McKinley orders a blockade of Cuban ports. The Spanish ship *Buena Vista* is captured by the U.S. gunboat *Nashville.* Congress passes the Volunteer Army Act, authorizing the First or Rough Riders under the command of Col. Leonard Wood and Lt. Col. Theodore Roosevelt. Spain and the United States officially declare war on one another.

April: *The Open Boat and Other Tales of Adventure.*

May: The U.S. Asiatic squadron defeats the Spanish fleet at the Battle of Manila Bay in the Philippines.

May: Crane, commissioned as war correspondent for Joseph Pulitzer's *New York World,* makes several brief voyages from Key West to gather information about naval operations with other correspondents, including Frank Norris.

June–Early July: U.S. forces invade Cuba, achieving important victories including the assault on San Juan Hill with the Rough Riders, led by Theodore Roosevelt, participating. Crane observes much of the action close at hand.

Late July: Crane briefly returns to New York, breaks with Pulitzer, signs with William Randolph Hearst's *New York Journal* to report the Puerto Rican campaign. U.S. troops invade Puerto Rico.

August: Crane returns to Cuba and enters Havana illegally. After hostilities end, Crane remains in Havana for several months, living in virtual seclusion. Cora, still in England, attempts to contact him but is unsuccessful.

December: Crane leaves Havana for New York and shortly thereafter leaves New York for England.

Henry James, *The Two Magics* (which includes "The Turn of the Screw"). Edward Noyes Westcott, *David Harum.*

1899

February: Stephen and Cora Crane move into Brede Place, a dilapidated, centuries-old country estate.

May: *War Is Kind.*

October: *Active Service.*

December: *The Monster and Other Stories.*

December: "The Ghost," a dramatic performance mostly written by Crane, though nominally written by Robert Barr, Joseph Conrad, George Gissing, H. Rider Haggard, Henry James, and H. G. Wells, among others, is performed during the Christmas festivities.

Spanish-American War ends.

Frank Norris (1870–1902), *McTeague.*

1900

April: Crane suffers two massive tubercular hemorrhages. Cora raises funds to seek a cure in Badenweiler, Germany.

June 5: Crane dies in Badenweiler.

November: Theodore Roosevelt is elected president.

Several of Crane's works are published posthumously: *Whilomville Stories* (1900), *Wounds in the Rain* (1900), *Great Battles of the World* (1901), *Last Words* (1902), and *The O'Ruddy* (completed by Robert Barr).

Theodore Dreiser (1871–1945), *Sister Carrie.* William Dean Howells, *Literary Friends and Acquaintance.* Jack London (1876–1916), *The Son of the Wolf.*

A Note on the Text

The period of American literary realism was a time when writers were willing to write what publishers were unwilling to publish. Few works better illustrate this truism than Stephen Crane's *Maggie*. Publishing the work himself, Crane avoided the editorial cleansing that unquestionably would have taken place had a commercial press accepted the work. The 1893 text closely represents the work as Crane intended it. Once he achieved fame with the publication of *The Red Badge of Courage* in 1895, that work's publisher, D. Appleton and Co., asked Crane for another work they might publish, and he offered them *Maggie*. Though Appleton was anxious to capitalize on the fame *Red Badge* created, the firm was unwilling to publish *Maggie* as Crane originally had written it. For the 1896 edition, they cleaned up the printing errors, yet they also cleaned up Crane's language and his imagery, deleting expletives and expurgating important descriptive passages that had been part of the 1893 *Maggie*. Despite the changes and deletions, the watered-down version helped further Crane's reputation, and the 1893 first edition text was forgotten. Reprints through the 1950s all were based on the 1896 Appleton edition. Not until Joseph Katz's 1966 article "The *Maggie* Nobody Knows" was the 1893 version recognized as the superior text. While several facsimile reprints of the 1893 edition followed Katz's article, inexpensive paperback editions continue to perpetuate the inferior 1896 text. Complicating matters further, the approach to textual editing that predominated in the late 1960s and early 1970s prompted the Virginia edition of *Maggie,* an eclectic text that combines elements from both the 1893 and the 1896 editions, to create a *Maggie* Stephen Crane never knew. The Virginia edition of *Maggie* was savagely reviewed after the volume appeared in 1969, and the eclectic approach to textual editing it exemplifies has since fallen into disfavor. The text of this volume is that of the 1893 first edition. Obvious typographical errors have been silently emended.

Maggie: A Girl of the Streets
(A Story of New York)

CHAPTER I

A very little boy stood upon a heap of gravel for the honor of Rum Alley. He was throwing stones at howling urchins from Devil's Row who were circling madly about the heap and pelting at him.

His infantile countenance was livid with fury. His small body was writhing in the delivery of great, crimson oaths.

"Run, Jimmie, run! Dey'll get yehs," screamed a retreating Rum Alley child.

"Naw," responded Jimmie with a valiant roar, "dese micks can't make me run."

Howls of renewed wrath went up from Devil's Row throats. Tattered gamins on the right made a furious assault on the gravel heap. On their small, convulsed faces there shone the grins of true assassins. As they charged, they threw stones and cursed in shrill chorus.

The little champion of Rum Alley stumbled precipitately down the other side. His coat had been torn to shreds in a scuffle, and his hat was gone. He had bruises on twenty parts of his body, and blood was dripping from a cut in his head. His wan features wore a look of a tiny, insane demon.

On the ground, children from Devil's Row closed in on their antagonist. He crooked his left arm defensively about his head and fought with cursing fury. The little boys ran to and fro, dodging, hurling stones and swearing in barbaric trebles.

From a window of an apartment house that upreared its form from amid squat, ignorant stables, there leaned a curious woman. Some laborers, unloading a scow at a dock at the river, paused for a moment and regarded the fight. The engineer of a passive tugboat hung lazily to a railing and watched. Over on the Island,[1] a worm of yellow convicts came from the shadow of a grey ominous building and crawled slowly along the river's bank.

A stone had smashed into Jimmie's mouth. Blood was bubbling over his chin and down upon his ragged shirt. Tears made furrows on his dirt-stained cheeks. His thin legs had begun to tremble and turn weak, causing his small body to reel. His roaring curses of the first part of the fight had changed to a blasphemous chatter.

In the yells of the whirling mob of Devil's Row children there were notes of joy like songs of triumphant savagery. The little boys seemed to leer gloatingly at the blood upon the other child's face.

Down the avenue came boastfully sauntering a lad of sixteen years, although the chronic sneer of an ideal manhood already sat upon his lips. His hat was tipped with an air of challenge over his eye. Between his teeth, a cigar stump was tilted at the angle of defiance. He walked with a certain swing of the shoulders which appalled the timid. He glanced over into the vacant lot in which the little raving boys from Devil's Row seethed about the shrieking and tearful child from Rum Alley.

"Gee!" he murmured with interest, "A scrap. Gee!"

He strode over to the cursing circle, swinging his shoulders in a manner which denoted that he held victory in his fists. He approached at the back of one of the most deeply engaged of the Devil's Row children.

"Ah, what deh hell," he said, and smote the deeply-engaged one on the back of the head. The little boy fell to the ground and gave a hoarse, tremendous howl. He scrambled to his feet, and perceiving, evidently, the size of his assailant, ran quickly off, shouting alarms. The entire Devil's Row party followed him. They came to a stand a

[1] *the Island:* Blackwell's Island (now Roosevelt Island), located in the East River, was acquired by New York City in 1828. During the nineteenth century, the city erected and maintained there a prison, an almshouse, a workhouse, and three hospitals: a lunatic asylum, a charity hospital, and a smallpox hospital.

short distance away and yelled taunting oaths at the boy with the chronic sneer. The latter, momentarily, paid no attention to them.

"What deh hell, Jimmie?" he asked of the small champion.

Jimmie wiped his blood-wet features with his sleeve.

"Well, it was dis way, Pete, see! I was goin' teh lick dat Riley kid and dey all pitched on me."

Some Rum Alley children now came forward. The party stood for a moment exchanging vainglorious remarks with Devil's Row. A few stones were thrown at long distances, and words of challenge passed between small warriors. Then the Rum Alley contingent turned slowly in the direction of their home street. They began to give, each to each, distorted versions of the fight. Causes of retreat in particular cases were magnified. Blows dealt in the fight were enlarged to catapultian power, and stones thrown were alleged to have hurtled with infinite accuracy. Valor grew strong again, and the little boys began to swear with great spirit.

"Ah, we blokies kin lick deh hull damn Row," said a child, swaggering.

Little Jimmie was striving to stanch the flow of blood from his cut lips. Scowling, he turned upon the speaker.

"Ah, where deh hell was yeh when I was doin' all deh fightin'?" he demanded. "Youse kids makes me tired."

"Ah, go ahn," replied the other argumentatively.

Jimmie replied with heavy contempt. "Ah, youse can't fight, Blue Billie! I kin lick yeh wid one han'."

"Ah, go ahn," replied Billie again.

"Ah," said Jimmie threateningly.

"Ah," said the other in the same tone.

They struck at each other, clinched, and rolled over on the cobble stones.

"Smash 'im, Jimmie, kick deh damn guts out of 'im," yelled Pete, the lad with the chronic sneer, in tones of delight.

The small combatants pounded and kicked, scratched and tore. They began to weep and their curses struggled in their throats with sobs. The other little boys clasped their hands and wriggled their legs in excitement. They formed a bobbing circle about the pair.

A tiny spectator was suddenly agitated.

"Cheese it, Jimmie, cheese it! Here comes yer fader," he yelled.

The circle of little boys instantly parted. They drew away and waited in ecstatic awe for that which was about to happen. The two little boys fighting in the modes of four thousand years ago, did not hear the warning.

Up the avenue there plodded slowly a man with sullen eyes. He was carrying a dinner pail and smoking an apple-wood pipe.

As he neared the spot where the little boys strove, he regarded them listlessly. But suddenly he roared an oath and advanced upon the rolling fighters.

"Here, you Jim, git up, now, while I belt yer life out, you damned disorderly brat."

He began to kick into the chaotic mass on the ground. The boy Billie felt a heavy boot strike his head. He made a furious effort and disentangled himself from Jimmie. He tottered away, damning.

Jimmie arose painfully from the ground and confronting his father, began to curse him. His parent kicked him. "Come home, now," he cried, "an' stop yer jawin', er I'll lam the everlasting head off yehs."

They departed. The man paced placidly along with the apple-wood emblem of serenity between his teeth. The boy followed a dozen feet in the rear. He swore luridly, for he felt that it was degradation for one who aimed to be some vague soldier, or a man of blood with a sort of sublime license, to be taken home by a father.

CHAPTER II

Eventually they entered into a dark region where, from a careening building, a dozen gruesome doorways gave up loads of babies to the street and the gutter. A wind of early autumn raised yellow dust from cobbles and swirled it against an hundred windows. Long streamers of garments fluttered from fire-escapes. In all unhandy places there were buckets, brooms, rags and bottles. In the street infants played or fought with other infants or sat stupidly in the way of vehicles. Formidable women, with uncombed hair and disordered dress, gossiped while leaning on railings, or screamed in frantic quarrels. Withered persons, in curious postures of submission to something, sat smoking pipes in obscure corners. A thousand odors of cooking food came forth to the street. The building quivered and creaked from the weight of humanity stamping about in its bowels.

A small ragged girl dragged a red, bawling infant along the crowded ways. He was hanging back, baby-like, bracing his wrinkled, bare legs.

The little girl cried out: "Ah, Tommie, come ahn. Dere's Jimmie and fader. Don't be a-pullin' me back."

She jerked the baby's arm impatiently. He fell on his face, roaring. With a second jerk she pulled him to his feet, and they went on. With the obstinacy of his order, he protested against being dragged in a chosen direction. He made heroic endeavors to keep on his legs, denounce his sister and consume a bit of orange peeling which he chewed between the times of his infantile orations.

As the sullen-eyed man, followed by the blood-covered boy, drew near, the little girl burst into reproachful cries. "Ah, Jimmie, youse bin fightin' agin."

The urchin swelled disdainfully.

"Ah, what deh hell, Mag. See?"

The little girl upbraided him, "Youse allus fightin', Jimmie, an' yeh knows it puts mudder out when yehs come home half dead, an' it's like we'll all get a poundin'."

She began to weep. The babe threw back his head and roared at his prospects.

"Ah, what deh hell!" cried Jimmie. "Shut up er I'll smack yer mout'. See?"

As his sister continued her lamentations, he suddenly swore and struck her. The little girl reeled and, recovering herself, burst into tears and quaveringly cursed him. As she slowly retreated her brother advanced dealing her cuffs. The father heard and turned about.

"Stop that, Jim, d'yeh hear? Leave yer sister alone on the street. It's like I can never beat any sense into yer damned wooden head."

The urchin raised his voice in defiance to his parent and continued his attacks. The babe bawled tremendously, protesting with great violence. During his sister's hasty manœuvres, he was dragged by the arm.

Finally the procession plunged into one of the gruesome doorways. They crawled up dark stairways and along cold, gloomy halls. At last the father pushed open a door and they entered a lighted room in which a large woman was rampant.

She stopped in a career from a seething stove to a pan-covered table. As the father and children filed in she peered at them.

"Eh, what? Been fightin' agin, by Gawd!" She threw herself upon Jimmie. The urchin tried to dart behind the others and in the scuffle the babe, Tommie, was knocked down. He protested with his usual vehemence, because they had bruised his tender shins against a table leg.

The mother's massive shoulders heaved with anger. Grasping the urchin by the neck and shoulder she shook him until he rattled. She dragged him to an unholy sink, and, soaking a rag in water, began to scrub his lacerated face with it. Jimmie screamed in pain and tried to twist his shoulders out of the clasp of the huge arms.

The babe sat on the floor watching the scene, his face in contortions like that of a woman at a tragedy. The father, with a newly-ladened pipe in his mouth, crouched on a backless chair near the stove. Jimmie's cries annoyed him. He turned about and bellowed at his wife:

"Let the damned kid alone for a minute, will yeh, Mary? Yer allus poundin' 'im. When I come nights I can't git no rest 'cause yer allus poundin' a kid. Let up, d'yeh hear? Don't be allus poundin' a kid."

The woman's operations on the urchin instantly increased in violence. At last she tossed him to a corner where he limply lay cursing and weeping.

The wife put her immense hands on her hips and with a chieftain-like stride approached her husband.

"Ho," she said, with a great grunt of contempt. "An' what in the devil are you stickin' your nose for?"

The babe crawled under the table and, turning, peered out cautiously. The ragged girl retreated and the urchin in the corner drew his legs carefully beneath him.

The man puffed his pipe calmly and put his great mudded boots on the back part of the stove.

"Go teh hell," he murmured, tranquilly.

The woman screamed and shook her fists before her husband's eyes. The rough yellow of her face and neck flared suddenly crimson. She began to howl.

He puffed imperturbably at his pipe for a time, but finally arose and began to look out at the window into the darkening chaos of back yards.

"You've been drinkin', Mary," he said. "You'd better let up on the bot', ol' woman, or you'll git done."

"You're a liar. I ain't had a drop," she roared in reply.

They had a lurid altercation, in which they damned each other's souls with frequence.

The babe was staring out from under the table, his small face working in his excitement.

The ragged girl went stealthily over to the corner where the urchin lay.

"Are yehs hurted much, Jimmie?" she whispered timidly.

"Not a damn bit! See?" growled the little boy.

"Will I wash deh blood?"

"Naw!"

"Will I" —

"When I catch dat Riley kid I'll break 'is face! Dat's right! See?"

He turned his face to the wall as if resolved to grimly bide his time.

In the quarrel between husband and wife, the woman was victor. The man grabbed his hat and rushed from the room, apparently determined upon a vengeful drunk. She followed to the door and thundered at him as he made his way down stairs.

She returned and stirred up the room until her children were bobbing about like bubbles.

"Git outa deh way," she persistently bawled, waving feet with their dishevelled shoes near the heads of her children. She shrouded herself, puffing and snorting, in a cloud of steam at the stove, and eventually extracted a frying-pan full of potatoes that hissed.

She flourished it. "Come teh yer suppers, now," she cried with sudden exasperation. "Hurry up, now, er I'll help yeh!"

The children scrambled hastily. With prodigious clatter they arranged themselves at table. The babe sat with his feet dangling high from a precarious infant chair and gorged his small stomach. Jimmie forced, with feverish rapidity, the grease-enveloped pieces between his wounded lips. Maggie, with side glances of fear of interruption, ate like a small pursued tigress.

The mother sat blinking at them. She delivered reproaches, swallowed potatoes and drank from a yellow-brown bottle. After a time her mood changed and she wept as she carried little Tommie into another room and laid him to sleep with his fists doubled in an old quilt of faded red and green grandeur. Then she came and moaned by the stove. She rocked to and fro upon a chair, shedding tears and crooning miserably to the two children about their "poor mother" and "yer fader, damn 'is soul."

The little girl plodded between the table and the chair with a dish-pan on it. She tottered on her small legs beneath burdens of dishes.

Jimmie sat nursing his various wounds. He cast furtive glances at his mother. His practised eye perceived her gradually emerge from a muddled mist of sentiment until her brain burned in drunken heat. He sat breathless.

Maggie broke a plate.

The mother started to her feet as if propelled.

"Good Gawd," she howled. Her eyes glittered on her child with sudden hatred. The fervent red of her face turned almost to purple. The little boy ran to the halls, shrieking like a monk in an earthquake.

He floundered about in darkness until he found the stairs. He stumbled, panic-stricken, to the next floor. An old woman opened a door. A light behind her threw a flare on the urchin's quivering face.

"Eh, Gawd, child, what is it dis time? Is yer fader beatin' yer mudder, or yer mudder beatin' yer fader?"

CHAPTER III

Jimmie and the old woman listened long in the hall. Above the muffled roar of conversation, the dismal wailings of babies at night, the thumping of feet in unseen corridors and rooms, mingled with the sound of varied hoarse shoutings in the street and the rattling of wheels over cobbles, they heard the screams of the child and the roars of the mother die away to a feeble moaning and a subdued bass muttering.

The old woman was a gnarled and leathery personage who could don, at will, an expression of great virtue. She possessed a small music-box capable of one tune, and a collection of "God bless yehs" pitched in assorted keys of fervency. Each day she took a position upon the stones of Fifth Avenue, where she crooked her legs under her and crouched immovable and hideous, like an idol. She received daily a small sum in pennies. It was contributed, for the most part, by persons who did not make their homes in that vicinity.

Once, when a lady had dropped her purse on the sidewalk, the gnarled woman had grabbed it and smuggled it with great dexterity beneath her cloak. When she was arrested she had cursed the lady into a partial swoon, and with her aged limbs, twisted from rheumatism, had almost kicked the stomach out of a huge policeman whose conduct upon that occasion she referred to when she said: "The police, damn 'em."

"Eh, Jimmie, it's cursed shame," she said. "Go, now, like a dear an' buy me a can,[1] an' if yer mudder raises 'ell all night yehs can sleep here."

Jimmie took a tendered tin-pail and seven pennies and departed. He passed into the side door of a saloon and went to the bar. Straining up on his toes he raised the pail and pennies as high as his arms would let

[1] *can:* As the next paragraph clarifies, the word *can* refers to a tin-pail filled with beer, also known as a growler. The reformers commonly complained about the ease with which children could obtain beer. See George Frederic Parsons, "The Saloon in Society" (pp. 174–82).

him. He saw two hands thrust down and take them. Directly the same hands let down the filled pail and he left.

In front of the gruesome doorway he met a lurching figure. It was his father, swaying about on uncertain legs.

"Give me deh can. See?" said the man, threateningly.

"Ah, come off! I got dis can fer dat ol' woman an' it 'ud be dirt teh swipe it. See?" cried Jimmie.

The father wrenched the pail from the urchin. He grasped it in both hands and lifted it to his mouth. He glued his lips to the under edge and tilted his head. His hairy throat swelled until it seemed to grow near his chin. There was a tremendous gulping movement and the beer was gone.

The man caught his breath and laughed. He hit his son on the head with the empty pail. As it rolled clanging into the street, Jimmie began to scream and kicked repeatedly at his father's shins.

"Look at deh dirt what yeh done me," he yelled. "Deh ol' woman 'ill be raisin' hell."

He retreated to the middle of the street, but the man did not pursue. He staggered toward the door.

"I'll club hell outa yeh when I ketch yeh," he shouted, and disappeared.

During the evening he had been standing against a bar drinking whiskies and declaring to all comers, confidentially: "My home reg'lar livin' hell! Damndes' place! Reg'lar hell! Why do I come an' drin' whisk' here thish way? 'Cause home reg'lar livin' hell!"

Jimmie waited a long time in the street and then crept warily up through the building. He passed with great caution the door of the gnarled woman, and finally stopped outside his home and listened.

He could hear his mother moving heavily about among the furniture of the room. She was chanting in a mournful voice, occasionally interjecting bursts of volcanic wrath at the father, who, Jimmie judged, had sunk down on the floor or in a corner.

"Why deh blazes don' chere try teh keep Jim from fightin'? I'll break yer jaw," she suddenly bellowed.

The man mumbled with drunken indifference. "Ah, wha' deh hell. W'a's odds? Wha' makes kick?"

"Because he tears 'is clothes, yeh damn fool," cried the woman in supreme wrath.

The husband seemed to become aroused. "Go teh hell," he thundered fiercely in reply. There was a crash against the door and something broke into clattering fragments. Jimmie partially suppressed a howl and darted down the stairway. Below he paused and listened. He heard howls and curses, groans and shrieks, confusingly in chorus as if a battle were raging. With all was the crash of splintering furniture. The eyes of the urchin glared in fear that one of them would discover him.

Curious faces appeared in door-ways, and whispered comments passed to and fro. "Ol' Johnson's raisin' hell agin."

Jimmie stood until the noises ceased and the other inhabitants of the tenement had all yawned and shut their doors. Then he crawled upstairs with the caution of an invader of a panther den. Sounds of labored breathing came through the broken door-panels. He pushed the door open and entered, quaking.

A glow from the fire threw red hues over the bare floor, the cracked and soiled plastering, and the overturned and broken furniture.

In the middle of the floor lay his mother asleep. In one corner of the room his father's limp body hung across the seat of a chair.

The urchin stole forward. He began to shiver in dread of awakening his parents. His mother's great chest was heaving painfully. Jimmie paused and looked down at her. Her face was inflamed and swollen from drinking. Her yellow brows shaded eye-lids that had grown blue. Her tangled hair tossed in waves over her forehead. Her mouth was set in the same lines of vindictive hatred that it had, perhaps, borne during the fight. Her bare, red arms were thrown out above her head in positions of exhaustion, something, mayhap, like those of a sated villain.

The urchin bended over his mother. He was fearful lest she should open her eyes, and the dread within him was so strong, that he could not forbear to stare, but hung as if fascinated over the woman's grim face.

Suddenly her eyes opened. The urchin found himself looking straight into that expression, which, it would seem, had the power to change his blood to salt. He howled piercingly and fell backward.

The woman floundered for a moment, tossed her arms about her head as if in combat, and again began to snore.

Jimmie crawled back in the shadows and waited. A noise in the next room had followed his cry at the discovery that his mother was awake. He grovelled in the gloom, the eyes from out his drawn face riveted upon the intervening door.

He heard it creak, and then the sound of a small voice came to him. "Jimmie! Jimmie! Are yehs dere?" it whispered. The urchin started. The thin, white face of his sister looked at him from the door-way of the other room. She crept to him across the floor.

The father had not moved, but lay in the same death-like sleep. The mother writhed in uneasy slumber, her chest wheezing as if she were in the agonies of strangulation. Out at the window a florid moon was peering over dark roofs, and in the distance the waters of a river glimmered pallidly.

The small frame of the ragged girl was quivering. Her features were haggard from weeping, and her eyes gleamed from fear. She grasped the urchin's arm in her little trembling hands and they huddled in a corner. The eyes of both were drawn, by some force, to stare at the woman's face, for they thought she need only to awake and all fiends would come from below.

They crouched until the ghost-mists of dawn appeared at the window, drawing close to the panes, and looking in at the prostrate, heaving body of the mother.

CHAPTER IV

The babe, Tommie, died. He went away in a white, insignificant coffin, his small waxen hand clutching a flower that the girl, Maggie, had stolen from an Italian.

She and Jimmie lived.

The inexperienced fibres of the boy's eyes were hardened at an early age. He became a young man of leather. He lived some red years without laboring. During that time his sneer became chronic. He studied human nature in the gutter, and found it no worse than he thought he had reason to believe it. He never conceived a respect for the world, because he had begun with no idols that it had smashed.

He clad his soul in armor by means of happening hilariously in at a mission church where a man composed his sermons of "yous." While they got warm at the stove, he told his hearers just where he calculated they stood with the Lord. Many of the sinners were impatient over the pictured depths of their degradation. They were waiting for soup-tickets.

A reader of words of wind-demons might have been able to see the portions of a dialogue pass to and fro between the exhorter and his hearers.

"You are damned," said the preacher. And the reader of sounds might have seen the reply go forth from the ragged people: "Where's our soup?"

Jimmie and a companion sat in a rear seat and commented upon the things that didn't concern them, with all the freedom of English gentlemen. When they grew thirsty and went out their minds confused the speaker with Christ.

Momentarily, Jimmie was sullen with thoughts of a hopeless altitude where grew fruit. His companion said that if he should ever meet God he would ask for a million dollars and a bottle of beer.

Jimmie's occupation for a long time was to stand on street-corners and watch the world go by, dreaming blood-red dreams at the passing of pretty women. He menaced mankind at the intersections of streets.

On the corners he was in life and of life. The world was going on and he was there to perceive it.

He maintained a belligerent attitude toward all well-dressed men. To him fine raiment was allied to weakness, and all good coats covered faint hearts. He and his order were kings, to a certain extent, over the men of untarnished clothes, because these latter dreaded, perhaps, to be either killed or laughed at.

Above all things he despised obvious Christians and ciphers with the chrysanthemums of aristocracy in their button-holes. He considered himself above both of these classes. He was afraid of neither the devil nor the leader of society.

When he had a dollar in his pocket his satisfaction with existence was the greatest thing in the world. So, eventually, he felt obliged to work. His father died and his mother's years were divided up into periods of thirty days.

He became a truck driver. He was given the charge of a pains-taking pair of horses and a large rattling truck. He invaded the turmoil and tumble of the down-town streets and learned to breathe maledictory defiance at the police who occasionally used to climb up, drag him from his perch and beat him.

In the lower part of the city he daily involved himself in hideous tangles. If he and his team chanced to be in the rear he preserved a demeanor of serenity, crossing his legs and bursting forth into yells when foot passengers took dangerous dives beneath the noses of his champing horses. He smoked his pipe calmly for he knew that his pay was marching on.

If in the front and the key-truck of chaos, he entered terrifically into the quarrel that was raging to and fro among the drivers on their high seats, and sometimes roared oaths and violently got himself arrested.

After a time his sneer grew so that it turned its glare upon all things. He became so sharp that he believed in nothing. To him the police were always actuated by malignant impulses and the rest of the world was composed, for the most part, of despicable creatures who were all trying to take advantage of him and with whom, in defense, he was obliged to quarrel on all possible occasions. He himself occupied a down-trodden position that had a private but distinct element of grandeur in its isolation.

The most complete cases of aggravated idiocy were, to his mind, rampant upon the front platforms of all of the street-cars. At first his tongue strove with these beings, but he eventually was superior. He became immured like an African cow. In him grew a majestic contempt for those strings of street cars that followed him like intent bugs.

He fell into the habit, when starting on a long journey, of fixing his eye on a high and distant object, commanding his horses to begin, and then going into a sort of a trance of observation. Multitudes of drivers might howl in his rear, and passengers might load him with opprobrium, he would not awaken until some blue policeman turned red and began to frenziedly tear bridles and beat the soft noses of the responsible horses.

When he paused to contemplate the attitude of the police toward himself and his fellows, he believed that they were the only men in the city who had no rights. When driving about, he felt that he was held liable by the police for anything that might occur in the streets, and was the common prey of all energetic officials. In revenge, he resolved never to move out of the way of anything, until formidable circumstances, or a much larger man than himself forced him to it.

Foot-passengers were mere pestering flies with an insane disregard for their legs and his convenience. He could not conceive their maniacal desires to cross the streets. Their madness smote him with eternal amazement. He was continually storming at them from his throne. He sat aloft and denounced their frantic leaps, plunges, dives and straddles.

When they would thrust at, or parry, the noses of his champing horses, making them swing their heads and move their feet, disturbing a solid dreamy repose, he swore at the men as fools, for he himself could perceive that Providence had caused it clearly to be written, that he and his team had the unalienable right to stand in the proper path of the sun chariot, and if they so minded, obstruct its mission or take a wheel off.

And, perhaps, if the god-driver had an ungovernable desire to step down, put up his flame-colored fists and manfully dispute the right of way, he would have probably been immediately opposed by a scowling mortal with two sets of very hard knuckles.

It is possible, perhaps, that this young man would have derided, in an axle-wide alley, the approach of a flying ferry boat. Yet he achieved a respect for a fire-engine. As one charged toward his truck, he would drive fearfully upon a sidewalk, threatening untold people with annihilation. When an engine would strike a mass of blocked trucks, splitting it into fragments, as a blow annihilates a cake of ice, Jimmie's team could usually be observed high and safe, with whole wheels, on the sidewalk. The fearful coming of the engine could break up the most intricate muddle of heavy vehicles at which the police had been swearing for the half of an hour.

A fire-engine was enshrined in his heart as an appalling thing that he loved with a distant dog-like devotion. They had been known to overturn street-cars. Those leaping horses, striking sparks from the cobbles in their forward lunge, were creatures to be ineffably admired. The clang of the gong pierced his breast like a noise of remembered war.

When Jimmie was a little boy, he began to be arrested. Before he reached a great age, he had a fair record.

He developed too great a tendency to climb down from his truck and fight with other drivers. He had been in quite a number of miscellaneous fights, and in some general barroom rows that had become known to the police. Once he had been arrested for assaulting a Chinaman. Two women in different parts of the city, and entirely unknown to each other, caused him considerable annoyance by breaking forth, simultaneously, at fateful intervals, into wailings about marriage and support and infants.

Nevertheless, he had, on a certain star-lit evening, said wonderingly and quite reverently: "Deh moon looks like hell, don't it?"

CHAPTER V

The girl, Maggie, blossomed in a mud puddle. She grew to be a most rare and wonderful production of a tenement district, a pretty girl.

None of the dirt of Rum Alley seemed to be in her veins. The philosophers up-stairs, down-stairs and on the same floor, puzzled over it.

When a child, playing and fighting with gamins in the street, dirt disguised her. Attired in tatters and grime, she went unseen.

There came a time, however, when the young men of the vicinity, said: "Dat Johnson goil is a puty good looker." About this period her brother remarked to her: "Mag, I'll tell yeh dis! See? Yeh've edder got teh go teh hell or go teh work!" Whereupon she went to work, having the feminine aversion of going to hell.

By a chance, she got a position in an establishment where they made collars and cuffs. She received a stool and a machine in a room where sat twenty girls of various shades of yellow discontent. She perched on the stool and treadled at her machine all day, turning out collars, the name of whose brand could be noted for its irrelevancy to anything in connection with collars. At night she returned home to her mother.

Jimmie grew large enough to take the vague position of head of the family. As incumbent of that office, he stumbled up-stairs late at night, as his father had done before him. He reeled about the room, swearing at his relations, or went to sleep on the floor.

The mother had gradually arisen to that degree of fame that she could bandy words with her acquaintances among the police-justices. Court-officials called her by her first name. When she appeared they pursued a course which had been theirs for months. They invariably grinned and cried out: "Hello, Mary, you here again?" Her grey head wagged in many a court. She always besieged the bench with voluble excuses, explanations, apologies and prayers. Her flaming face and rolling eyes were a sort of familiar sight on the island. She measured time by means of sprees, and was eternally swollen and dishevelled.

One day the young man, Pete, who as a lad had smitten the Devil's Row urchin in the back of the head and put to flight the antagonists of his friend, Jimmie, strutted upon the scene. He met Jimmie one day on the street, promised to take him to a boxing match in Williamsburg,[1] and called for him in the evening.

Maggie observed Pete.

He sat on a table in the Johnson home and dangled his checked legs with an enticing nonchalance. His hair was curled down over his forehead in an oiled bang. His rather pugged nose seemed to revolt from contact with a bristling moustache of short, wire-like hairs. His blue double-breasted coat, edged with black braid, buttoned close to a red puff tie, and his patent-leather shoes looked like murder-fitted weapons.

His mannerisms stamped him as a man who had a correct sense of his personal superiority. There was valor and contempt for circum-

[1] *Williamsburg:* Located across the East River from Manhattan, Williamsburgh was incorporated as a village in 1827 and as a city in 1851. Williamsburg (without the *h*) was consolidated with Brooklyn in 1855. It became and remained a fashionable suburb for the well-to-do among people of Austrian, German, and Irish descent until the Williamsburg Bridge opened in 1903.

stances in the glance of his eye. He waved his hands like a man of the world, who dismisses religion and philosophy, and says "Fudge." He had certainly seen everything and with each curl of his lip, he declared that it amounted to nothing. Maggie thought he must be a very elegant and graceful bartender.

He was telling tales to Jimmie.

Maggie watched him furtively, with half-closed eyes, lit with a vague interest.

"Hully gee! Dey makes me tired," he said. "Mos' e'ry day some farmer comes in an' tries teh run deh shop. See? But dey gits t'rowed right out! I jolt dem right out in deh street before dey knows where dey is! See?"

"Sure," said Jimmie.

"Dere was a mug come in deh place deh odder day wid an idear he wus goin' teh own deh place! Hully gee, he wus goin' teh own deh place! I see he had a still on an' I didn' wanna giv'im no stuff, so I says: 'Git deh hell outa here an' don' make no trouble,' I says like dat! See? 'Git deh hell outa here an' don' make no trouble'; like dat. 'Git deh hell outa here,' I says. See?"

Jimmie nodded understandingly. Over his features played an eager desire to state the amount of his valor in a similar crisis, but the narrator proceeded.

"Well, deh blokie he says: 'T'hell wid it! I ain' lookin' for no scrap,' he says (See?) 'but' he says, 'I'm 'spectable cit'zen an' I wanna drink an' purtydamnsoon, too.' See? 'Deh hell,' I says. Like dat! 'Deh hell,' I says. See? 'Don' make no trouble,' I says. Like dat. 'Don' make no trouble.' See? Den deh mug he squared off an' said he was fine as silk wid his dukes (See?) an' he wanned a drink damnquick. Dat's what he said. See?"

"Sure," repeated Jimmie.

Pete continued. "Say, I jes' jumped deh bar an' deh way I plunked dat blokie was great. See? Dat's right! In deh jaw! See? Hully gee, he t'rowed a spittoon true deh front windee. Say, I taut I'd drop dead. But deh boss, he comes in after an' he says, 'Pete, yehs done jes' right! Yeh've gota keep order an' it's all right.' See? 'It's all right,' he says. Dat's what he said."

The two held a technical discussion.

"Dat bloke was a dandy," said Pete, in conclusion, "but he had'n' oughta made no trouble. Dat's what I says teh dem: 'Don' come in here an' make no trouble,' I says, like dat. 'Don' make no trouble.' See."

As Jimmie and his friend exchanged tales descriptive of their prowess, Maggie leaned back in the shadow. Her eyes dwelt wonderingly and rather wistfully upon Pete's face. The broken furniture, grimey walls, and general disorder and dirt of her home of a sudden appeared before her and began to take a potential aspect. Pete's aristocratic person looked as if it might soil. She looked keenly at him, occasionally, wondering if he was feeling contempt. But Pete seemed to be enveloped in reminiscence.

"Hully gee," said he, "dose mugs can't phase me. Dey knows I kin wipe up deh street wid any tree of dem."

When he said, "Ah, what deh hell," his voice was burdened with disdain for the inevitable and contempt for anything that fate might compel him to endure.

Maggie perceived that here was the beau ideal of a man. Her dim thoughts were often searching for far away lands where, as God says, the little hills[2] sing together in the morning. Under the trees of her dream-gardens there had always walked a lover.

CHAPTER VI

Pete took note of Maggie.

"Say, Mag, I'm stuck on yer shape. It's outa sight," he said, parenthetically, with an affable grin.

As he became aware that she was listening closely, he grew still more eloquent in his descriptions of various happenings in his career. It appeared that he was invincible in fights.

"Why," he said, referring to a man with whom he had had a misunderstanding, "dat mug scrapped like a damn dago. Dat's right. He was dead easy. See? He tau't he was a scrapper! But he foun' out diff'ent! Hully gee."

He walked to and fro in the small room, which seemed then to grow even smaller and unfit to hold his dignity, the attribute of a supreme warrior. That swing of the shoulders that had frozen the timid when he was but a lad had increased with his growth and education at the ratio of ten to one. It, combined with the sneer upon his mouth, told mankind that there was nothing in space which could

[2] *little hills:* See Psalms 65.12.

appall him. Maggie marvelled at him and surrounded him with greatness. She vaguely tried to calculate the altitude of the pinnacle from which he must have looked down upon her.

"I met a chump deh odder day way up in deh city," he said. "I was goin' teh see a frien' of mine. When I was a-crossin' deh street deh chump runned plump inteh me, an' den he turns aroun' an' says, 'Yer insolen' ruffin,' he says, like dat. 'Oh, gee,' I says, 'oh, gee, go teh hell and git off deh eart',' I says, like dat. See? 'Go teh hell an' git off deh eart',' like dat. Den deh blokie he got wild. He says I was a contempt'ble scoun'el, er something like dat, an' he says I was doom' teh everlastin' pe'dition an' all like dat. 'Gee,' I says, 'gee! Deh hell I am,' I says. 'Deh hell I am,' like dat. An' den I slugged 'im. See?"

With Jimmie in his company, Pete departed in a sort of a blaze of glory from the Johnson home. Maggie, leaning from the window, watched him as he walked down the street.

Here was a formidable man who disdained the strength of a world full of fists. Here was one who had contempt for brass-clothed power; one whose knuckles could defiantly ring against the granite of law. He was a knight.

The two men went from under the glimmering street-lamp and passed into shadows.

Turning, Maggie contemplated the dark, dust-stained walls, and the scant and crude furniture of her home. A clock, in a splintered and battered oblong box of varnished wood, she suddenly regarded as an abomination. She noted that it ticked raspingly. The almost vanished flowers in the carpet-pattern, she conceived to be newly hideous. Some faint attempts she had made with blue ribbon, to freshen the appearance of a dingy curtain, she now saw to be piteous.

She wondered what Pete dined on.

She reflected upon the collar and cuff factory. It began to appear to her mind as a dreary place of endless grinding. Pete's elegant occupation brought him, no doubt, into contact with people who had money and manners. It was probable that he had a large acquaintance of pretty girls. He must have great sums of money to spend.

To her the earth was composed of hardships and insults. She felt instant admiration for a man who openly defied it. She thought that if the grim angel of death should clutch his heart, Pete would shrug his shoulders and say: "Oh, ev'ryt'ing goes."

She anticipated that he would come again shortly. She spent some of her week's pay in the purchase of flowered cretonne for a lambrequin.[1] She made it with infinite care and hung it to the slightly-careening mantel, over the stove, in the kitchen. She studied it with painful anxiety from different points in the room. She wanted it to look well on Sunday night when, perhaps, Jimmie's friend would come. On Sunday night, however, Pete did not appear.

Afterward the girl looked at it with a sense of humiliation. She was now convinced that Pete was superior to admiration for lambrequins.

A few evenings later Pete entered with fascinating innovations in his apparel. As she had seen him twice and he had different suits on each time, Maggie had a dim impression that his wardrobe was prodigiously extensive.

"Say, Mag," he said, "put on yer bes' duds Friday night an' I'll take yehs teh deh show. See?"

He spent a few moments in flourishing his clothes and then vanished, without having glanced at the lambrequin.

Over the eternal collars and cuffs in the factory Maggie spent the most of three days in making imaginary sketches of Pete and his daily environment. She imagined some half dozen women in love with him and thought he must lean dangerously toward an indefinite one, whom she pictured with great charms of person, but with an altogether contemptible disposition.

She thought he must live in a blare of pleasure. He had friends, and people who were afraid of him.

She saw the golden glitter of the place where Pete was to take her. An entertainment of many hues and many melodies where she was afraid she might appear small and mouse-colored.

Her mother drank whiskey all Friday morning. With lurid face and tossing hair she cursed and destroyed furniture all Friday afternoon. When Maggie came home at half-past six her mother lay asleep amidst the wreck of chairs and a table. Fragments of various household utensils were scattered about the floor. She had vented some phase of drunken fury upon the lambrequin. It lay in a bedraggled heap in the corner.

"Hah," she snorted, sitting up suddenly, "where deh hell yeh been? Why deh hell don' yeh come home earlier? Been loafin' 'round deh streets. Yer gettin' teh be a reg'lar devil."

[1] *cretonne for a lambrequin:* Cretonne is a colorfully printed, heavy-duty cotton or linen fabric normally used for curtains and chair coverings. A lambrequin is a short curtain or piece of drapery, often with a scalloped lower edge, suspended for ornament from a mantelshelf.

When Pete arrived Maggie, in a worn black dress, was waiting for him in the midst of a floor strewn with wreckage. The curtain at the window had been pulled by a heavy hand and hung by one tack, dangling to and fro in the draft through the cracks at the sash. The knots of blue ribbons appeared like violated flowers. The fire in the stove had gone out. The displaced lids and open doors showed heaps of sullen grey ashes. The remnants of a meal, ghastly, like dead flesh, lay in a corner. Maggie's red mother, stretched on the floor, blasphemed and gave her daughter a bad name.

CHAPTER VII

An orchestra of yellow silk women and bald-headed men on an elevated stage near the centre of a great green-hued hall,[1] played a popular waltz. The place was crowded with people grouped about little tables. A battalion of waiters slid among the throng, carrying trays of beer glasses and making change from the inexhaustible vaults of their trousers pockets. Little boys, in the costumes of French chefs, paraded up and down the irregular aisles vending fancy cakes. There was a low rumble of conversation and a subdued clinking of glasses. Clouds of tobacco smoke rolled and wavered high in air about the dull gilt of the chandeliers.

The vast crowd had an air throughout of having just quitted labor. Men with calloused hands and attired in garments that showed the wear of an endless trudge for a living, smoked their pipes contentedly and spent five, ten, or perhaps fifteen cents for beer. There was a mere sprinkling of kid-gloved men who smoked cigars purchased elsewhere. The great body of the crowd was composed of people who showed that all day they strove with their hands. Quiet Germans, with maybe their wives and two or three children, sat listening to the music, with the expressions of happy cows. An occasional party of sailors from a war-ship, their faces pictures of sturdy health, spent the earlier hours of the evening at the small round tables. Very infrequent tipsy men, swollen with the value of their opinions, engaged their companions in earnest and confidential conversation. In the balcony, and here

[1] *green-hued hall:* Crane refers to a beer garden here. For a fuller description, see James D. McCabe, "The Beer Gardens" (pp. 170–73).

and there below, shone the impassive faces of women. The nationalities of the Bowery[2] beamed upon the stage from all directions.

Pete aggressively walked up a side aisle and took seats with Maggie at a table beneath the balcony.

"Two beehs!"

Leaning back he regarded with eyes of superiority the scene before them. This attitude affected Maggie strongly. A man who could regard such a sight with indifference must be accustomed to very great things.

It was obvious that Pete had been to this place many times before, and was very familiar with it. A knowledge of this fact made Maggie feel little and new.

He was extremely gracious and attentive. He displayed the consideration of a cultured gentleman who knew what was due.

"Say, what deh hell? Bring deh lady a big glass! What deh hell use is dat pony?"

"Don't be fresh, now," said the waiter, with some warmth, as he departed.

"Ah, git off deh eart'," said Pete, after the other's retreating form.

Maggie perceived that Pete brought forth all his elegance and all his knowledge of high-class customs for her benefit. Her heart warmed as she reflected upon his condescension.

The orchestra of yellow silk women and bald-headed men gave vent to a few bars of anticipatory music and a girl, in a pink dress with short skirts, galloped upon the stage. She smiled upon the throng as if in acknowledgment of a warm welcome, and began to walk to and fro, making profuse gesticulations and singing, in brazen soprano tones, a song, the words of which were inaudible. When she broke into the swift rattling measures of a chorus some half tipsy men near the stage joined in the rollicking refrain and glasses were pounded rhythmically upon the tables. People leaned forward to watch her and to try to catch the words of the song. When she vanished there were long rollings of applause.

Obedient to more anticipatory bars, she reappeared amidst the half-suppressed cheering of the tipsy men. The orchestra plunged into dance music and the laces of the dancer fluttered and flew in the glare

[2] *Bowery:* A street in lower Manhattan that stretches for about one mile from Chatham Square to Cooper Square. After the Third Avenue elevated line was placed over it, the Bowery became a run-down place peopled by immigrants and derelicts and containing cheap lodging houses, seedy saloons, and brothels. For good contemporary descriptions of the Bowery, see David Graham Phillips, "The Bowery at Night"; and Julian Ralph, "The Bowery" (pp. 145–65).

of gas jets. She divulged the fact that she was attired in some half dozen skirts. It was patent that any one of them would have proved adequate for the purpose for which skirts are intended. An occasional man bent forward, intent upon the pink stockings. Maggie wondered at the splendor of the costume and lost herself in calculations of the cost of the silks and laces.

The dancer's smile of stereotyped enthusiasm was turned for ten minutes upon the faces of her audience. In the finale she fell into some of those grotesque attitudes which were at the time popular among the dancers in the theatres up-town, giving to the Bowery public the phantasies of the aristocratic theatre-going public, at reduced rates.

"Say, Pete," said Maggie, leaning forward, "dis is great."

"Sure," said Pete, with proper complacence.

A ventriloquist followed the dancer. He held two fantastic dolls on his knees. He made them sing mournful ditties and say funny things about geography and Ireland.

"Do dose little men talk?" asked Maggie.

"Naw," said Pete, "it's some damn fake. See?"

Two girls, on the bills as sisters, came forth and sang a duet that is heard occasionally at concerts given under church auspices. They supplemented it with a dance which of course can never be seen at concerts given under church auspices.

After the duettists had retired, a woman of debatable age sang a negro melody. The chorus necessitated some grotesque waddlings supposed to be an imitation of a plantation darkey, under the influence, probably, of music and the moon. The audience was just enthusiastic enough over it to have her return and sing a sorrowful lay, whose lines told of a mother's love and a sweetheart who waited and a young man who was lost at sea under the most harrowing circumstances. From the faces of a score or so in the crowd, the self-contained look faded. Many heads were bent forward with eagerness and sympathy. As the last distressing sentiment of the piece was brought forth, it was greeted by that kind of applause which rings as sincere.

As a final effort, the singer rendered some verses which described a vision of Britain being annihilated by America, and Ireland bursting her bonds.[3] A carefully prepared crisis was reached in the last line of

[3] *vision . . . bonds:* American political animosity toward Great Britain lingered through much of the nineteenth century. It was not until after 1895 with the Great Rapprochement that the two countries established their lasting political alliance. See Bradford Perkins, *The Great Rapprochement: England and the United States, 1895–1914* (New York: Athenaeum, 1968).

the last verse, where the singer threw out her arms and cried, "The star-spangled banner." Instantly a great cheer swelled from the throats of the assemblage of the masses. There was a heavy rumble of booted feet thumping the floor. Eyes gleamed with sudden fire, and calloused hands waved frantically in the air.

After a few moments' rest, the orchestra played crashingly, and a small fat man burst out upon the stage. He began to roar a song and stamp back and forth before the foot-lights, wildly waving a glossy silk hat and throwing leers, or smiles, broadcast. He made his face into fantastic grimaces until he looked like a pictured devil on a Japanese kite. The crowd laughed gleefully. His short, fat legs were never still a moment. He shouted and roared and bobbed his shock of red wig until the audience broke out in excited applause.

Pete did not pay much attention to the progress of events upon the stage. He was drinking beer and watching Maggie.

Her cheeks were blushing with excitement and her eyes were glistening. She drew deep breaths of pleasure. No thoughts of the atmosphere of the collar and cuff factory came to her.

When the orchestra crashed finally, they jostled their way to the sidewalk with the crowd. Pete took Maggie's arm and pushed a way for her, offering to fight with a man or two.

They reached Maggie's home at a late hour and stood for a moment in front of the gruesome doorway.

"Say, Mag," said Pete, "give us a kiss for takin' yeh teh deh show, will yer?"

Maggie laughed, as if startled, and drew away from him.

"Naw, Pete," she said, "dat wasn't in it."

"Ah, what deh hell?" urged Pete.

The girl retreated nervously.

"Ah, what deh hell?" repeated he.

Maggie darted into the hall, and up the stairs. She turned and smiled at him, then disappeared.

Pete walked slowly down the street. He had something of an astonished expression upon his features. He paused under a lamp-post and breathed a low breath of suprise.

"Gawd," he said, "I wonner if I've been played fer a duffer."

CHAPTER VIII

As thoughts of Pete came to Maggie's mind, she began to have an intense dislike for all of her dresses.

"What deh hell ails yeh? What makes yeh be allus fixin' and fussin'? Good Gawd," her mother would frequently roar at her.

She began to note, with more interest, the well-dressed women she met on the avenues. She envied elegance and soft palms. She craved those adornments of person which she saw every day on the street, conceiving them to be allies of vast importance to women.

Studying faces, she thought many of the women and girls she chanced to meet, smiled with serenity as though forever cherished and watched over by those they loved.

The air in the collar and cuff establishment strangled her. She knew she was gradually and surely shrivelling in the hot, stuffy room. The begrimed windows rattled incessantly from the passing of elevated trains.[1] The place was filled with a whirl of noises and odors.

She wondered as she regarded some of the grizzled women in the room, mere mechanical contrivances sewing seams and grinding out, with heads bended over their work, tales of imagined or real girl-hood happiness, past drunks, the baby at home, and unpaid wages. She speculated how long her youth would endure. She began to see the bloom upon her cheeks as valuable.

She imagined herself, in an exasperating future, as a scrawny woman with an eternal grievance. Too, she thought Pete to be a very fastidious person concerning the appearance of women.

She felt she would love to see somebody entangle their fingers in the oily beard of the fat foreigner who owned the establishment. He was a detestable creature. He wore white socks with low shoes.

He sat all day delivering orations, in the depths of a cushioned chair. His pocket-book deprived them of the power of retort.

[1] *elevated trains:* The first elevated railways or el trains in New York City were constructed from 1867–70. By the early 1890s, other elevated lines crisscrossed Manhattan from South Ferry to 155th Street. The noisy, dirty railways darkened and polluted streets, made buildings shake, and continually disturbed anyone living at the third-floor level along the elevated routes. In *Manhattan Transfer* (1925), John Dos Passos would make effective literary use of the el train. As Dos Passos emphasized, the el trains served as a pervasive reminder of the extent to which industrial technology controlled man's life. Small wonder that in the film *King Kong* (1933), one of Kong's first acts upon escaping his captivity is to rip up an elevated line.

"What een hell do you sink I pie fife dolla a week for?[2] Play? No, py damn!"

Maggie was anxious for a friend to whom she could talk about Pete. She would have liked to discuss his admirable mannerisms with a reliable mutual friend. At home, she found her mother often drunk and always raving.

It seems that the world had treated this woman very badly, and she took a deep revenge upon such portions of it as came within her reach. She broke furniture as if she were at last getting her rights. She swelled with virtuous indignation as she carried the lighter articles of household use, one by one under the shadows of the three gilt balls, where Hebrews chained them with chains of interest.

Jimmie came when he was obliged to by circumstances over which he had no control. His well-trained legs brought him staggering home and put him to bed some nights when he would rather have gone elsewhere.

Swaggering Pete loomed like a golden sun to Maggie. He took her to a dime museum[3] where rows of meek freaks astonished her. She contemplated their deformities with awe and thought them a sort of chosen tribe.

Pete, raking his brains for amusement, discovered the Central Park Menagerie and the Museum of Arts.[4] Sunday afternoons would sometimes find them at these places. Pete did not appear to be particularly interested in what he saw. He stood around looking heavy, while Maggie giggled in glee.

Once at the Menagerie he went into a trance of admiration before the spectacle of a very small monkey threatening to thrash a cageful

[2] *fife dolla a week:* Maggie's earnings are just slightly under the average $5.50 per week that women working in New York City shirt factories made in 1888. See *Fourth Annual Report of the Commissioner of Labor, 1888* (Washington: GPO, 1889), 507.

[3] *dime museum:* Seedy, run-for-profit exhibits of wonder, unbelievable occurrences, and oddities of nature which were mostly located along the Bowery. For a good contemporary description of the dime museums, see David Graham Phillips, "The Bowery at Night" (pp. 159–65).

[4] *Central Park Menagerie and the Museum of Arts:* Central Park, the first landscaped public park in the United States, was opened to the public in 1859. The Menagerie or Zoo moved into permanent quarters in 1871 and became the park's most popular attraction, though the bourgeois park visitors and those living in the fashionable neighborhoods adjacent to the park disapproved of the lower classes that the Menagerie attracted. In 1880, the Metropolitan Museum of Art opened its permanent location on Fifth Avenue adjacent to Central Park. Besides its collections and exhibits, the museum organized lectures, school programs, and craft workshops during its first two decades.

because one of them had pulled his tail and he had not wheeled about quickly enough to discover who did it. Ever after Pete knew that monkey by sight and winked at him, trying to induce him to fight with other and larger monkeys.

At the Museum, Maggie said, "Dis is outa sight."

"Oh hell," said Pete, "wait till next summer an' I'll take yehs to a picnic."

While the girl wandered in the vaulted rooms, Pete occupied himself in returning stony stare for stony stare, the appalling scrutiny of the watch-dogs of the treasures. Occasionally he would remark in loud tones: "Dat jay[5] has got glass eyes," and sentences of the sort. When he tired of this amusement he would go to the mummies and moralize over them.

Usually he submitted with silent dignity to all which he had to go through, but, at times, he was goaded into comment.

"What deh hell," he demanded once. "Look at all dese little jugs! Hundred jugs in a row! Ten rows in a case an' 'bout a t'ousand cases! What deh blazes use is dem?"

Evenings during the week he took her to see plays in which the brain-clutching heroine was rescued from the palatial home of her guardian, who is cruelly after her bonds, by the hero with the beautiful sentiments. The latter spent most of his time out at soak in pale-green snow storms, busy with a nickel-plated revolver, rescuing aged strangers from villains.

Maggie lost herself in sympathy with the wanderers swooning in snow storms beneath happy-hued church windows. And a choir within singing "Joy to the World." To Maggie and the rest of the audience this was transcendental realism. Joy always within, and they, like the actor, inevitably without. Viewing it, they hugged themselves in ecstatic pity of their imagined or real condition.

The girl thought the arrogance and granite-heartedness of the magnate of the play was very accurately drawn. She echoed the maledictions that the occupants of the gallery showered on this individual when his lines compelled him to expose his extreme selfishness.

Shady persons in the audience revolted from the pictured villainy of the drama. With untiring zeal they hissed vice and applauded virtue. Unmistakably bad men evinced an apparently sincere admiration for virtue.

[5] *jay:* An ignorant, rustic, or unsophisticated person.

The loud gallery was overwhelmingly with the unfortunate and the oppressed. They encouraged the struggling hero with cries, and jeered the villain, hooting and calling attention to his whiskers. When anybody died in the pale-green snow storms, the gallery mourned. They sought out the painted misery and hugged it as akin.

In the hero's erratic march from poverty in the first act, to wealth and triumph in the final one, in which he forgives all the enemies that he has left, he was assisted by the gallery, which applauded his generous and noble sentiments and confounded the speeches of his opponents by making irrelevant but very sharp remarks. Those actors who were cursed with villainy parts were confronted at every turn by the gallery. If one of them rendered lines containing the most subtle distinctions between right and wrong, the gallery was immediately aware if the actor meant wickedness, and denounced him accordingly.

The last act was a triumph for the hero, poor and of the masses, the representative of the audience, over the villain and the rich man, his pockets stuffed with bonds, his heart packed with tyrannical purposes, imperturbable amid suffering.

Maggie always departed with raised spirits from the showing places of the melodrama. She rejoiced at the way in which the poor and virtuous eventually surmounted the wealthy and wicked. The theater made her think. She wondered if the culture and refinement she had seen imitated, perhaps grotesquely, by the heroine on the stage, could be acquired by a girl who lived in a tenement house and worked in a shirt factory.

CHAPTER IX

A group of urchins were intent upon the side door of a saloon. Expectancy gleamed from their eyes. They were twisting their fingers in excitement.

"Here she comes," yelled one of them suddenly.

The group of urchins burst instantly asunder and its individual fragments were spread in a wide, respectable half circle about the point of interest. The saloon door opened with a crash, and the figure of a woman appeared upon the threshold. Her gray hair fell in knotted masses about her shoulders. Her face was crimsoned and wet with perspiration. Her eyes had a rolling glare.

"Not a damn cent more of me money will yehs ever get, not a damn cent. I spent me money here fer t'ree years an' now yehs tells me yeh'll sell me no more stuff! T'hell wid yeh, Johnnie Murckre! 'Disturbance?' Disturbance be damned! T'hell wid yeh, Johnnie — "

The door received a kick of exasperation from within and the woman lurched heavily out on the sidewalk.

The gamins in the half circle became violently agitated. They began to dance about and hoot and yell and jeer. Wide dirty grins spread over each face.

The woman made a furious dash at a particularly outrageous cluster of little boys. They laughed delightedly and scampered off a short distance, calling out over their shoulders to her. She stood tottering on the curbstone and thundered at them.

"Yeh devil's kids," she howled, shaking red fists. The little boys whooped in glee. As she started up the street they fell in behind and marched uproariously. Occasionally she wheeled about and made charges on them. They ran nimbly out of reach and taunted her.

In the frame of a gruesome doorway she stood for a moment cursing them. Her hair straggled, giving her crimson features a look of insanity. Her great fists quivered as she shook them madly in the air.

The urchins made terrific noises until she turned and disappeared. Then they filed quietly in the way they had come.

The woman floundered about in the lower hall of the tenement house and finally stumbled up the stairs. On an upper hall a door was opened and a collection of heads peered curiously out, watching her. With a wrathful snort the woman confronted the door, but it was slammed hastily in her face and the key was turned.

She stood for a few minutes, delivering a frenzied challenge at the panels.

"Come out in deh hall, Mary Murphy, damn yeh, if yehs want a row. Come ahn, yeh overgrown terrier, come ahn."

She began to kick the door with her great feet. She shrilly defied the universe to appear and do battle. Her cursing trebles brought heads from all doors save the one she threatened. Her eyes glared in every direction. The air was full of her tossing fists.

"Come ahn, deh hull damn gang of yehs, come ahn," she roared at the spectators. An oath or two, cat-calls, jeers and bits of facetious advice were given in reply. Missiles clattered about her feet.

"What deh hell's deh matter wid yeh?" said a voice in the gathered gloom, and Jimmie came forward. He carried a tin dinner-pail in his hand and under his arm a brown truckman's apron done in a bundle. "What deh hell's wrong?" he demanded.

"Come out, all of yehs, come out," his mother was howling. "Come ahn an' I'll stamp yer damn brains under me feet."

"Shet yer face, an' come home, yeh damned old fool," roared Jimmie at her. She strided up to him and twirled her fingers in his face. Her eyes were darting flames of unreasoning rage and her frame trembled with eagerness for a fight.

"T'hell wid yehs! An' who deh hell are yehs? I ain't givin' a snap of me fingers fer yehs," she bawled at him. She turned her huge back in tremendous disdain and climbed the stairs to the next floor.

Jimmie followed, cursing blackly. At the top of the flight he seized his mother's arm and started to drag her toward the door of their room.

"Come home, damn yeh," he gritted between his teeth.

"Take yer hands off me! Take yer hands off me," shrieked his mother.

She raised her arm and whirled her great fist at her son's face. Jimmie dodged his head and the blow struck him in the back of the neck. "Damn yeh," gritted he again. He threw out his left hand and writhed his fingers about her middle arm. The mother and the son began to sway and struggle like gladiators.

"Whoop!" said the Rum Alley tenement house. The hall filled with interested spectators.

"Hi, ol' lady, dat was a dandy!"

"T'ree to one on deh red!"

"Ah, stop yer dam scrappin'!"

The door of the Johnson home opened and Maggie looked out. Jimmie made a supreme cursing effort and hurled his mother into the room. He quickly followed and closed the door. The Rum Alley tenement swore disappointedly and retired.

The mother slowly gathered herself up from the floor. Her eyes glittered menacingly upon her children.

"Here, now," said Jimmie, "we've had enough of dis. Sit down, an' don' make no trouble."

He grasped her arm, and twisting it, forced her into a creaking chair.

"Keep yer hands off me," roared his mother again.

"Damn yer ol' hide," yelled Jimmie, madly. Maggie shrieked and ran into the other room. To her there came the sound of a storm of crashes and curses. There was a great final thump and Jimmie's voice cried: "Dere damn yeh, stay still." Maggie opened the door now, and went warily out. "Oh, Jimmie."

He was leaning against the wall and swearing. Blood stood upon bruises on his knotty fore-arms where they had scraped against the floor or the walls in the scuffle. The mother lay screeching on the floor, the tears running down her furrowed face.

Maggie, standing in the middle of the room, gazed about her. The usual upheaval of the tables and chairs had taken place. Crockery was strewn broadcast in fragments. The stove had been disturbed on its legs, and now leaned idiotically to one side. A pail had been upset and water spread in all directions.

The door opened and Pete appeared. He shrugged his shoulders. "Oh, Gawd," he observed.

He walked over to Maggie and whispered in her ear. "Ah, what deh hell, Mag? Come ahn and we'll have a hell of a time."

The mother in the corner upreared her head and shook her tangled locks.

"Teh hell wid him and you," she said, glowering at her daughter in the gloom. Her eyes seemed to burn balefully. "Yeh've gone teh deh devil, Mag Johnson, yehs knows yehs have gone teh deh devil. Yer a disgrace teh yer people, damn yeh. An' now, git out an' go ahn wid dat doe-faced jude of yours. Go teh hell wid him, damn yeh, an' a good riddance. Go teh hell an' see how yeh likes it."

Maggie gazed long at her mother.

"Go teh hell now, an' see how yeh likes it. Git out. I won't have sech as yehs in me house! Get out, d'yeh hear! Damn yeh, git out!"

The girl began to tremble.

At this instant Pete came forward. "Oh, what deh hell, Mag, see," whispered he softly in her ear. "Dis all blows over. See? Deh ol' woman 'ill be all right in deh mornin'. Come ahn out wid me! We'll have a hell of a time."

The woman on the floor cursed. Jimmie was intent upon his bruised fore-arms. The girl cast a glance about the room filled with a chaotic mass of debris, and at the red, writhing body of her mother.

"Go teh hell an' good riddance."

She went.

CHAPTER X

Jimmie had an idea it wasn't common courtesy for a friend to come to one's home and ruin one's sister. But he was not sure how much Pete knew about the rules of politeness.

The following night he returned home from work at rather a late hour in the evening. In passing through the halls he came upon the gnarled and leathery old woman who possessed the music box. She was grinning in the dim light that drifted through dust-stained panes. She beckoned to him with a smudged fore-finger.

"Ah, Jimmie, what do yehs tink I got onto las' night. It was deh funnies' ting I ever saw," she cried, coming close to him and leering. She was trembling with eagerness to tell her tale. "I was by me door las' night when yer sister and her jude feller came in late, oh, very late. An' she, the dear, she was a-cryin' as if her heart would break, she was. It was deh funnies' ting I ever saw. An' right out here by me door she asked him did he love her, did he. An' she was a-cryin' as if her heart would break, poor t'ing. An' him, I could see by deh way what he said it dat she had been askin' orften, he says: 'Oh, hell, yes,' he says, says he, 'Oh, hell, yes.'"

Storm-clouds swept over Jimmie's face, but he turned from the leathery old woman and plodded on up stairs.

"Oh, hell, yes," called she after him. She laughed a laugh that was like a prophetic croak. "'Oh, hell, yes,' he says, says he, 'Oh, hell, yes.'"

There was no one in at home. The rooms showed that attempts had been made at tidying them. Parts of the wreckage of the day before had been repaired by an unskilful hand. A chair or two and the table, stood uncertainly upon legs. The floor had been newly swept. Too, the blue ribbons had been restored to the curtains, and the lambrequin, with its immense sheaves of yellow wheat and red roses of equal size, had been returned, in a worn and sorry state, to its position at the mantel. Maggie's jacket and hat were gone from the nail behind the door.

Jimmie walked to the window and began to look through the blurred glass. It occurred to him to vaguely wonder, for an instant, if some of the women of his acquaintance had brothers.

Suddenly, however, he began to swear.

"But he was me frien'! I brought 'im here! Dat's deh hell of it!"

He fumed about the room, his anger gradually rising to the furious pitch.

"I'll kill deh jay! Dat's what I'll do! I'll kill deh jay!"

He clutched his hat and sprang toward the door. But it opened and his mother's great form blocked the passage.

"What deh hell's deh matter wid yeh?" exclaimed she, coming into the rooms.

Jimmie gave vent to a sardonic curse and then laughed heavily.

"Well, Maggie's gone teh deh devil! Dat's what! See?"

"Eh?" said his mother.

"Maggie's gone teh deh devil! Are yehs deaf?" roared Jimmie, impatiently.

"Deh hell she has," murmured the mother, astounded.

Jimmie grunted, and then began to stare out at the window. His mother sat down in a chair, but a moment later sprang erect and delivered a maddened whirl of oaths. Her son turned to look at her as she reeled and swayed in the middle of the room, her fierce face convulsed with passion, her blotched arms raised high in imprecation.

"May Gawd curse her forever," she shrieked. "May she eat nothin' but stones and deh dirt in deh street. May she sleep in deh gutter an' never see deh sun shine agin. Deh damn — "

"Here, now," said her son. "Take a drop on yourself."

The mother raised lamenting eyes to the ceiling.

"She's deh devil's own chil', Jimmie," she whispered. "Ah, who would tink such a bad girl could grow up in our fambly, Jimmie, me son. Many deh hour I've spent in talk wid dat girl an' tol' her if she ever went on deh streets I'd see her damned. An' after all her bringin' up an' what I tol' her and talked wid her, she goes teh deh bad, like a duck teh water."

The tears rolled down her furrowed face. Her hands trembled.

"An' den when dat Sadie MacMallister next door to us was sent teh deh devil by dat feller what worked in deh soap-factory, didn't I tell our Mag dat if she — "

"Ah, dat's anudder story," interrupted the brother. "Of course, dat Sadie was nice an' all dat — but — see — it ain't dessame as if — well, Maggie was diff'ent — see — she was diff'ent."

He was trying to formulate a theory that he had always unconsciously held, that all sisters, excepting his own, could advisedly be ruined.

He suddenly broke out again. "I'll go t'ump hell outa deh mug what did her deh harm. I'll kill 'im! He tinks he kin scrap, but when he gits me a-chasin' 'im he'll fin' out where he's wrong, deh damned duffer. I'll wipe up deh street wid 'im."

In a fury he plunged out of the doorway. As he vanished the mother raised her head and lifted both hands, entreating.

"May Gawd curse her forever," she cried.

In the darkness of the hallway Jimmie discerned a knot of women talking volubly. When he strode by they paid no attention to him.

"She allus was a bold thing," he heard one of them cry in an eager voice. "Dere wasn't a feller come teh deh house but she'd try teh mash 'im. My Annie says deh shameless t'ing tried teh ketch her feller, her own feller, what we useter know his fader."

"I could a' tol' yehs dis two years ago," said a woman, in a key of triumph. "Yessir, it was over two years ago dat I says teh my ol' man, I says, 'Dat Johnson girl ain't straight,' I says. 'Oh, hell,' he says. 'Oh, hell.' 'Dat's all right,' I says, 'but I know what I knows,' I says, 'an' it' 'ill come out later. You wait an' see,' I says, 'you see.'"

"Anybody what had eyes could see dat dere was somethin' wrong wid dat girl. I didn't like her actions."

On the street Jimmie met a friend. "What deh hell?" asked the latter.

Jimmie explained. "An' I'll 'tump 'im till he can't stand."

"Oh, what deh hell," said the friend. "What's deh use! Yeh'll git pulled in! Everybody 'ill be onto it! An' ten plunks![1] Gee!"

Jimmie was determined. "He t'inks he kin scrap, but he'll fin' out diff'ent."

"Gee," remonstrated the friend. "What deh hell?"

CHAPTER XI

On a corner a glass-fronted building shed a yellow glare upon the pavements. The open mouth of a saloon called seductively to passengers to enter and annihilate sorrow or create rage.

The interior of the place was papered in olive and bronze tints of imitation leather. A shining bar of counterfeit massiveness extended down the side of the room. Behind it a great mahogany-appearing sideboard reached the ceiling. Upon its shelves rested pyramids of shimmering glasses that were never disturbed. Mirrors set in the face of the sideboard multiplied them. Lemons, oranges and paper napkins, arranged with mathematical precision, sat among the glasses. Many-hued decanters of liquor perched at regular intervals on the lower shelves. A nickel-plated cash register occupied a position in the exact center of the general effect. The elementary senses of it all seemed to be opulence and geometrical accuracy.

Across from the bar a smaller counter held a collection of plates upon which swarmed frayed fragments of crackers, slices of boiled

[1] *plunks!:* Dollars.

ham, dishevelled bits of cheese, and pickles swimming in vinegar. An odor of grasping, begrimed hands and munching mouths pervaded.

Pete, in a white jacket, was behind the bar bending expectantly toward a quiet stranger. "A beeh," said the man. Pete drew a foam-topped glassful and set it dripping upon the bar.

At this moment the light bamboo doors at the entrance swung open and crashed against the siding. Jimmie and a companion entered. They swaggered unsteadily but belligerently toward the bar and looked at Pete with bleared and blinking eyes.

"Gin," said Jimmie.

"Gin," said the companion.

Pete slid a bottle and two glasses along the bar. He bended his head sideways as he assiduously polished away with a napkin at the gleaming wood. He had a look of watchfulness upon his features.

Jimmie and his companion kept their eyes upon the bartender and conversed loudly in tones of contempt.

"He's a dindy masher, ain't he, by Gawd?" laughed Jimmie.

"Oh, hell, yes," said the companion, sneering widely. "He's great, he is. Git onto deh mug on deh blokie. Dat's enough to make a feller turn hand-springs in 'is sleep."

The quiet stranger moved himself and his glass a trifle further away and maintained an attitude of oblivion.

"Gee! ain't he hot stuff!"

"Git onto his shape! Great Gawd!"

"Hey," cried Jimmie, in tones of command. Pete came along slowly, with a sullen dropping of the under lip.

"Well," he growled, "what's eatin' yehs?"

"Gin," said Jimmie.

"Gin," said the companion.

As Pete confronted them with the bottle and the glasses, they laughed in his face. Jimmie's companion, evidently overcome with merriment, pointed a grimy forefinger in Pete's direction.

"Say, Jimmie," demanded he, "what deh hell is dat behind deh bar?"

"Damned if I knows," replied Jimmie. They laughed loudly. Pete put down a bottle with a bang and turned a formidable face toward them. He disclosed his teeth and his shoulders heaved restlessly.

"You fellers can't guy me," he said. "Drink yer stuff an' git out an' don' make no trouble."

Instantly the laughter faded from the faces of the two men and expressions of offended dignity immediately came.

"Who deh hell has said anyt'ing teh you," cried they in the same breath.

The quiet stranger looked at the door calculatingly.

"Ah, come off," said Pete to the two men. "Don't pick me up for no jay. Drink yer rum an' git out an' don' make no trouble."

"Oh, deh hell," airily cried Jimmie.

"Oh, deh hell," airily repeated his companion.

"We goes when we git ready! See!" continued Jimmie.

"Well," said Pete in a threatening voice, "don' make no trouble."

Jimmie suddenly leaned forward with his head on one side. He snarled like a wild animal.

"Well, what if we does? See?" said he.

Dark blood flushed into Pete's face, and he shot a lurid glance at Jimmie.

"Well, den we'll see who's deh bes' man, you or me," he said.

The quiet stranger moved modestly toward the door.

Jimmie began to swell with valor.

"Don' pick me up fer no tenderfoot. When yeh tackles me yeh tackles one of deh bes' men in deh city. See? I'm a scrapper, I am. Ain't dat right, Billie?"

"Sure, Mike," responded his companion in tones of conviction.

"Oh, hell," said Pete, easily. "Go fall on yerself."

The two men again began to laugh.

"What deh hell is dat talkin?" cried the companion.

"Damned if I knows," replied Jimmie with exaggerated contempt.

Pete made a furious gesture. "Git outa here now, an' don' make no trouble. See? Youse fellers er lookin' fer a scrap an' it's damn likely yeh'll fin' one if yeh keeps on shootin' off yer mout's. I know yehs! See? I kin lick better men dan yehs ever saw in yer lifes. Dat's right! See? Don' pick me up fer no stuff er yeh might be jolted out in deh street before yeh knows where yeh is. When I comes from behind dis bar, I t'rows yehs boat inteh deh street. See?"

"Oh, hell," cried the two men in chorus.

The glare of a panther came into Pete's eyes. "Dat's what I said! Unnerstan'?"

He came through a passage at the end of the bar and swelled down upon the two men. They stepped promptly forward and crowded close to him.

They bristled like three roosters. They moved their heads pugnaciously and kept their shoulders braced. The nervous muscles about each mouth twitched with a forced smile of mockery.

"Well, what deh hell yer goin' teh do?" gritted Jimmie.

Pete stepped warily back, waving his hands before him to keep the men from coming too near.

"Well, what deh hell yer goin' teh do?" repeated Jimmie's ally. They kept close to him, taunting and leering. They strove to make him attempt the initial blow.

"Keep back, now! Don' crowd me," ominously said Pete.

Again they chorused in contempt. "Oh, hell!"

In a small, tossing group, the three men edged for positions like frigates contemplating battle.

"Well, why deh hell don' yeh try teh t'row us out?" cried Jimmie and his ally with copious sneers.

The bravery of bull-dogs sat upon the faces of the men. Their clenched fists moved like eager weapons.

The allied two jostled the bartender's elbows, glaring at him with feverish eyes and forcing him toward the wall.

Suddenly Pete swore redly. The flash of action gleamed from his eyes. He threw back his arm and aimed a tremendous, lightning-like blow at Jimmie's face. His foot swung a step forward and the weight of his body was behind his fist. Jimmie ducked his head, Bowery-like, with the quickness of a cat. The fierce, answering blows of him and his ally crushed on Pete's bowed head.

The quiet stranger vanished.

The arms of the combatants whirled in the air like flails. The faces of the men, at first flushed to flame-colored anger, now began to fade to the pallor of warriors in the blood and heat of a battle. Their lips curled back and stretched tightly over the gums in ghoul-like grins. Through their white, gripped teeth struggled hoarse whisperings of oaths. Their eyes glittered with murderous fire.

Each head was huddled between its owner's shoulders, and arms were swinging with marvelous rapidity. Feet scraped to and fro with a loud scratching sound upon the sanded floor. Blows left crimson blotches upon pale skin. The curses of the first quarter minute of the fight died away. The breaths of the fighters came wheezingly from their lips and the three chests were straining and heaving. Pete at intervals gave vent to low, labored hisses, that sounded like a desire to kill. Jimmie's ally gibbered at times like a wounded maniac. Jimmie was silent, fighting with the face of a sacrificial priest. The rage of fear shone in all their eyes and their blood-colored fists swirled.

At a tottering moment a blow from Pete's hand struck the ally and he crashed to the floor. He wriggled instantly to his feet and grasping the quiet stranger's beer glass from the bar, hurled it at Pete's head.

High on the wall it burst like a bomb, shivering fragments flying in all directions. Then missiles came to every man's hand. The place had heretofore appeared free of things to throw, but suddenly glass and bottles went singing through the air. They were thrown point blank at bobbing heads. The pyramid of shimmering glasses, that had never been disturbed, changed to cascades as heavy bottles were flung into them. Mirrors splintered to nothing.

The three frothing creatures on the floor buried themselves in a frenzy for blood. There followed in the wake of missiles and fists some unknown prayers, perhaps for death.

The quiet stranger had sprawled very pyrotechnically out on the sidewalk. A laugh ran up and down the avenue for the half of a block.

"Dey've trowed a bloke inteh deh street."

People heard the sound of breaking glass and shuffling feet within the saloon and came running. A small group, bending down to look under the bamboo doors, watching the fall of glass, and three pairs of violent legs, changed in a moment to a crowd.

A policeman came charging down the sidewalk and bounced through the doors into the saloon. The crowd bended and surged in absorbing anxiety to see.

Jimmie caught first sight of the on-coming interruption. On his feet he had the same regard for a policeman that, when on his truck, he had for a fire engine. He howled and ran for the side door.

The officer made a terrific advance, club in hand. One comprehensive sweep of the long night stick threw the ally to the floor and forced Pete to a corner. With his disengaged hand he made a furious effort at Jimmie's coat-tails. Then he regained his balance and paused.

"Well, well, you are a pair of pictures. What in hell yeh been up to?"

Jimmie, with his face drenched in blood, escaped up a side street, pursued a short distance by some of the more law-loving, or excited individuals of the crowd.

Later, from a corner safely dark, he saw the policeman, the ally and the bartender emerge from the saloon. Pete locked the doors and then followed up the avenue in the rear of the crowd-encompassed policeman and his charge.

On first thoughts Jimmie, with his heart throbbing at battle heat, started to go desperately to the rescue of his friend, but he halted.

"Ah, what deh hell?" he demanded of himself.

CHAPTER XII

In a hall of irregular shape sat Pete and Maggie drinking beer. A submissive orchestra dictated to by a spectacled man with frowsy hair and a dress suit, industriously followed the bobs of his head and the waves of his baton. A ballad singer, in a dress of flaming scarlet, sang in the inevitable voice of brass. When she vanished, men seated at the tables near the front applauded loudly, pounding the polished wood with their beer glasses. She returned attired in less gown, and sang again. She received another enthusiastic encore. She reappeared in still less gown and danced. The deafening rumble of glasses and clapping of hands that followed her exit indicated an overwhelming desire to have her come on for the fourth time, but the curiosity of the audience was not gratified.

Maggie was pale. From her eyes had been plucked all look of self-reliance. She leaned with a dependent air toward her companion. She was timid, as if fearing his anger or displeasure. She seemed to beseech tenderness of him.

Pete's air of distinguished valor had grown upon him until it threatened stupendous dimensions. He was infinitely gracious to the girl. It was apparent to her that his condescension was a marvel.

He could appear to strut even while sitting still and he showed that he was a lion of lordly characteristics by the air with which he spat.

With Maggie gazing at him wonderingly, he took pride in commanding the waiters who were, however, indifferent or deaf.

"Hi, you, git a russle on yehs! What deh hell yeh's lookin' at? Two more beehs, d'yeh hear?"

He leaned back and critically regarded the person of a girl with a straw-colored wig who upon the stage was flinging her heels in somewhat awkward imitation of a well-known danseuse.

At times Maggie told Pete long confidential tales of her former home life, dwelling upon the escapades of the other members of the family and the difficulties she had to combat in order to obtain a degree of comfort. He responded in tones of philanthropy. He pressed her arm with an air of reassuring proprietorship.

"Dey was damn jays," he said, denouncing the mother and brother.

The sound of the music which, by the efforts of the frowsy-headed leader, drifted to her ears through the smoke-filled atmosphere, made the girl dream. She thought of her former Rum Alley environment and turned to regard Pete's strong protecting fists. She thought of the collar and cuff manufactory and the eternal moan of the proprietor: "What een hell do you sink I pie fife dolla a week for? Play? No, py damn."

She contemplated Pete's man-subduing eyes and noted that wealth and prosperity was indicated by his clothes. She imagined a future, rose-tinted, because of its distance from all that she previously had experienced.

As to the present she perceived only vague reasons to be miserable. Her life was Pete's and she considered him worthy of the charge. She would be disturbed by no particular apprehensions, so long as Pete adored her as he now said he did. She did not feel like a bad woman. To her knowledge she had never seen any better.

At times men at other tables regarded the girl furtively. Pete, aware of it, nodded at her and grinned. He felt proud.

"Mag, yer a bloomin' good-looker," he remarked, studying her face through the haze. The men made Maggie fear, but she blushed at Pete's words as it became apparent to her that she was the apple of his eye.

Grey-headed men, wonderfully pathetic in their dissipation, stared at her through clouds. Smooth cheeked boys, some of them with faces of stone and mouths of sin, not nearly so pathetic as the grey heads, tried to find the girl's eyes in the smoke wreaths. Maggie considered she was not what they thought her. She confined her glances to Pete and the stage.

The orchestra played negro melodies and a versatile drummer pounded, whacked, clattered and scratched on a dozen machines to make noise.

Those glances of the men, shot at Maggie from under half-closed lids, made her tremble. She thought them all to be worse men than Pete.

"Come, let's go," she said.

As they went out Maggie perceived two women seated at a table with some men. They were painted and their cheeks had lost their roundness. As she passed them the girl, with a shrinking movement, drew back her skirts.

CHAPTER XIII

Jimmie did not return home for a number of days after the fight with Pete in the saloon. When he did, he approached with extreme caution.

He found his mother raving. Maggie had not returned home. The parent continually wondered how her daughter could come to such a pass. She had never considered Maggie as a pearl dropped unstained into Rum Alley from Heaven, but she could not conceive how it was possible for her daughter to fall so low as to bring disgrace upon her family. She was terrific in denunciation of the girl's wickedness.

The fact that the neighbors talked of it, maddened her. When women came in, and in the course of their conversation casually asked, "Where's Maggie dese days?" the mother shook her fuzzy head at them and appalled them with curses. Cunning hints inviting confidence she rebuffed with violence.

"An' wid all deh bringin' up she had, how could she?" moaningly she asked of her son. "Wid all deh talkin' wid her I did an' deh t'ings I tol' her to remember? When a girl is bringed up deh way I bringed up Maggie, how kin she go teh deh devil?"

Jimmie was transfixed by these questions. He could not conceive how under the circumstances his mother's daughter and his sister could have been so wicked.

His mother took a drink from a squdgy bottle that sat on the table. She continued her lament.

"She had a bad heart, dat girl did, Jimmie. She was wicked teh deh heart an' we never knowed it."

Jimmie nodded, admitting the fact.

"We lived in deh same house wid her an' I brought her up an' we never knowed how bad she was."

Jimmie nodded again.

"Wid a home like dis an' a mudder like me, she went teh deh bad," cried the mother, raising her eyes.

One day, Jimmie came home, sat down in a chair and began to wriggle about with a new and strange nervousness. At last he spoke shamefacedly.

"Well, look-a-here, dis t'ing queers us! See? We're queered! An' maybe it 'ud be better if I — well, I t'ink I kin look 'er up an' — maybe it 'ud be better if I fetched her home an — "

The mother started from her chair and broke forth into a storm of passionate anger.

"What! Let 'er come an' sleep under deh same roof wid her mudder agin! Oh, yes, I will, won't I? Sure? Shame on yehs, Jimmie Johnson, fer sayin' such a t'ing teh yer own mudder — teh yer own mudder! Little did I tink when yehs was a babby playin' about me feet dat ye'd grow up teh say sech a t'ing teh yer mudder — yer own mudder. I never taut — "

Sobs choked her and interrupted her reproaches.

"Dere ain't nottin teh raise sech hell about," said Jimmie. "I on'y says it 'ud be better if we keep dis t'ing dark, see? It queers us! See?"

His mother laughed a laugh that seemed to ring through the city and be echoed and re-echoed by countless other laughs. "Oh, yes, I will, won't I! Sure!"

"Well, yeh must take me fer a damn fool," said Jimmie, indignant at his mother for mocking him. "I didn't say we'd make 'er inteh a little tin angel, ner nottin, but deh way it is now she can queer us! Don' che see?"

"Aye, she'll git tired of deh life atter a while an' den she'll wanna be a-comin' home, won' she, deh beast! I'll let 'er in den, won' I?"

"Well, I didn' mean none of dis prod'gal bus'ness anyway," explained Jimmie.

"It wasn't no prod'gal dauter, yeh damn fool," said the mother. "It was prod'gal son, anyhow."

"I know dat," said Jimmie.

For a time they sat in silence. The mother's eyes gloated on a scene her imagination could call before her. Her lips were set in a vindictive smile.

"Aye, she'll cry, won' she, an' carry on, an' tell how Pete, or some odder feller, beats 'er an' she'll say she's sorry an' all dat an' she ain't happy, she ain't, an' she wants to come home agin, she does."

With grim humor, the mother imitated the possible wailing notes of the daughter's voice.

"Den I'll take 'er in, won't I, deh beast. She kin cry 'er two eyes out on deh stones of deh street before I'll dirty deh place wid her. She abused an' ill-treated her own mudder — her own mudder what loved her an' she'll never git anodder chance dis side of hell."

Jimmie thought he had a great idea of women's frailty, but he could not understand why any of his kin should be victims.

"Damn her," he fervidly said.

Again he wondered vaguely if some of the women of his acquaintance had brothers. Nevertheless, his mind did not for an instant confuse himself with those brothers nor his sister with theirs. After the mother had, with great difficulty, suppressed the neighbors, she went among them and proclaimed her grief. "May Gawd forgive dat girl," was her continual cry. To attentive ears she recited the whole length and breadth of her woes.

"I bringed 'er up deh way a dauter oughta be bringed up an' dis is how she served me! She went teh deh devil deh first chance she got! May Gawd forgive her."

When arrested for drunkenness she used the story of her daughter's downfall with telling effect upon the police justices. Finally one of them said to her, peering down over his spectacles: "Mary, the records of this and other courts show that you are the mother of forty-two daughters who have been ruined. The case is unparalleled in the annals of this court, and this court thinks — "

The mother went through life shedding large tears of sorrow. Her red face was a picture of agony.

Of course Jimmie publicly damned his sister that he might appear on a higher social plane. But, arguing with himself, stumbling about in ways that he knew not, he, once, almost came to a conclusion that his sister would have been more firmly good had she better known why. However, he felt that he could not hold such a view. He threw it hastily aside.

CHAPTER XIV

In a hilarious hall there were twenty-eight tables and twenty-eight women[1] and a crowd of smoking men. Valiant noise was made on a stage at the end of the hall by an orchestra composed of men who looked as if they had just happened in. Soiled waiters ran to and fro, swooping down like hawks on the unwary in the throng; clattering along the aisles with trays covered with glasses; stumbling over women's skirts and charging two prices for everything but beer, all with a swiftness that blurred the view of the cocoanut palms and dusty monstrosities painted upon the walls of the room. A bouncer, with an immense load of business upon his hands, plunged about in the crowd, dragging bashful strangers to prominent chairs, ordering waiters here and there and quarreling furiously with men who wanted to sing with the orchestra.

The usual smoke cloud was present, but so dense that heads and arms seemed entangled in it. The rumble of conversation was replaced by a roar. Plenteous oaths heaved through the air. The room rang with the shrill voices of women bubbling o'er with drink-laughter. The chief element in the music of the orchestra was speed. The musicians played

[1] *twenty-eight tables and twenty-eight women:* One woman, employed by the establishment, sits at each table to coax male patrons into sitting with her and buying multiple expensive drinks. See J. W. Buel, *Metropolitan Life Unveiled* (p. 193).

in intent fury. A woman was singing and smiling upon the stage, but no one took notice of her. The rate at which the piano, cornet and violins were going, seemed to impart wildness to the half-drunken crowd. Beer glasses were emptied at a gulp and conversation became a rapid chatter. The smoke eddied and swirled like a shadowy river hurrying toward some unseen falls. Pete and Maggie entered the hall and took chairs at a table near the door. The woman who was seated there made an attempt to occupy Pete's attention and, failing, went away.

Three weeks had passed since the girl had left home. The air of spaniel-like dependence had been magnified and showed its direct effect in the peculiar off-handedness and ease of Pete's ways toward her.

She followed Pete's eyes with hers, anticipating with smiles gracious looks from him.

A woman of brilliance and audacity, accompanied by a mere boy, came into the place and took seats near them.

At once Pete sprang to his feet, his face beaming with glad surprise.

"By Gawd, there's Nellie," he cried.

He went over to the table and held out an eager hand to the woman.

"Why, hello, Pete, me boy, how are you," said she, giving him her fingers.

Maggie took instant note of the woman. She perceived that her black dress fitted her to perfection. Her linen collar and cuffs were spotless. Tan gloves were stretched over her well-shaped hands. A hat of a prevailing fashion perched jauntily upon her dark hair. She wore no jewelry and was painted with no apparent paint. She looked clear-eyed through the stares of the men.

"Sit down, and call your lady-friend over," she said cordially to Pete. At his beckoning Maggie came and sat between Pete and the mere boy.

"I thought yeh were gone away fer good," began Pete, at once. "When did yeh git back? How did dat Buff'lo bus'ness turn out?"

The woman shrugged her shoulders. "Well, he didn't have as many stamps[2] as he tried to make out, so I shook him, that's all."

"Well, I'm glad teh see yehs back in deh city," said Pete, with awkward gallantry.

He and the woman entered into a long conversation, exchanging reminiscences of days together. Maggie sat still, unable to formulate

[2] *stamps:* Money.

an intelligent sentence upon the conversation and painfully aware of it.

She saw Pete's eyes sparkle as he gazed upon the handsome stranger. He listened smilingly to all she said. The woman was familiar with all his affairs, asked him about mutual friends, and knew the amount of his salary.

She paid no attention to Maggie, looking toward her once or twice and apparently seeing the wall beyond.

The mere boy was sulky. In the beginning he had welcomed with acclamations the additions.

"Let's all have a drink! What'll you take, Nell? And you, Miss what's-your-name. Have a drink, Mr. ——— , you, I mean."

He had shown a sprightly desire to do the talking for the company and tell all about his family. In a loud voice he declaimed on various topics. He assumed a patronizing air toward Pete. As Maggie was silent, he paid no attention to her. He made a great show of lavishing wealth upon the woman of brilliance and audacity.

"Do keep still, Freddie! You gibber like an ape, dear," said the woman to him. She turned away and devoted her attention to Pete.

"We'll have many a good time together again, eh?"

"Sure, Mike," said Pete, enthusiastic at once.

"Say," whispered she, leaning forward, "let's go over to Billie's and have a heluva time."

"Well, it's dis way! See?" said Pete. "I got dis lady frien' here."

"Oh, t'hell with her," argued the woman.

Pete appeared disturbed.

"All right," said she, nodding her head at him. "All right for you! We'll see the next time you ask me to go anywheres with you."

Pete squirmed.

"Say," he said, beseechingly, "come wid me a minit an' I'll tell yer why."

The woman waved her hand.

"Oh, that's all right, you needn't explain, you know. You wouldn't come merely because you wouldn't come, that's all there is of it."

To Pete's visible distress she turned to the mere boy, bringing him speedily from a terrific rage. He had been debating whether it would be the part of a man to pick a quarrel with Pete, or would he be justified in striking him savagely with his beer glass without warning. But he recovered himself when the woman turned to renew her smilings. He beamed upon her with an expression that was somewhat tipsy and inexpressibly tender.

"Say, shake that Bowery jay," requested he, in a loud whisper.

"Freddie, you are so droll," she replied.

Pete reached forward and touched the woman on the arm.

"Come out a minit while I tells yeh why I can't go wid yer. Yer doin' me dirt, Nell! I never taut ye'd do me dirt, Nell. Come on, will yer?" He spoke in tones of injury.

"Why, I don't see why I should be interested in your explanations," said the woman, with a coldness that seemed to reduce Pete to a pulp.

His eyes pleaded with her. "Come out a minit while I tells yeh."

The woman nodded slightly at Maggie and the mere boy, "Scuse me."

The mere boy interrupted his loving smile and turned a shriveling glare upon Pete. His boyish countenance flushed and he spoke, in a whine, to the woman:

"Oh, I say, Nellie, this ain't a square deal, you know. You aren't goin' to leave me and go off with that duffer, are you? I should think — "

"Why, you dear boy, of course I'm not," cried the woman, affectionately. She bended over and whispered in his ear. He smiled again and settled in his chair as if resolved to wait patiently.

As the woman walked down between the rows of tables, Pete was at her shoulder talking earnestly, apparently in explanation. The woman waved her hands with studied airs of indifference. The doors swung behind them, leaving Maggie and the mere boy seated at the table.

Maggie was dazed. She could dimly perceive that something stupendous had happened. She wondered why Pete saw fit to remonstrate with the woman, pleading for forgiveness with his eyes. She thought she noted an air of submission about her leonine Pete. She was astounded.

The mere boy occupied himself with cock-tails and a cigar. He was tranquilly silent for half an hour. Then he bestirred himself and spoke.

"Well," he said, sighing, "I knew this was the way it would be." There was another stillness. The mere boy seemed to be musing.

"She was pulling m'leg. That's the whole amount of it," he said, suddenly. "It's a bloomin' shame the way that girl does. Why, I've spent over two dollars in drinks to-night. And she goes off with that plug-ugly who looks as if he had been hit in the face with a coin-die. I call it rocky treatment for a fellah like me. Here, waiter, bring me a cock-tail and make it damned strong."

Maggie made no reply. She was watching the doors. "It's a mean piece of business," complained the mere boy. He explained to her how amazing it was that anybody should treat him in such a manner. "But

I'll get square with her, you bet. She won't get far ahead of yours truly, you know," he added, winking. "I'll tell her plainly that it was bloomin' mean business. And she won't come it over me with any of her 'now-Freddie-dears.' She thinks my name is Freddie, you know, but of course it ain't. I always tell these people some name like that, because if they got onto your right name they might use it sometime. Understand? Oh, they don't fool me much."

Maggie was paying no attention, being intent upon the doors. The mere boy relapsed into a period of gloom, during which he exterminated a number of cock-tails with a determined air, as if replying defiantly to fate. He occasionally broke forth into sentences composed of invectives joined together in a long string.

The girl was still staring at the doors. After a time the mere boy began to see cobwebs just in front of his nose. He spurred himself into being agreeable and insisted upon her having a charlotte-russe and a glass of beer.

"They's gone," he remarked, "they's gone." He looked at her through the smoke wreaths. "Shay, lil' girl, we mightish well make bes' of it. You ain't such bad lookin' girl, y'know. Not half bad. Can't come up to Nell, though. No, can't do it! Well, I should shay not! Nell fine-lookin' girl! F — i — n — ine. You look damn bad longsider her, but by y'self ain't so bad. Have to do anyhow. Nell gone. O'ny you left. Not half bad, though."

Maggie stood up.

"I'm going home," she said.

The mere boy started.

"Eh? What? Home," he cried, struck with amazement. "I beg pardon, did hear say home?"

"I'm going home," she repeated.

"Great Gawd, what hava struck," demanded the mere boy of himself, stupefied.

In a semi-comatose state he conducted her on board an up-town car, ostentatiously paid her fare, leered kindly at her through the rear window and fell off the steps.

CHAPTER XV

A forlorn woman went along a lighted avenue. The street was filled with people desperately bound on missions. An endless crowd darted at the elevated station stairs and the horse cars were thronged with owners of bundles.

The pace of the forlorn woman was slow. She was apparently searching for some one. She loitered near the doors of saloons and watched men emerge from them. She scanned furtively the faces in the rushing stream of pedestrians. Hurrying men, bent on catching some boat or train, jostled her elbows, failing to notice her, their thoughts fixed on distant dinners.

The forlorn woman had a peculiar face. Her smile was no smile. But when in repose her features had a shadowy look that was like a sardonic grin, as if some one had sketched with cruel fore-finger indelible lines about her mouth.

Jimmie came strolling up the avenue. The woman encountered him with an aggrieved air.

"Oh, Jimmie, I've been lookin' all over fer yehs — ," she began.

Jimmie made an impatient gesture and quickened his pace.

"Ah, don't bodder me! Good Gawd!" he said, with the savageness of a man whose life is pestered.

The woman followed him along the sidewalk in somewhat the manner of a suppliant.

"But, Jimmie," she said, "yehs told me ye'd — "

Jimmie turned upon her fiercely as if resolved to make a last stand for comfort and peace.

"Say, fer Gawd's sake, Hattie, don' foller me from one end of deh city teh deh odder. Let up, will yehs! Give me a minute's res', can't yehs? Yehs makes me tired, allus taggin' me. See? Ain' yehs got no sense? Do yehs want people teh get onto me? Go chase yourself, fer Gawd's sake."

The woman stepped closer and laid her fingers on his arm. "But, look-a here — "

Jimmie snarled. "Oh, go teh hell."

He darted into the front door of a convenient saloon and a moment later came out into the shadows that surrounded the side door. On the brilliantly lighted avenue he perceived the forlorn woman dodging about like a scout. Jimmie laughed with an air of relief and went away.

When he arrived home he found his mother clamoring. Maggie had returned. She stood shivering beneath the torrent of her mother's wrath.

"Well, I'm damned," said Jimmie in greeting.

His mother, tottering about the room, pointed a quivering forefinger.

"Lookut her, Jimmie, lookut her. Dere's yer sister, boy. Dere's yer sister. Lookut her! Lookut her!"

She screamed in scoffing laughter.

The girl stood in the middle of the room. She edged about as if unable to find a place on the floor to put her feet.

"Ha, ha, ha," bellowed the mother. "Dere she stands! Ain' she purty? Lookut her! Ain' she sweet, deh beast? Lookut her! Ha, ha, lookut her!"

She lurched forward and put her red and seamed hands upon her daughter's face. She bent down and peered keenly up into the eyes of the girl.

"Oh, she's jes' dessame as she ever was, ain' she? She's her mudder's purty darlin' yit, ain' she? Lookut her, Jimmie! Come here, fer Gawd's sake, and lookut her."

The loud, tremendous sneering of the mother brought the denizens of the Rum Alley tenement to their doors. Women came in the hallways. Children scurried to and fro.

"What's up? Dat Johnson party on anudder tear?"

"Naw! Young Mag's come home!"

"Deh hell yeh say?"

Through the open doors curious eyes stared in at Maggie. Children ventured into the room and ogled her, as if they formed the front row at a theatre. Women, without, bended toward each other and whispered, nodding their heads with airs of profound philosophy. A baby, overcome with curiosity concerning this object at which all were looking, sidled forward and touched her dress, cautiously, as if investigating a red-hot stove. Its mother's voice rang out like a warning trumpet. She rushed forward and grabbed her child, casting a terrible look of indignation at the girl.

Maggie's mother paced to and fro, addressing the doorful of eyes, expounding like a glib showman at a museum. Her voice rang through the building.

"Dere she stands," she cried, wheeling suddenly and pointing with dramatic finger. "Dere she stands! Lookut her! Ain' she a dindy? An' she was so good as to come home teh her mudder, she was! Ain' she a beaut'? Ain' she a dindy? Fer Gawd's sake!"

The jeering cries ended in another burst of shrill laughter.

The girl seemed to awaken. "Jimmie — "

He drew hastily back from her.

"Well, now, yer a hell of a t'ing, ain' yeh?" he said, his lips curling in scorn. Radiant virtue sat upon his brow and his repelling hands expressed horror of contamination.

Maggie turned and went.

The crowd at the door fell back precipitately. A baby falling down in front of the door, wrenched a scream like a wounded animal from its mother. Another woman sprang forward and picked it up, with a chivalrous air, as if rescuing a human being from an oncoming express train.

As the girl passed down through the hall, she went before open doors framing more eyes strangely microscopic, and sending broad beams of inquisitive light into the darkness of her path. On the second floor she met the gnarled old woman who possessed the music box.

"So," she cried, "'ere yehs are back again, are yehs? An' dey've kicked yehs out? Well, come in an' stay wid me teh-night. I ain' got no moral standin'."

From above came an unceasing babble of tongues, over all of which rang the mother's derisive laughter.

CHAPTER XVI

Pete did not consider that he had ruined Maggie. If he had thought that her soul could never smile again, he would have believed the mother and brother, who were pyrotechnic over the affair, to be responsible for it.

Besides, in his world, souls did not insist upon being able to smile. "What deh hell?"

He felt a trifle entangled. It distressed him. Revelations and scenes might bring upon him the wrath of the owner of the saloon, who insisted upon respectability of an advanced type.

"What deh hell do dey wanna' raise such a smoke about it fer?" demanded he of himself, disgusted with the attitude of the family. He saw no necessity for anyone's losing their equilibrium merely because their sister or their daughter had stayed away from home.

Searching about in his mind for possible reasons for their conduct, he came upon the conclusion that Maggie's motives were correct, but that the two others wished to snare him. He felt pursued.

The woman of brilliance and audacity whom he had met in the hilarious hall showed a disposition to ridicule him.

"A little pale thing with no spirit," she said. "Did you note the expression of her eyes? There was something in them about pumpkin pie and virtue. That is a peculiar way the left corner of her mouth has of twitching, isn't it? Dear, dear, my cloud-compelling Pete, what are you coming to?"

Pete asserted at once that he never was very much interested in the girl. The woman interrupted him, laughing.

"Oh, it's not of the slightest consequence to me, my dear young man. You needn't draw maps for my benefit. Why should I be concerned about it?"

But Pete continued with his explanations. If he was laughed at for his tastes in women, he felt obliged to say that they were only temporary or indifferent ones.

The morning after Maggie had departed from home, Pete stood behind the bar. He was immaculate in white jacket and apron and his hair was plastered over his brow with infinite correctness. No customers were in the place. Pete was twisting his napkined fist slowly in a beer glass, softly whistling to himself and occasionally holding the object of his attention between his eyes and a few weak beams of sunlight that had found their way over the thick screens and into the shaded room.

With lingering thoughts of the woman of brilliance and audacity, the bartender raised his head and stared through the varying cracks between the swaying bamboo doors. Suddenly the whistling pucker faded from his lips. He saw Maggie walking slowly past. He gave a great start, fearing for the previously-mentioned eminent respectability of the place.

He threw a swift, nervous glance about him, all at once feeling guilty. No one was in the room.

He went hastily over to the side door. Opening it and looking out, he perceived Maggie standing, as if undecided, on the corner. She was searching the place with her eyes.

As she turned her face toward him Pete beckoned to her hurriedly, intent upon returning with speed to a position behind the bar and to the atmosphere of respectability upon which the proprietor insisted.

Maggie came to him, the anxious look disappearing from her face and a smile wreathing her lips.

"Oh, Pete — ," she began brightly.

The bartender made a violent gesture of impatience.

"Oh, my Gawd," cried he, vehemently. "What deh hell do yeh wanna hang aroun' here fer? Do yeh wanna git me inteh trouble?" he demanded with an air of injury.

Astonishment swept over the girl's features. "Why, Pete! yehs tol' me — "

Pete glanced profound irritation. His countenance reddened with the anger of a man whose respectability is being threatened.

"Say, yehs makes me tired. See? What deh hell deh yeh wanna tag aroun' atter me fer? Yeh'll git me inteh trouble wid deh ol' man an' dey'll be hell teh pay! If he sees a woman roun' here he'll go crazy an' I'll lose me job! See? Ain' yehs got no sense? Don' be allus bodderin' me. See? Yer brudder come in here an' raised hell an' deh ol' man hada put up fer it! An' now I'm done! See? I'm done."

The girl's eyes stared into his face. "Pete, don' yeh remem — "

"Oh, hell," interrupted Pete, anticipating.

The girl seemed to have a struggle with herself. She was apparently bewildered and could not find speech. Finally she asked in a low voice: "But where kin I go?"

The question exasperated Pete beyond the powers of endurance. It was a direct attempt to give him some responsibility in a matter that did not concern him. In his indignation he volunteered information.

"Oh, go teh hell," cried he. He slammed the door furiously and returned, with an air of relief, to his respectability.

Maggie went away.

She wandered aimlessly for several blocks. She stopped once and asked aloud a question of herself: "Who?"

A man who was passing near her shoulder, humorously took the questioning word as intended for him.

"Eh? What? Who? Nobody! I didn't say anything," he laughingly said, and continued his way.

Soon the girl discovered that if she walked with such apparent aimlessness, some men looked at her with calculating eyes. She quickened her step, frightened. As a protection, she adopted a demeanor of intentness as if going somewhere.

After a time she left rattling avenues and passed between rows of houses with sternness and stolidity stamped upon their features. She hung her head for she felt their eyes grimly upon her.

Suddenly she came upon a stout gentleman in a silk hat and a chaste black coat, whose decorous row of buttons reached from his chin to his knees.[1] The girl had heard of the Grace of God and she decided to approach this man.

His beaming, chubby face was a picture of benevolence and kindheartedness. His eyes shone good-will.

[1] *stout gentleman . . . knees:* A clergyman. Here, Crane echoes Edgar Fawcett's description of the indifferent clergyman in "The Woes of the New York Working-Girl" (pp. 241–42).

But as the girl timidly accosted him, he gave a convulsive movement and saved his respectability by a vigorous side-step. He did not risk it to save a soul. For how was he to know that there was a soul before him that needed saving?

CHAPTER XVII

Upon a wet evening, several months after the last chapter, two interminable rows of cars, pulled by slipping horses, jangled along a prominent side-street. A dozen cabs, with coat-enshrouded drivers, clattered to and fro. Electric lights, whirring softly, shed a blurred radiance. A flower dealer, his feet tapping impatiently, his nose and his wares glistening with rain-drops, stood behind an array of roses and chrysanthemums. Two or three theatres emptied a crowd upon the storm-swept pavements. Men pulled their hats over their eye-brows and raised their collars to their ears. Women shrugged impatient shoulders in their warm cloaks and stopped to arrange their skirts for a walk through the storm. People having been comparatively silent for two hours burst into a roar of conversation, their hearts still kindling from the glowings of the stage.

The pavements became tossing seas of umbrellas. Men stepped forth to hail cabs or cars, raising their fingers in varied forms of polite request or imperative demand. An endless procession wended toward elevated stations. An atmosphere of pleasure and prosperity seemed to hang over the throng, born, perhaps, of good clothes and of having just emerged from a place of forgetfulness.

In the mingled light and gloom of an adjacent park, a handful of wet wanderers, in attitudes of chronic dejection, was scattered among the benches.

A girl of the painted cohorts of the city went along the street. She threw changing glances at men who passed her, giving smiling invitations to men of rural or untaught pattern and usually seeming sedately unconscious of the men with a metropolitan seal upon their faces.

Crossing glittering avenues, she went into the throng emerging from the places of forgetfulness. She hurried forward through the crowd as if intent upon reaching a distant home, bending forward in her handsome cloak, daintily lifting her skirts and picking for her well-shod feet the dryer spots upon the pavements.

The restless doors of saloons, clashing to and fro, disclosed animated rows of men before bars and hurrying barkeepers.

A concert hall gave to the street faint sounds of swift, machine-like music, as if a group of phantom musicians were hastening.

A tall young man, smoking a cigarette with a sublime air, strolled near the girl. He had on evening dress, a moustache, a chrysanthemum, and a look of ennui, all of which he kept carefully under his eye. Seeing the girl walk on as if such a young man as he was not in existence, he looked back transfixed with interest. He stared glassily for a moment, but gave a slight convulsive start when he discerned that she was neither new, Parisian, nor theatrical. He wheeled about hastily and turned his stare into the air, like a sailor with a search-light.

A stout gentleman, with pompous and philanthropic whiskers, went stolidly by, the broad of his back sneering at the girl.

A belated man in business clothes, and in haste to catch a car, bounced against her shoulder. "Hi, there, Mary, I beg your pardon! Brace up, old girl." He grasped her arm to steady her, and then was away running down the middle of the street.

The girl walked on out of the realm of restaurants and saloons. She passed more glittering avenues and went into darker blocks than those where the crowd travelled.

A young man in light overcoat and derby hat received a glance shot keenly from the eyes of the girl. He stopped and looked at her, thrusting his hands in his pockets and making a mocking smile curl his lips. "Come, now, old lady," he said, "you don't mean to tell me that you sized me up for a farmer?"

A laboring man marched along with bundles under his arms. To her remarks, he replied: "It's a fine evenin', ain't it?"

She smiled squarely into the face of a boy who was hurrying by with his hand buried in his overcoat, his blonde locks bobbing on his youthful temples, and a cheery smile of unconcern upon his lips. He turned his head and smiled back at her, waving his hands.

"Not this eve — some other eve!"

A drunken man, reeling in her pathway, began to roar at her. "I ain' ga no money, dammit," he shouted, in a dismal voice. He lurched on up the street, wailing to himself, "Dammit, I ain' ga no money. Damn ba' luck. Ain' ga no more money."

The girl went into gloomy districts near the river, where the tall black factories shut in the street and only occasional broad beams of light fell across the pavements from saloons. In front of one of these places, from whence came the sound of a violin vigorously scraped, the patter of feet on boards and the ring of loud laughter, there stood a man with blotched features.

"Ah, there," said the girl.

"I've got a date," said the man.

Further on in the darkness she met a ragged being with shifting, blood-shot eyes and grimey hands. "Ah, what deh hell? Tink I'm a millionaire?"

She went into the blackness of the final block. The shutters of the tall buildings were closed like grim lips. The structures seemed to have eyes that looked over her, beyond her, at other things. Afar off the lights of the avenues glittered as if from an impossible distance. Street car bells jingled with a sound of merriment.

When almost to the river the girl saw a great figure. On going forward she perceived it to be a huge fat man in torn and greasy garments. His grey hair straggled down over his forehead. His small, bleared eyes, sparkling from amidst great rolls of red fat, swept eagerly over the girl's upturned face. He laughed, his brown, disordered teeth gleaming under a grey, grizzled moustache from which beer-drops dripped. His whole body gently quivered and shook like that of a dead jelly fish. Chuckling and leering, he followed the girl of the crimson legions.[1]

At their feet the river appeared a deathly black hue. Some hidden factory sent up a yellow glare, that lit for a moment the waters lapping oilily against timbers. The varied sounds of life, made joyous by distance and seeming unapproachableness, came faintly and died away to a silence.

CHAPTER XVIII

In a partitioned-off section of a saloon sat a man with a half dozen women, gleefully laughing, hovering about him. The man had arrived at that stage of drunkenness where affection is felt for the universe.

"I'm good f'ler, girls," he said, convincingly. "I'm damn good f'ler. An'body treats me right, I allus trea's zem right! See?"

The women nodded their heads approvingly. "To be sure," they cried in hearty chorus. "You're the kind of a man we like, Pete. You're outa sight! What yeh goin' to buy this time, dear?"

"An'thin' yehs wants, damn it," said the man in an abandonment of good will. His countenance shone with the true spirit of benevolence.

[1] This paragraph was deleted in the 1896 edition of *Maggie*.

He was in the proper mode of missionaries. He would have fraternized with obscure Hottentots.[1] And above all, he was overwhelmed in tenderness for his friends, who were all illustrious.

"An'thing yehs wants, damn it," repeated he, waving his hands with beneficent recklessness. "I'm good f'ler, girls, an' if an'body treats me right I — here," called he through an open door to a waiter, "bring girls drinks, damn it. What 'ill yehs have, girls? An'thing yehs want, damn it!"

The waiter glanced in with the disgusted look of the man who serves intoxicants for the man who takes too much of them. He nodded his head shortly at the order from each individual, and went.

"Damn it," said the man, "we're havin' heluva time. I like you girls! Damn'd if I don't! Yer right sort! See?"

He spoke at length and with feeling, concerning the excellencies of his assembled friends.

"Don' try pull man's leg, but have a heluva time! Das right! Das way teh do! Now, if I sawght yehs tryin' work me fer drinks, wouldn' buy damn t'ing! But yer right sort, damn it! Yehs know how ter treat a f'ler, an' I stays by yehs 'til spen' las' cent! Das right! I'm good f'ler an' I knows when an'body treats me right!"

Between the times of the arrival and departure of the waiter, the man discoursed to the women on the tender regard he felt for all living things. He laid stress upon the purity of his motives in all dealings with men in the world and spoke of the fervor of his friendship for those who were amiable. Tears welled slowly from his eyes. His voice quavered when he spoke to them.

Once when the waiter was about to depart with an empty tray, the man drew a coin from his pocket and held it forth.

"Here," said he, quite magnificently, "here's quar'."

The waiter kept his hands on his tray.

"I don' want yer money," he said.

The other put forth the coin with tearful insistence.

"Here, damn it," cried he, "tak't! Yer damn goo' f'ler an' I wan' yehs tak't!"

"Come, come, now," said the waiter, with the sullen air of a man who is forced into giving advice. "Put yer mon in yer pocket! Yer loaded an' yehs on'y makes a damn fool of yerself."

[1] *Hottentots:* Specifically, the term refers to an African tribe near the Cape of Good Hope, yet it generally was used, as it is here, to describe any savages.

As the latter passed out of the door the man turned pathetically to the women.

"He don' know I'm damn goo' f'ler," cried he, dismally.

"Never you mind, Pete, dear," said a woman of brilliance and audacity, laying her hand with great affection upon his arm. "Never you mind, old boy! We'll stay by you, dear!"

"Das ri'," cried the man, his face lighting up at the soothing tones of the woman's voice. "Das ri', I'm damn goo' f'ler an' w'en anyone trea's me ri', I treats zem ri'! Shee!"

"Sure!" cried the women. "And we're not goin' back on you, old man."

The man turned appealing eyes to the woman of brilliance and audacity. He felt that if he could be convicted of a contemptible action he would die.

"Shay, Nell, damn it, I allus trea's yehs shquare, didn' I? I allus been goo' f'ler wi' yehs, ain't I, Nell?"

"Sure you have, Pete," assented the woman. She delivered an oration to her companions. "Yessir, that's a fact. Pete's a square fellah, he is. He never goes back on a friend. He's the right kind an' we stay by him, don't we, girls?"

"Sure," they exclaimed. Looking lovingly at him they raised their glasses and drank his health.

"Girlsh," said the man, beseechingly, "I allus trea's yehs ri', didn' I? I'm goo' f'ler, ain' I, girlsh?"

"Sure," again they chorused.

"Well," said he finally, "le's have nozzer drink, zen."

"That's right," hailed a woman, "that's right. Yer no bloomin' jay! Yer spends yer money like a man. Dat's right."

The man pounded the table with his quivering fists.

"Yessir," he cried, with deep earnestness, as if someone disputed him. "I'm damn goo' f'ler, an' w'en anyone trea's me ri', I allus trea's — le's have nozzer drink."

He began to beat the wood with his glass.

"Shay," howled he, growing suddenly impatient. As the waiter did not then come, the man swelled with wrath.

"Shay," howled he again.

The waiter appeared at the door.

"Bringsh drinksh," said the man.

The waiter disappeared with the orders.

"Zat f'ler dam fool," cried the man. "He insul' me! I'm ge'man! Can' stan' be insul'! I'm goin' lickim when comes!"

"No, no," cried the women, crowding about and trying to subdue him. "He's all right! He didn't mean anything! Let it go! He's a good fellah!"

"Din' he insul' me?" asked the man earnestly.

"No," said they. "Of course he didn't! He's all right!"

"Sure he didn' insul' me?" demanded the man, with deep anxiety in his voice.

"No, no! We know him! He's a good fellah. He didn't mean anything."

"Well, zen," said the man, resolutely, "I'm go' 'pol'gize!"

When the waiter came, the man struggled to the middle of the floor.

"Girlsh shed you insul' me! I shay damn lie! I 'pol'gize!"

"All right," said the waiter.

The man sat down. He felt a sleepy but strong desire to straighten things out and have a perfect understanding with everybody.

"Nell, I allus trea's yeh shquare, din' I? Yeh likes me, don' yehs, Nell? I'm goo' f'ler?"

"Sure," said the woman of brilliance and audacity.

"Yeh knows I'm stuck on yehs, don' yehs, Nell?"

"Sure," she repeated, carelessly.

Overwhelmed by a spasm of drunken adoration, he drew two or three bills from his pocket, and, with the trembling fingers of an offering priest, laid them on the table before the woman.

"Yehs knows, damn it, yehs kin have all got, 'cause I'm stuck on yehs, Nell, damn't, I — I'm stuck on yehs, Nell — buy drinksh — damn't — we're havin' heluva time — w'en anyone trea's me ri' — I — damn't, Nell — we're havin' heluva — time."

Shortly he went to sleep with his swollen face fallen forward on his chest.

The women drank and laughed, not heeding the slumbering man in the corner. Finally he lurched forward and fell groaning to the floor.

The women screamed in disgust and drew back their skirts.

"Come ahn," cried one, starting up angrily, "let's get out of here."

The woman of brilliance and audacity stayed behind, taking up the bills and stuffing them into a deep, irregularly-shaped pocket. A guttural snore from the recumbent man caused her to turn and look down at him.

She laughed. "What a damn fool," she said, and went.

The smoke from the lamps settled heavily down in the little compartment, obscuring the way out. The smell of oil, stifling in its intensity, pervaded the air. The wine from an overturned glass dripped softly down upon the blotches on the man's neck.

CHAPTER XIX

In a room a woman sat at a table eating like a fat monk in a picture.

A soiled, unshaven man pushed open the door and entered.

"Well," said he, "Mag's dead."

"What?" said the woman, her mouth filled with bread.

"Mag's dead," repeated the man.

"Deh hell she is," said the woman. She continued her meal. When she finished her coffee she began to weep.

"I kin remember when her two feet was no bigger dan yer tumb, and she weared worsted boots," moaned she.

"Well, whata dat?" said the man.

"I kin remember when she weared worsted boots," she cried.

The neighbors began to gather in the hall, staring in at the weeping woman as if watching the contortions of a dying dog. A dozen women entered and lamented with her. Under their busy hands the rooms took on that appalling appearance of neatness and order with which death is greeted.

Suddenly the door opened and a woman in a black gown rushed in with outstretched arms. "Ah, poor Mary," she cried, and tenderly embraced the moaning one.

"Ah, what ter'ble affliction is dis," continued she. Her vocabulary was derived from mission churches. "Me poor Mary, how I feel fer yehs! Ah, what a ter'ble affliction is a disobed'ent chile."

Her good, motherly face was wet with tears. She trembled in eagerness to express her sympathy. The mourner sat with bowed head, rocking her body heavily to and fro, and crying out in a high, strained voice that sounded like a dirge on some forlorn pipe.

"I kin remember when she weared worsted boots an' her two feets was no bigger dan yer tumb an' she weared worsted boots, Miss Smith," she cried, raising her streaming eyes.

"Ah, me poor Mary," sobbed the woman in black. With low, coddling cries, she sank on her knees by the mourner's chair, and put her arms about her. The other women began to groan in different keys.

"Yer poor misguided chil' is gone now, Mary, an' let us hope it's fer deh bes'. Yeh'll fergive her now, Mary, won't yehs, dear, all her disobed'ence? All her tankless behavior to her mudder an' all her badness? She's gone where her ter'ble sins will be judged."

The woman in black raised her face and paused. The inevitable sunlight came streaming in at the windows and shed a ghastly cheerfulness upon the faded hues of the room. Two or three of the spectators were sniffling, and one was loudly weeping. The mourner arose and staggered into the other room. In a moment she emerged with a pair of faded baby shoes held in the hollow of her hand.

"I kin remember when she used to wear dem," cried she. The women burst anew into cries as if they had all been stabbed. The mourner turned to the soiled and unshaven man.

"Jimmie, boy, go git yer sister! Go git yer sister an' we'll put deh boots on her feets!"

"Dey won't fit her now, yeh damn fool," said the man.

"Go git yer sister, Jimmie," shrieked the woman, confronting him fiercely.

The man swore sullenly. He went over to a corner and slowly began to put on his coat. He took his hat and went out, with a dragging, reluctant step.

The woman in black came forward and again besought the mourner.

"Yeh'll fergive her, Mary! Yeh'll fergive yer bad, bad chil'! Her life was a curse an' her days were black an' yeh'll fergive yer bad girl? She's gone where her sins will be judged."

"She's gone where her sins will be judged," cried the other women, like a choir at a funeral.

"Deh Lord gives and deh Lord takes away," said the woman in black, raising her eyes to the sunbeams.

"Deh Lord gives and deh Lord takes away," responded the others.

"Yeh'll fergive her, Mary!" pleaded the woman in black. The mourner essayed to speak but her voice gave way. She shook her great shoulders frantically, in an agony of grief. Hot tears seemed to scald her quivering face. Finally her voice came and arose like a scream of pain.

"Oh, yes, I'll fergive her! I'll fergive her!"

Part Two

Maggie: A Girl of the Streets (A Story of New York) Cultural Contexts

Childe Hassam, *Fifth Avenue in Winter,* ca. 1890. Courtesy of the Carnegie Museum of Art, Pittsburgh.

1

In Darkest New York

The stone fight that begins *Maggie* occurs between young Jimmie's gang and some Irish children or, to use Jimmie's word, "micks." As an adult, Jimmie continues to get into fights and is arrested for assaulting a "Chinaman." Pete also has a reputation for his fighting ability, which he exploits to impress Maggie. Describing one fight, Pete uses a derogatory term for an Italian and says his opponent "scrapped like a damn dago" (52). Though New York's ethnic population is not a prominent theme in *Maggie,* it is an omnipresent one. These three derisive references cast the immigrants in an adversarial role. Jimmie and Pete represent the Native-born, Anglo-Saxon population, who must fight their various battles in an effort to assert their superiority over others. These instances of physical violence emphasize the ethnic tensions that significantly contributed to the urban tenement problem during the late nineteenth century and that set the tone for Stephen Crane's novel. The violence in *Maggie* — both the street fights and the domestic violence within the "gruesome doorways" — does more than simply illustrate the tension-filled urban life, for it suggests that tenement dwellers are in the process of degenerating to a more primitive state of existence. The attitude underlying Crane's savage imagery was not unusual for the times. A good portion of the middle-class American reading public held similar beliefs toward the urban poor. No words better reflect their attitude toward New York's tenement district in the 1890s than the phrase, "in darkest New York."

After the explorer Henry M. Stanley described his latest adventures within the African interior in his book, *In Darkest Africa* (1890), the phrase became proverbial, and was applied to anything foreign, different, unfamiliar, or mysterious. From the viewpoint of New York's middle class, the poor tenement dwellers — Russian, Eastern European, Italian, Irish, and German — lived "in darkest New York." Not only does the phrase embody the alien quality of the tenement population, it also suggests that tenement dwellers lived in a semibarbarous state, a step below the rest of the city's population on the evolutionary scale. Describing summertime in the tenements, Harry P. Mawson applied Stanley's racist characterizations of Africans as barbarous savages to New York's urban poor and suggested that missionary efforts would be better directed toward the near-savages of New York (see p. 105). Mawson's argument embodies the idea of paternalism, which casts civilized society into a fatherly role and the less civilized into the role of children who need a parent's protection, an idea that underlies several of the essays in this section.

Though intended to alleviate urban overcrowding, the tenement buildings became the place where overcrowding occurred, for they could not keep pace with the influx of people, both immigrants from overseas and migrants from rural America. The tenements were targeted by legislators and reformers. Alice Wellington Rollins, one of the most vocal reformers, compared the late-nineteenth-century tenement problem to slavery, for both oppressed underprivileged segments of the American population, and both gave middle-class, reform-minded people a cause to rally around. Many efforts to improve the tenements did little to alleviate their inherent problems, however. Even the most well-intended reformers could not help but be a little repulsed by their foreign quality. In "Tenement-House Morality," a fairly sensitive treatment of the problem of urban overcrowding, the Reverend James O. S. Huntington likened the tenement to "an Eastern caravansary" and "the steerage of an emigrant vessel" (see p. 122).

The following essays are divided into two groups. The first contains essays describing life in the tenements while the second group includes articles describing the amusements available to tenement dwellers, places where people could go to escape the squalor of their everyday lives. The essays in the first group are by reform-minded authors. Several of the authors in the second group are more cosmopolitan observers who describe the urban landscape with a chuckle and a grin — but not all — George Frederic Parsons, for example, found nothing to laugh at when it came to saloons. Overall, the following essays provide a sample of the kinds of articles that appeared in the middle-class newspapers and magazines of Crane's day and thus indicate the bourgeois fashion for reading about how the other half lived.

Jacob Riis deserves credit for popularizing the phrase "the other half." After his *How the Other Half Lives* appeared in 1890, Riis continued to study the poor and to publish his findings. His next book, *The Children of the Poor,* though less original than the earlier work, is, perhaps, more touching for its concentration on children. Sentimentality remained a dominant mode in the day's popular fiction; for nonfiction, it provided a useful rhetorical strategy. *The Children of the Poor* reveals that Jimmie's violent behavior in *Maggie* was neither unique nor unusual. Not all of the writers painted bleak pictures, however. Among the reformers represented below, William T. Elsing is the most objective; his essay is neither denunciatory nor polemical. Though he notes problems with the tenements, he emphasizes the fact that they did contain many wholesome, upright citizens. Elsing refused to accept the theory that the environment of the tenement necessarily leads to the moral degradation of its inhabitants.

Although the streets of New York provided the only place children could escape the oppressive tenement interiors, they hardly allowed them to escape darkest New York. After 1878, the year the Third Avenue elevated railroad began operating, the phrase literally was appropriate for some parts of the Lower East Side. This elevated line ran up the Bowery, casting shadows on the street, its shops, and indeed, every street-facing apartment below the third story. Much of the adult tenement population spent their daylight hours at work. The streets they saw after dark differed significantly from their daytime appearance. At night, the Bowery was garishly lighted throughout its length. The practical application of electricity and Charles Brush's invention of the arc light allowed the pawn shops, rooming houses, and store fronts along the Bowery to advertise themselves in large illuminated signs.

When Maggie and Pete begin going out, they visit several local attractions. Many dime museums — exhibits of wonder, unbelievable occurrences, and oddities of nature — were located along the Bowery, as Julian Ralph and David Graham Phillips describe. The Central Park Menagerie or Zoo moved into permanent quarters in 1871 and became the park's most popular attraction, though those living in the nearby fashionable neighborhoods disapproved of the lower classes that the Menagerie attracted. While the Metropolitan Museum of Art or the "Met," as it is known today, may seem too upscale for the likes of Maggie and Pete, after opening its permanent location on Fifth Avenue adjacent to Central Park, the museum organized lectures, school programs, and craft workshops designed to attract the lower classes. In the summertime, outdoor concerts held in local neighborhood parks provided free entertainment for the tenement dwellers, as Crane describes in "Where 'De Gang' Hears the Band Play," a sketch that anticipates *Maggie* (see pp. 165–69).

Pete and Maggie also visit the beer gardens along the Bowery, one of the many different types of drinking establishments available. In *Lights and Shadows of New York Life,* James D. McCabe provides an excellent description of the beer gardens of New York. Inspired by the German originals, these establishments, with their cavernous interiors, were relatively pleasant and innocuous. Husbands could bring wives without concern, and groups of friends could gather for an evening of conviviality. Saloons, on the other hand, were small, inexpensive neighborhood bars; every tenement building had one, and there were numerous others located nearby. Unlike the beer garden, a saloon was not a place a man took his wife. Rather, it was a place where men could escape the squalor of their home lives. As Parsons explains in his essay, the saloon, more than other drinking establishments, represented the biggest threat to home and family life. The local grocery shops also sold beer and liquor, contributing to the burgeoning problem of female alcoholism in the tenements, as the *New-York Tribune*'s investigative report shows. Another type of drinking establishment, known euphemistically as the concert hall, was a place where women were employed for the purpose of coaxing drinks from male patrons. Concert halls often served as fronts for other types of vice, mainly prostitution, which took place in its private rooms or off the premises.

One type of concert hall was the black-and-tan, a place where black men associated with white women and black women with white men. In *Metropolitan Life Unveiled,* J. W. Buel enjoys the sights of the Bowery with detached amusement until he encounters a black-and-tan, which he views with abomination. The existence of black-and-tans confirmed the worst fears of mainstream America. The idea of black America and white America coming together in such intimacy was more than even the most reform-minded people of the day, Jacob Riis included, were willing to accept. In *How the Other Half Lives,* Riis wrote, "The border-land where the white and black races meet in common debauch, the aptly-named black-and-tan saloon, has never been debatable ground from a moral stand-point. It has always been the worst of the desperately bad. Than this commingling of the utterly depraved of both sexes, white and black, on such ground, there can be no greater abomination" (161–62). Truly, many believed, this was darkest America.

Stanley's phrase influenced the literary discourse, as well. At least one critic found it useful for describing *Maggie*. The *Boston Courier* asserted that in *Maggie* "the story of Darkest America has been told in the most realistic way by Stephen Crane. In all the work he has ventured upon, he has rendered the seamy side of modern existence, the real life of the slums, with a force and actuality of description that has not been equalled by any depiction of low life" (quoted in *Log* 188).

Tenement Life

HARRY P. MAWSON

"A Hot Wave among the Poor"

Harry P. Mawson (*b.* 1853) contributed to magazines, wrote campaign literature for William McKinley and a play based on Charles Dickens's *Old Curiosity Shop*, and, most important, corroborated with J. W. Buel on *Leslie's Official History of the Spanish-American War* (Washington, 1899). The following article comes from *Harper's Weekly* 36 (August 20, 1892): 814.

When this new hemisphere was formed, an all-seeing Providence placed a great city, like this New York of ours, close to the great waterway that separates it from the old. Were it not for the ocean breeze which in the hottest weather finds its way at some period in the twenty-four hours among the hovels and tenements of this great city, our poor would die by hundreds in the streets, as we read of in those plague-stricken cities of antiquity. Take a look with me at a few figures. The district on the east side of New York below Fourteenth Street presents now such an anomaly of contrast to what the honest burghers of New Amsterdam expected of their settlement in the New World, that few indeed of us who stop to consider the *raison d'être* of this district would dare to claim that this is yet a "new country," when all the characteristics of the oldest of the Old World have taken root in our midst. In the district referred to over 51 per cent of the dwellings contain 20 persons to a dwelling. In New York there are 312,766 families living in 81,825 dwellings, while in Philadelphia there are 205,135 families living in 187,052 dwellings, an average of only 5.60, while the above figures make New York's average 18.50 persons to a dwelling. And it is this terrible east side that raises New York's percentage so high, and makes this district one of the most densely populated in the world.

Think of this district during a heated term, when the thermometer reaches the 100 line, and the percentage of humidity follows fast in its footsteps. If you wish to realize what New York really is, take a trip through Hester, Oak, Orchard, Rivington, Cherry, etc., on this east

Jacob A. Riis, *An Old Rear-Tenement in Roosevelt Street* (1890). Courtesy of the Museum of the City of New York.

side, or Sullivan and West, etc., on the other side of the town. See the people in the doorways and windows of their wretched homes. What a mockery of the word! See them clustering in a small spot beneath some awning or shade from an overhanging pile of lumber, where the cruel sun does not strike in; at night or after sundown, standing upon the corners looking for the breeze, fairly panting for breath, gasping for life. In the winter-time you see these poor creatures huddled about the itinerant coal-wagon, with buckets, baskets, or the dress held up, to buy their scanty supply of the "black diamonds." July and August see the same vender peddling out small blocks of the "frozen liquid" to the same poor, half-starved wretches. In winter these people gather about the end of the cart, with shawls over their heads; for the buyers are always women, although occasionally children, their faces drawn and pinched with the cold. Now they are bare of head, and often, too, of foot, their faces wan and white, with a look of despair upon them that seems almost ready to break out into a cry for help. It is curious to see with what feverish anxiety they watch the ice-man chop off their "nickel's worth," how it is hurried into basket or bucket or wrapped in the folds of their wretched and, alas! often filthy gown. Once they have secured the ice, they hurry down some alley or cellar or stairway into the seemingly endless blackness of a hallway to some tenement hovel, hugging the crystal to their bosoms like some dear friend. A grade lower yet are those who patronize the free ice-wagon of the Moderation Society, and whose income is so uncertain that not even a nickel can be spared to help keep cool.

At night those who can, take to the roofs and sleep there, in preference to the fetid atmosphere of the tenement closet, without light and without air, by apology called a room. And when it rains, even then they stick it out upon the roof and get soaking wet — anything in preference to the atmosphere in the terrors below.

Walk through these districts during a hot wave and take note of the babies. There upon the steps of this pestilent-breeding building is a group of half a dozen children, the infant of the flock held in an elder sister's arms, not over five years old, and all so deathly white and emaciated that involuntarily one almost shudders. As I stood taking in the group, a little girl's voice exclaimed, "It's Jimmy Callahan as is dead — him as wuz took up las' Christmas for stealin'."

And as I looked across the street a woman was attaching to the door-bell a dirty piece of white crape. I turned to the group on the step; a deathlike silence had fallen upon it; they were staring intently at the signal of death, as if just then a glimmering of their own dreadful situation had reached their dulled brains, and they were mentally wondering how soon their own turn might come.

. . . What to do with this conglomerate mass of sizzling humanity is a problem as great as any that presents itself to the humanitarian side of our metropolitan life. Another almost ineradicable difficulty in dealing with the question of what to do for the improvident poor during the summer months is the ignorance and suspiciousness of the foreign population, which forms so large a percentage in these overcrowded districts; dirt, vice, and squalor travel as their close attendants, and when a hot wave makes its appearance these people seem to revel in their surroundings. It is here where the death-rate takes a jump when the thermometer climbs to a hundred. Riding one night in an elevated train, a party of Italians sat opposite to me — father, mother, and three small children, the youngest a baby in arms. Evidently they had been on some excursion, and this was the winding up; the aforesaid baby was in the act of commencing on a huge banana, and looked so wan and puny that it seemed an effort for it to eat. Finally I interfered, and made them understand I wanted to buy that banana, which I did for a dime, finding its way into the palm of the eldest hopeful. "What they need in this town is a Society for the Education of Mothers," I said to a friend as the guard yelled "Eigh' fus'." As we passed out, the baby had another banana in its dirty little paws — a contingency I had not counted upon.

Naturally there is much organized relief. St. John's Guild and kindred societies, the Fresh Air Funds of the *Tribune, World,* and *Life,* do an unfathomable amount of good. Three dollars sent to *Life* will send a child to the country for a week. Many a child is saved for future usefulness, perhaps, and the ties of family life, no matter how wretched or squalid, are preserved through these funds and societies. But in spite of all these aids the misery and suffering caused by a hot wave among the poor are untold and untellable.

Perhaps one remedy would be to tear down these pestilent rookeries and make city parks of their sites: this might work a hardship to some, but it would have the effect of scattering these people, of breaking up that clannishness which so distinguishes those arch allies, poverty and vice, and of giving more lungs and more breathing-spots to our city. And yet how difficult to minister to the wants and needs of the people! Passing Manhattan Park the other day, a couple of men, carters they seemed to be, were holding a desultory conversation.

"I don't believe in these here parks; there's Central Park, and the Riverside Drive, and another one 'bove; got too darned many parks, I believe," said one.

Now how is it possible to legislate for the good of such a barbarian? We read of "Darkest Africa" and of other barbarous and semi-civilized countries, where much missionary work is done to help save the souls of these multicolored savages. Spend the money these excursions cost upon excursions to help our poor escape from the evils of our tenement system during our hot waves. Help them to see a little of God's country, of the ocean and its cooling waters. Think of the "white savages" in "Darkest New York," and of what they suffer in the killing heat; give them free ice and plenty of it, free fruit, and free medical attendance; teach them sanitary laws that mean life and health. There is crying need right here in New York for teachers who can show our own people how to keep body and soul together all the year round.

ALICE WELLINGTON ROLLINS

"The New Uncle Tom's Cabin"

"The Tenement-House Problem"

Born into an established Boston family and well educated, Alice Wellington Rollins (1847–1897) began writing not long after her marriage to Daniel M. Rollins. She published several volumes of verse and some children's books, compiled collections of aphorisms in the manner of John Bartlett, and contributed to the day's leading periodicals. The following two essays appeared in *Forum* 4 (October 1887): 221–27; 5 (April 1888): 207–15. Rollins's heartfelt regard for the plight of tenement dwellers expressed in these two essays rises above the commonplace sentimentality of her verse. Her comparison to Harriet Beecher Stowe's *Uncle Tom's Cabin* (1852) shows that work's ongoing usefulness in the cause of reform. Rollins's concerns prompted her most important book and only novel, *Uncle Tom's Tenement* (1888).

"The New Uncle Tom's Cabin"

We all remember the old one: "The cabin of Uncle Tom was a small log building close adjoining to 'the house,' as the negro *par excellence* designates his master's dwelling. In front it had a neat garden patch, where every summer strawberries, raspberries, and a variety of fruits and vegetables flourished under careful training." It was small, very small as a dwelling-place for Uncle Tom and his wife and the row of little woolly-heads. Inside, its one general apartment was even more limited in space than Boffin's Bower;[1] though, like the bower, a strip of

[1] *Boffin's Bower:* See Charles Dickens, *Our Mutual Friend* (1864–65), ch. 5.

flowery carpet marked off one corner for a drawing-room, while the spot where vegetation ceased was covered by a table that indicated the dining-room, and still another corner was distinguished as the kitchen by a goodly cooking-stove, redolent of griddle-cakes. It was a very small place for so many people to live in, and, to add to their discomfort, they were slaves.

Nous avons changé tout cela.[2] That is, to some extent. The cabin is now at the North, instead of the South. Uncle Tom is white, not colored. And the cabin is very much larger, forty or fifty times as large. It is so large now that we no longer call it a cabin, but a tenement, possibly because ten persons live in the space which only one ought to occupy. For, unfortunately, there are more Uncle Toms to occupy the larger space: fifty or sixty times as many Uncle Toms. So each one gets no more space to himself than his southern slave brother had. There is the same one little room for drawing-room, dining-room, and kitchen; alas! it is sometimes also sleeping-room and laundry. For the northern Aunt Chloe[3] cannot move her tubs out into the fresh air, or send the children rollicking out under the sky, over the grass, all over the sunny, wide plantation. Land is expensive in New York; we cannot afford space around the cabin. It is not, as in the old days, "adjoining to 'the house'" of its owner. Oh, no! the owner lives — Uncle Tom does not know where he lives; somewhere three or four miles off up-town, likely. Or, maybe, as in the olden days the master was a man and Uncle Tom a chattel, so now, perhaps, it is Uncle Tom that is the man and his master a chattel; that is, the master is, perhaps, "an estate," an enormous, wealthy estate, with heirs scattered here and there, who hire an agent as their southern brothers hired an overseer, irresponsible, unsympathetic, caring only to please his patrons by showing a large balance of profit. And the poorer the tenement the larger the balance. No repairs, no janitor, no supervision to pay for; accommodation so wretched that only the very wretched, who will expect to be crowded and miserable, will apply for it. Oh, landlord or "estate," too busy to collect your own rents, be not too indolent to require of your agent a strict account when he brings you twenty per cent. instead of six! You would quickly bring him to book if he were suddenly to hand you six instead of twenty. But the time to question him is when it is twenty.

[2] *Nous avons changé tout cela:* We have changed all that.
[3] *Aunt Chloe:* The wife of Uncle Tom in Hariet Beecher Stowe's *Uncle Tom's Cabin* (1852).

No, the cabin is no longer near "the house;" we cannot afford even the space to build the cabins next each other, even in close rows. We must build them as we do the elevated roads, over the streets, over each other, story after story, behind each other, literally even *in* each other, for I believe it is matter of history that four families sometimes live in the four corners of one room, and get on in tolerable peace till one of them begins to take boarders. If there are bedrooms, they are little more than closets, dark, ill-ventilated, crowded. In the improved tenements that pay six per cent. there is a general laundry for every twelve families. Here there is always hot water, without the necessity for having the heat of a fire and the misery of steaming clothes, all day and every day. When your agent hands you twenty per cent. income on your tenement investments, look him in the eye and ask where he has located the laundry. You will find that no woman in the house can have hot water without making her own fire; it will be a mercy if there is even any water at all above the first floor. It may be August, and her sick child may be wailing in the corner of a room at the temperature of 115°; but she must not let the fire go out, nor stop her washing to attend to her child; if she does, there will be no food in the house to-morrow for those who are well. But at least, you think, she might put a bit of ice in the child's fevered mouth now and then. Ice? She would look at you as her sister of the French Revolution would have looked at the *grande dame* who suggested that if the poor had no bread they ought to be satisfied with cake.[4] She has no ice; if she had ice, she has no refrigerator; if she had a refrigerator she has nowhere to put it. How long would ice keep in a refrigerator standing next the stove? And there is no other room; even if there were space in the wretched passage-way, the passage is just as hot, and the neighbors would steal the ice.

In front of Uncle Tom's cabin, at the South, in summer, there were "strawberries and raspberries and a variety of fruits and vegetables." So there are in front of the tenement at the North, strawberries and raspberries, oranges and bananas, pears and pineapples, lettuce and squash, beans and cherries and grapes. But they are in carts. They are the refuse, brought down into the poorer streets after they have become unsalable to better customers. They are already beginning to turn black. Poor as they are, however, longing eyes are turned to them

[4] *grande dame . . . cake:* The French Queen Marie Antoinette (1755–1793) was notorious for her insensitivity to the plight of the underprivileged.

from tenement windows. Cheap as they are, those who want them have to hesitate. Yonder a child of six plucks at its mother's gown and begs for a banana: "It's only a penny, mammy, an' 'tain't very rotten!" The mother hesitates; there is a penny on the table, but she had meant it to get a little milk for the child; but maybe the milk is not much better in quality than the fruit. Let her take the penny; the mother cannot even stop work long enough to see that she eats only the part that "ain't very rotten."

The one feature of southern slavery, that the slave could be bought and sold, of course outweighs every other, and makes that sort of slavery the most accursed on the face of the earth. But with that single exception, granting frankly that it is even the greatest exception, the condition of the slaves of New York is a hundred times worse than that of the southern slave. The southern slave had a "chance;" there was a chance that he might have a good master. The tenement-house slave has no chance; for those who have begun that degradation there is no hope of better days. Uncle Tom, in the Shelby household, had nothing to fear but that sudden loss of wealth might work for him a change of masters; but Uncle Tom of the tenement has to fear his master's growing rich; has to fear that the tinkle of increasing ducats in his pocket will make him rejoice in and increase the extortion that thrives so well on tenants who have forgotten to expect rights. The southern Uncle Tom had his ways of earning; of laying aside, if only penny by penny, something. The northern Uncle Tom never can save; there is not enough from day to day to live even decently; and one break in a day's work, one hour of sickness in father, mother, or child, leaves a gap that it seems as if never again they could bridge over. They never can "catch up." The southern Tom had his family perhaps torn from his arms, never again to lay eyes on daughter or son. Think you that the northern Tom can keep his family about him? As the slave Eliza clasped her boy to her breast and ran away that she might not have to give him up, so slips from the tenement house every day some Eliza with a child in her arms, not to keep him, but to give him up; to lay him in the crib at the foundling hospital, whence he will be taken by others, to be known hereafter not as "Johnny," but as "No. 22,716;" or to give him to benevolent agents[5] who will find him a home at the West. Saddest reflection of all is to think of the callousness that misery

[5] *benevolent agents:* A reference to the efforts of the Children's Aid Society, founded in 1853 by Charles Loring Brace, which found foster homes in the Midwest for New York children.

produces. Of all the pathetic stories you can tell of Little Nell or Oliver Twist,[6] none so impresses me with the horrors of abject poverty as I am impressed on hearing a mother, when I tell her I am glad her sick baby is better, say calmly: "Lor', yes! Warn't it a mercy it didn't die! It's just orful to have children dyin' now, when it costs so high to bury 'em."

And if the callousness to suffering is so terrible, what shall be said of the callousness to vice? You will remind me of that hideous phase of southern slavery which brought with every daughter born to Uncle Tom a certain sorrow. But Tom's daughter of the tenement is worse off. Hating her wretched home, and seeking diversion elsewhere, she wanders to the dance-house of her own accord. She learns to like it. She is not a helpless victim, ruined by men; she is a voluntary agent, ruining men. Sometimes she is driven by poverty to a life of shame that she abhors; that is terrible enough; but more often she is driven to like a life of shame. The southern slave girl was sinned against; the northern tenement girl sins. You shudder at the thought of a mother with a daughter ruined; will you not shudder at the thought of a mother with a daughter ruining? Will you not shudder still more at the thought of a mother in whom natural affection is so dulled that she does not care what happens to her daughter? You agonize all day over the tidings you must carry to a mother respectable above the average, that her boy has stolen a sum of money; and when at last you stammer out the words that you fear will kill her, she only straightens herself with dignity, and exclaims: "Well, he ain't brought none of that money here; we ain't seen a red cent of it, have we, Martha?" You bear to another mother the tidings that her son is in jail, and she only says: "Ain't that a shame! D'yer think he'll manage to get off?"

The southern slave had no anxiety for food, shelter, or clothing. Ask the very poor what that means. It means that the defender of slavery was probably right as far as he went when he said that Virginia, before the war, was utterly free from the crimes that spring from poverty. The crimes that spring from poverty are probably two-thirds of all the crimes on the earth. The southern slave was kept in ignorance; the northern slave is kept in ignorance in sight of knowledge. There are schools for his children; there are even laws that his children

[6] *Little Nell or Oliver Twist:* Little Nell is the heroine of Charles Dickens's *The Old Curiosity Shop* (1841), and Oliver Twist is, of course, the title character of Dickens's *Oliver Twist* (1837–38).

must go to school; but they can't. They have no clothes, and as soon as they can do anything they must either earn or steal. The southern slave was wronged, stultified, abused, bought and sold; the northern slaves are wronged, stultified, abused, and made vicious. Uncle Tom had to see wife or child mercilessly beaten; the northern Tom learns to beat wife and child himself.

The southern planter acknowledges to-day that one of the worst evils of slavery was the reaction on the aristocracy. Do you suppose there is no reaction on the rich from the tenement-house slavery? One hesitates to speak of retribution; it is so degrading an idea that we cannot do the right thing till we find out that if we don't we shall suffer. Let us not call it retribution, but reaction. Remember the pestilence breeding in those dens where they are eating bananas that "ain't very rotten." Remember that the sewer pipes connect Murray Hill[7] with some of the worst of tenements. Remember that when your soiled linen is sent to a washerwoman you know not where it goes or whence it comes back to you; you can only be sure that if it is snowy, it has braved what the water-lily does that manages to come white to the surface. When you go to the intelligence office[8] for a wet-nurse, or, worse still, for a nurse to care for your older growing children night and day, remember the haunts they come from, the tastes and habits they bring with them. Remember to what are being driven the girls that will not, or cannot, "live out," and so live in degradation that drags down with them — you know not whom.

But now you will say, granting all this, what can be done about it? Do you advocate going back to slavery? Is it not poverty that is responsible for the home, instead of the home for the poverty? Why do you call these people slaves? Are we responsible? What do you ask of us — that we take a few millions out of our pockets and hand over some nice tenements to them as a charity? Surely it is the employer, not the landlord, who ought to reform. With better wages, a division of profits, the poor could have better homes.

No, I do not advocate slavery; I advocate the abolition of all slavery. Having somewhat roughly removed the mote from our brother's eye, let us now gracefully remove the beam from our own. It is the home that causes the poverty; it is the tenement that is the root of the

[7] *Murray Hill:* A fashionable neighborhood on the East Side of Manhattan.

[8] *intelligence office:* Employment office. For a good contemporary description, see "In Employment Offices: Trials of Women Who Look There for Work" (pp. 244–50).

evil; debasing, unfitting its inmates to be either good or competent citizens. Reform the employer, and exact from him higher wages, before you have reformed the landlord, and you will only have the tenants staying along in the same old places, with their rents raised. We are not responsible for the poverty, but we are for taking advantage of the poverty. We are not to give these people homes as a charity; we are to take money for what we give them; but having taken money, we are to give them their money's worth. If a man pays you $1.50 a week, you are to give him two rooms and a scullery, with separate sink and closet. This you can do, and still draw a six per cent. dividend on your investment. We are responsible for the state of things inasmuch as we leave no one responsible. The philanthropist lays the blame on the city, the city on the landlord, the landlord — where is he? He is a myth; he may be a man, he may be an estate, he may be a church organization. There are even rumors that sometimes he preaches from a pulpit. We do not know that this is so, but we ought to know that it is not so. There is an obsolete law about the name of the landlord being posted inside the door of every tenement; but even if this were carried out it would do little good. Few people enter those doors to see those names except the miserable inmates and the agent who collects the rents. The southern planter was known to some extent as a good master or a bad master; the tenement owner is not known at all. You cannot even point the finger of scorn at him. We know the good tenement owners, but not the bad. While they can so securely defy the laws that exist, requiring so many cubic feet of air for each individual, of what use to plead for more laws? Charity? no; philanthropy pays six per cent. But, unfortunately, abuse pays twenty per cent., and the hidden landlord is safe even from scorn.

"The Tenement-House Problem"

The tenement problem includes the question of what shall be done for the very lowest class of the degraded poor, and reformers are at once met with the objection, "Why, these people are not suffering as you suppose they are. You project yourself, with your tastes and habits, into their environment, and fancy what you would suffer in such surroundings. But they are not suffering like that. They will not care enough about your better tenements to move into them, even at the rents they pay now. They are wedded to filth and misery. They don't want you to do anything for them."

The embarrassing part of this objection is that it is true. The degraded poor are not suffering from a keen sense of degradation; they do not desire either your sympathy or your interference. Attempt to deny this, and proceed loftily to relieve an oppressed class suffering bitterly and ready to worship you as their benefactor, and you will be discouraged at the very first step. Of course, this is not asserting that no poor people are suffering. There is a large number of poor people who do suffer, but they are not the degraded creatures of the very lowest class. They are, as a rule, *nouveaux pauvres,* not absolutely poor perhaps, but poor in their own estimation because they are a little poorer than they were yesterday. The *nouveaux pauvres* are suffering; but it is *les misérables* whom we are chiefly considering in the tenement problem, and the most awful part of the misery of *les misérables* is that their misery does not bring with it any great sense of degradation.

No, we may not assert that the degraded poor are anxious to be improved. But that should not discourage effort. On the contrary it should intensify it a hundred-fold. Do you suppose that Philanthropy will shut her purse, and turn away her eyes, and go quietly home again, because you tell her the tenement poor are satisfied to be left as they are? I think she will open her eyes a little wider, and say with increased emphasis, "You tell me that they are satisfied as they are. What, then, have you been doing to these people for generations past, that they have learned to be satisfied with so little? How did they become acquainted with the filth and misery to which you say they are now wedded?"

The assertion that it is the tenants that determine the condition of a tenement, not the landlord or the architect, is in a sense justified. There are, in one quarter of New York, two tenements built alike, in the same neighborhood, and yielding the same rents. One is a den unfit to live in, the other is a decent and respectable house. The difference is due to the habits of the tenants. But analyze this, and what have you? You have a class of tenants made what they are by their original surroundings. True, we are not at fault for those original surroundings. But this does not excuse the landlord. He may not be compelling these people to live so, but at any rate he is allowing them to live so. We ascribe the evils of tenement life largely to the close herding together of so many human beings; but objectors will at once remind us that out on the western prairie, where a man builds his own house, and can put in it as many rooms as he pleases, and can have all the land he wants, you will still find what is one of the great evils of the tenement question in crowded cities: a man and his wife and eleven children living, cooking, eating, washing, and sleeping in one room. But what does this prove? Merely that the tenement evil spreads beyond its native *habitat*. "But if you drive the degraded poor out of the bad tenements," say the objectors, "they will only go off to herd in worse places." That is the point under discussion; there should be no "worse places." We are supposed to be dealing with the tenement problem at its lowest terms.

I have said that Philanthropy will not be discouraged by the objection that the tenants do not care for reform; but the true court of appeal to decide the matter is not even Philanthropy. The popular idea of the tenement problem is this: "There are in the community a class of people very, very poor, who can afford to live only in the most squalid surroundings; now cannot we club together and arrange to do something for them?" That is not the problem at all. The problem is this: "There are in the community a class of people who, it is true, are poor, yet who are paying enough to have decent, even comfortable, surroundings. But they are not getting their money's worth. They are crowded into dens where they are getting used to dirt — mental, physical, and moral. Shall this thing be allowed to go on?" This is a question not of benevolence, but of self-preservation. It means not merely discomfort to the suffering few, but danger to the state.

The tenement problem is not a question of what tenants desire, or even of what it would be nice for them to have; it is a moral situation, to be considered without any regard to what the tenants may think about it. Our own children do not want to have their faces washed, but we wash them. No child, even with the finest pedigree, the most favorable surroundings, the best examples, is born into this world with any overwhelming desire to be clean. Until the lad is ten or twelve, we must insist on washing his face; for three or four years more, we must see that he does it himself. By the time our boy is sixteen, however, the task will be done. The *vis inertiæ* of cleanliness will have been established.

Nothing is more astonishing, in investigating the slums, than the discovery of the enormous prices the poor are paying for the most wretched accommodations. One man boasts that he draws 33 per cent. on his tenement investments. Mr. Alfred White's experiments[9] with improved tenements have been carried on for ten years, and have

[9] *Mr. Alfred White's experiments:* Alfred Tredway White (1846–1921) devoted much of his life to building affordable, model tenements for New York's workers and campaigned hard to improve existing tenements.

been made in the city which is third in size in the United States, so that he has certainly had to grapple with all the problems presented by a large city; and he states that for $1.50 a week you can give tenants two light, airy rooms, with separate sink, scullery, and arrangements for coal, and draw six per cent. on your investment; yet you will find families paying $6 a week for two rooms, with right to use the hallway for some of their "things;" and in the same house a woman with three children paying $2 a week for one room in the basement, where she cooks, eats, and does washing for a living, with a dark closet and one bed where she and the three children sleep. In a semicircle of sheds occupied by rag-pickers one woman pays $1 a week for the end of one shed.

More than half the population of the city of New York live in tenement houses. There are 30,000 of these tenements, 2,000 of them reported in the official statistics as "very bad." In one block on the east side there are as many people as you would find in a country village stretching over several hundred acres of land. Between two avenues and two streets in the same district are 3,000 or 4,000 souls. This in itself is not objectionable; for the same space, if built up with "apartment houses," such as we see in other quarters of the city, might afford to a much larger number of persons even luxurious privacy; but these tenements are only ordinary houses. In some rooms you will, in the daytime, see mattresses piled up till they touch the ceiling; at night, when the "boarders" stream in from their day's work, these mattresses are taken down and spread over the floor, touching each other. Forty-five people sometimes sleep in one room.

You will return from your first visit to the slums with two very strong impressions: one, of the utter hopelessness of trying to do anything; the other, of the necessity for doing something immediately, lest the heavens fall. Perhaps you have evolved in your *boudoir* some beautiful scheme of amelioration; it has occurred to you that if ten rich men of the city could be persuaded to give $100,000 apiece, not as a charity, but as an investment, to build ten tenements, each to accommodate seventy families, it would be a great and glorious thing. But, as you stand in the "Bend" in Mulberry Street[10] and gaze about you, it will be to say in despair, "$1,000,000, ten tenements, seven hundred families! Of what possible use to plan such an infinitesimal oasis of

[10] *the "Bend" in Mulberry Street:* In *How the Other Half Lives*, ch. 6, Jacob Riis remarked, "Where Mulberry Street crooks like an elbow within hail of the old depravity of the Five Points, is 'the Bend,' foul core of New York's slums" (96).

relief in this universe of misery and degradation?" You have never seen people so hived before. Above you, below you, behind you, in front of you, to the right, to the left, in the rear, in the distance, crowded against each other, behind each other, above each other, are human beings. They swarm on the sidewalks, they are entering and issuing from the doorways, they lean out of the windows. You have always supposed that in the homes of the very poor you would be filled with pity for the hard work you would be seeing them do: women bending over washtubs or ironing-tables, cobblers cobbling, tailors sewing, seamstresses running machines, tinkers mending, children weeping bitterly as they, too, turn a machine, or try to make a shoe; everybody toiling for dear life, for a mouthful of bread, too busy to look up, even, as you pass. But the very first impression made upon you in the slums is that of a horrible leisure. What are these people doing? Nothing. What do they want to do? Nothing. What are they capable of doing? Nothing. What do they want you to do for them? Nothing. What can you do for them? Nothing.

And finally the great question, What is the remedy? Does the outcry against tenements mean that there should be no tenements? Not at all. The problem of homes for the poor in great cities can be solved only by tenements. Indeed, there is not the slightest objection to a home with others in an improved tenement. There are certain advantages in one large building: it is more comfortable in winter, and in summer such improvements as a general laundry prevent much discomfort in the individual living rooms. Let there be a good building, in a decent locality, with a conscientious landlord and a janitor, and a thousand souls may live under one roof comfortably and in decency. The janitor is quite as essential as the conscientious landlord. He must be there day and night, to see that forty-five boarders do not straggle in to spend the night with the family on the third floor, and to enforce the rules which at first your tenants will not care to obey.

The obvious query is, "Why do not these people move into the suburbs, or the country, where they can have so much better homes at so little cost?" But you may as well face at once this difficulty in the case: that the poor will not go into the country. There are limits to what we can dictate. We can say they shall not live as they do in the Mulberry Bend, but we cannot say they shall live in Harlem[11] or at Staten Island.

[11] *Harlem:* Harlem remained largely unsettled until the elevated railways were extended to 155th street during the 1880s, at which time it became a predominately German, middle-class neighborhood. African Americans did not begin moving there until the 1890s.

So great is their dread of the country that the children sent out by the fresh-air funds[12] are almost invariably wretchedly homesick for the slums for several days. A little girl taken from one of the poorest quarters of the city to spend the summer with a family out of town, where she was given the best of food, every comfort of clothing and shelter, plenty of strawberries, flowers, drives, games, and picnics, was utterly broken down by homesickness; and finally, when pressed to state what she had had in the city that she missed at Deerfield, confessed, between her sobs, "Oh, ma'am, we could sit on the stoops an' talk to folks!" But there is this element of hope: in less than a week the reaction sets in, and the children begin to love the country. Hundreds of them come back to exclaim, "As soon as I'm a man, I'm going to live in the country!"

A visionary reformer will say, "Make these poor people richer, and then they can have better homes." Nothing could be more fallacious. You cannot make a man rich by giving him money. It is no use to give him privileges till you have taught him appreciation of them. Cases of injustice, of oppression, there will inevitably be; yet, looking at the class of people who go forth from these tenements to their work every morning, you feel reasonably sure that workmen such as these are receiving all that they are capable of earning. Moreover, as we have seen, the people are not occupying these wretched quarters because they cannot afford to pay for better; they are paying now enough to have good accommodation. Philanthropy would like to believe that these people, if their wages were raised, would go home to their wretched surroundings and say cheerfully, "Come, Mary, pack up! I'm to have a dollar a week more now, and we can afford a better place." But we all know they would not, even were the "better place" to be had. The woman paying $2 a week for a basement room, and the family paying $6 a week for two rooms and the hallway on the second floor, could afford now to move into Alfred White's improved tenements at $1.50 a week, without Mr. White's giving them anything in the way of chairty. But the tenants will not move: first, because they do not know enough, or care enough, to move, being, as the anti-reformers say, "wedded to filth and misery;" and secondly, because there are not enough improved tenements to go round.

[12] *fresh-air funds:* Begun in June 1877 by clergyman Willard Parsons, the Fresh Air Fund provided summer vacations for New York City children. Later in 1877, Parsons turned the management of the fund over to the *New-York Tribune*, which relinquished control in the 1960s. Following the *Tribune*, other New York newspapers established fresh air funds. A Fresh Air Fund continues today.

The second remedy suggested is that of the socialists, share and share alike. Nothing could be more fallacious. Their cry practically is this: "It is a horrible thing to be poor; therefore let us all be poor." If the new order of things could make us all rich, it might be worth considering; but it is perfectly understood, even by its advocates, that the only result would be to make us all poor. Happily we do not need to linger over the discussion; every sane individual knows that perfect equality of men in possessions, endowments, and condition will never come about.

The third remedy proposed is that of those who rely upon the government, who make herculean exertions to get new laws through the legislature, more inspectors under the Board of Health, and a million of dollars for the establishment of small parks. This is all very well. I have myself often wondered why the city government, so generous in building hospitals and prisons and reform schools, should not take the tenements in hand. "Fall ill, or break your leg," says the city, "and I will give you doctors, nurses, comfortable bed, food, and everything you need. Commit a crime, and I will put you into a nice, clean prison where you will have food, clothing, and shelter without any expense. But so long as you can manage to keep well and virtuous, you may go to the — tenements for all I care!" It is well to get from the government all you can, but it is not desirable to depend solely upon government for reform.

No, nothing will, in reform, take the place of individual conscientious landlordism. Individualism seems minute, but it is mighty. Let in the light of day upon the landlordism of the slums, as you have let it in upon Mormonism and other hateful things that prefer darkness rather than light. The landlord is not to be a philanthropist, willing to sacrifice himself for the good of others; he is to be an intelligent capitalist, putting in his money purely as an investment, and philanthropic only to the degree of being satisfied with six per cent. returns, of hiring a janitor to be on hand day and night, of being his own agent, or keeping a sharp look-out on the one he may have to employ, and of urging his wife to collect the rents.

But individual landlordism need not necessarily be confined to individual persons. Individual corporations can become landlords. Why should not some of the insurance companies, that complain of being unable to find suitable investments for their immense funds, take hold of the tenement question? A life-insurance company of Boston, complaining of the low rates of interest obtainable, announce that they never expect over five per cent., and find it difficult at times to get four. The great decrease in rates of interest has been made the excuse for not giving members all the privileges which they once enjoyed. The risk of a tenement investment evidently cannot be any greater than the risk of other investments has proved.

Half of the trouble is caused by the willful cruelty, but half by the thoughtlessness, of the landlords. A wise writer has said recently, "Often you don't need to say to a man, '*Why* do you do so?' If you can show him *what* he is doing, it is often enough to rouse him to reform." I have faith enough in human nature to believe that if we could organize a procession of landlords, and compel them to walk through the tenement districts, they would begin the reform themselves. Half of them do not know what they are doing; they trust the care of their property to agents, whose interest it is not to trouble them with demands for repairs or any lessening of income.

And of course there is one other important factor: education. We say we can do nothing for the degraded occupants of the tenements because they are wedded to filth and misery, and we cannot educate their grandmothers. But there are grandmothers whom we can educate. Children of to-day are the grandmothers of the generations yet unborn. We can educate them.

But by education I mean something more than the development of intelligence or the cultivation of ideas: I mean the training in habits. And I also mean something more than the education of the poor: I mean also the education of the rich; that no boy who is to be a future millionaire shall grow up without a sense of his trusteeship. Flood your public schools with knowledge, and compel the children to come to them; yet so long as you let them go back at night to see and hear and learn the things they are seeing and hearing and learning at the places they call their homes, so long are you pouring fresh, pure, sparkling water into a sieve that empties in the gutter; so long are you trying to build a palace of pure white marble on supports of rotten wood. Nay, more: you are doing what is perhaps not only fruitless, but fatal; you are lighting a torch of intelligence that may end in setting fire to your own homes. Anarchy is not misery wedded to stupidity, as we are fond of fancying; it is misery touched with just too little intelligence. Waken the intelligence to rebel against results before you have taught the capacity for altering conditions, and woe be unto you in the struggle! You may say that intelligence ought to conquer conditions, and it will, in time; but you may well dread the contest if you do not do your share toward preventing the battle. Educate the grandmothers; but educate them to something more than ideas. Throw all of the English grammars and half of the Latin ones into the ash-barrel, and introduce in their places manual-labor, cooking, and sewing classes, that shall teach the young not only how to do things neatly, but to care about doing them neatly.

Every generation has its own wrongs to right, its pet grievance. The Abolitionists of the last generation are already jealous of our asserting that we can possibly have as big a grievance as theirs. Let it, then, be our aim and glory to settle our grievance before it becomes quite so big as theirs, and to settle it without a war. No adjustment of the difficulty between capital and labor, or of the tenement problem, can be either relied on or admired, that depends on any supercilious charity from the rich on the one hand, or on establishing their perfect equality with the poor on the other. It is not when we see another mother buying a handsomer overcoat for her boy than we can afford that the envy and bitterness come; it is when we see her buying a warmer one than we can afford that we begin to hate her. The poor are willing for you to be rich, and for themselves to work; but make it hard for them to get work — and may God have mercy upon your souls as well as theirs!

JAMES O. S. HUNTINGTON

"Tenement-House Morality"

James O. S. Huntington (1854–1935), born in Boston, received his A.B. from Harvard in 1875, and studied theology at St. Andrew's Divinity School. He was ordained a deacon in 1878 and advanced to the Episcopalian priesthood two years later. In 1884, he founded the Order of the Holy Cross, the oldest monastic order of the Episcopal Church in the United States. For the first ten years, its members dedicated themselves to mission work in New York's Lower East Side. "Tenement-House Morality" reflects Huntington's experience living and working among the city's poor. The text is from *Forum* 3 (1887): 513–22, a periodical devoted to bipartisan discussions of contemporary issues facing the United States and the World. For the most part, Huntington's copious notes have been excluded.

Some time ago a lad came back to me, after making his confession, and asked, in a troubled tone: "Father, must I confess what that man says at the shop?" That, it seems to me, is a fair example of the effect not only of the shops where tenement-house people work, but of the streets where they walk and the buildings in which they live. Here was a boy with strong impulses toward goodness, trying and struggling to do right and to keep himself pure, hating the blasphemy and obscenity which he heard from those around him, and yet compelled for so many hours each day to breathe an atmosphere foul with moral corruption that he had come to feel that the sin about him was somehow his own, and that he needed cleansing from others' guilt as if he were himself defiled. That this is the case in many shops where children work, is clear from their own pathetic acknowledgment. "How can we be good," they cry, "when we have to hear such talk all day?" Or, as the older ones say, in yet sadder tones: "When I first went to the factory I thought I couldn't stand it; then I got used to it; now I say the same things myself." Would that the evil stopped short at *words!*

But it is not of shops that I have to speak now, but of a more sacred place, of that which must ever be the source from which the life of society flows forth — of the homes of our working people. And I solemnly aver that the tenement-house system surrounds the poor in their very families with just such corrupting influences as those found in the factories and shops; yes, and with yet more deadly moral contagion. How can it be otherwise? Take one block in a tenement-house district. It will measure 700 by 200 feet. On all four sides are rows of tenements four or five stories high. Behind one-third of the houses in these rows are rear houses, with smaller rooms, darker and dirtier passages, backed often by another rear-house, a brewery, a stable, or a factory. Altogether there are 1,736 rooms. In these rooms live 2,076 souls, divided into 460 families; thus, on the average, each family of five persons occupies three rooms. The population of some parts of New York is 290,000 to the square mile: the most densely populated part of London has 170,000. Of course in many cases the family is larger (some of the very poorest people take lodgers), and in a number of cases we have found fourteen or fifteen grown persons occupying two rooms, or even one. And then many of these "rooms" are hardly more than closets, and dark closets at that. Almost all the bedrooms measure only seven feet by nine, and have but one door and one window. The door leads into the apartment that serves as kitchen, parlor, sitting-room, laundry, and workshop, and the window opens on a dark stairway, up which the moisture from the cellar and the sewer-gas from the drains are continually rising. One-fifth of these rooms, too, are in basements below the level of the street, and nearly half of even the outer rooms open into courts only twenty feet wide, in which there are usually several wooden privies for the use of the fifteen or twenty families in the front and rear houses.

I know that these statistics will give but a faint conception of the density of the population to any except those who have gone in and out of the houses day and night for months, if not years; but most people, by a little effort of the imagination, can form some sort of an idea how impossible it is for dwellers in tenement blocks to get out of the sight and sound of their neighbors, whose names are often unknown, but whose voices and foot-steps are as familiar as those of their own room-mates. At all seasons of the year the inhabitants of a tenement-house must meet one another in the entries (sometimes less than three feet wide), on the stairs, at the sink (there is but one on each floor); must see into one another's rooms as each person goes in and out; must use the roof, the doorway, the yard, in common. But when the summer heats are on, and men and women crowd together on the top of the house waiting for a breeze to come; when men will sit all night on a seat in the park to escape the closeness of a room where a fire has been burning all day (not for cooking, but to heat the irons for the laundry or the tailor's shop); when every window must stand open to let in what little air there is; then it may be seen that privacy in a tenement-house is not much more possible than in an Eastern caravansary or in the steerage of an emigrant vessel. At such a time every loud word spoken reaches the ears of scores of people. From one room come the harsh tones of a husband and wife in the heat of a "family quarrel," oaths and imprecations ringing out on the fetid air; from another window come the shouts and frantic laughter of men and women (God pity them!) trying to drown their misery in liquor from the gin-mill on the corner; while from the roof of a neighboring house come the words of a ribald song flung out shamelessly to all within hearing, whether they choose or not. And, as if this were not debasing enough, in many of these blocks every other house has, on the ground floor, a saloon or rum-shop, from which the smell of alcohol issues at all times; where the monotonous click of balls on the pool table sounds till after midnight, when it gives place to the howls of drunken men turned out on the street; and past the door of which, often open into the entry, every person, every child, in the house must pass to and from his room.

And who are the people that crowd these tenements? Perhaps it will be thought that the very badness of the condition of such places shows that the people are all "filthy and debased creatures," and that, therefore, very little can be done or need be done for them. Men will be inclined to dismiss the whole matter with a shrug of the shoulders and an impatient sigh. "It is all very dreadful, no doubt, but there will always be base, corrupt people; they naturally herd together, they create their own misery; if you root them out of one locality they will simply transfer themselves and their brutality and vice to some other." No doubt there are such people in tenement-houses, but that they represent the great body of the tenement-house population I entirely deny. Side by side with these poor outcasts of humanity are hard-working men and women who are leading lives of heroic purity and nobility. They are fighting, at fearful odds, to keep themselves and their children from the filth and pollution all about them. It is in their name that I plead; and not for their sake only, but for that great middle class of those who are not determinedly vicious, and yet are not striving with such desperate resolution as these others after goodness and truth — those who would gladly do right, but lack the courage to rise above the mass of simple low-living and coarseness around them. Surely the case of these people is pitiful enough. They are pressed together under conditions which make it well-nigh impossible for them to help themselves or one another. The bad almost inevitably drag down the good; and the good have not the chance to lift up the bad. Remember that the tenement population of most of our cities is a heterogeneous mixture of all the races and nationalities of the globe. There is no place in such a conglomeration for the public spirit and popular sentiment that so often exercise a restraining and elevating influence. There is no standard of morality. Human nature is left to do pretty nearly what it likes, and the lower passions are not slow to assert themselves.

This is all the more the case that so many of these people are emigrants. They have come from the villages of England, Germany, Russia, where they were under the constraint of a certain conventional morality, backed up by a strong and vigilant, even if a despotic, government that made it often easier to do right than to do wrong. Here they are jumbled together in utter disorder, Prussians, Bohemians, Swiss, Scotch, Chinese, Italians, Turks, Jews, and Christians, black and white; a restless, seething mass of human beings, unable to talk together, unable to think together, able only, under some overmastering passion, to act together. In a city like New York may be found representatives of almost "every epoch of history and every locality of the world." One scholar says that in New York, he has heard eighty-four languages and distinct dialects spoken. The signs alone in the crowded parts of the city show the cosmopolitan character of the population. Is it not evident that in such a chaotic state of things, with the reins of government held very loosely, every one, man, woman, boy, and girl, must actually live in an atmosphere of defilement night and day; not merely going into it, as in the case of work in a shop, and then coming back into pure and elevating surroundings, but breathing in the polluted air with every breath? Why, the very tones of the voices that I have heard from my room in a tenement-house brought with them a sense of moral contamination. Even bodily cleanliness is almost impossible. Bath-rooms are unknown in tenement-houses, and the public baths, open only a few months of the year, often afford but fresh opportunities for vice. In most families what little washing is done must be done in the presence of others, and often all the water used must be carried up three or four flights of narrow winding stairs.

Of course sickness and death have their own horrors and their own depraving influences. What little privacy may be possible for the well is often denied to the sick, who, to get any air at all, must lie in the room used by the whole family for almost every purpose. Many of the diseases are infectious, but isolation is impossible, and therefore almost every child suffers from scarlet fever, measles, chicken-pox, and diphtheria, and often bears the results through life. And death, from its frequency, and the coarseness that surrounds it, loses, if not all its terrors, at least its dignity, and is regarded as one of the many disagreeable accidents of life, hardly worthy even of idle curiosity. The corpse lies for two days in the room where the family eats, works, and often sleeps.

But this by no means exhausts the abominations of the system of tenement-house life. As I have said, it is only by an effort quite beyond the powers of many people that grown men and women can resist the lowering influences about them. What, then, must be the lot of the children? They must not only hear all that older people hear, and see all that they see, at an age when every such sight and sound leaves its impression, but they are practically forced into acquaintanceship with the other dwellers in the tenement which their elders can avoid. Many mothers do try to keep their children in their own rooms, but as the children grow up this is increasingly difficult, and at length impossible. Once beyond the mother's supervision, the child inevitably becomes one of a group of children representing, perhaps, almost all the nationalities and religions of which the population consists. This group of children finds its playground in the dirty street in front of the block, or in the dirty yard, half filled with privies, behind. Here and there is a yard where turf has been laid, and a few flowers coaxed to grow; but there, of course, is no room for children. When it rains the children play in the cellars, sailing their boats on the water that often stands there, or wading ankle-deep in it. Wherever they play they are without any real oversight. The fathers are at their work, or in the saloon; the mothers are working wearily at the sewing-machine or the wash-tub, too driven to stop and watch their children, even if they can see them from the window. Think of what possibilities of moral contagion lie in such associations, amid such surroundings. Think how horribly ruinous the presence of one older bad child can be. As a fact, I could not here relate what I know to be the effects of such companionship; I could not even describe the games at which they play.

But suppose that a child passes with some degree of safety through the period of mere unconscious and, even in tenement-houses, light-hearted childhood; suppose the child has not been afflicted by many of the disorders — granulated eyelids, scrofula, rickets, heart disease — so shockingly prevalent among these children, what then awaits these boys and girls? As life begins to open, and the desire for a little of the brightness and happiness of the world makes itself felt, what is the scene that confronts them? A wilderness of ignorance, poverty, and crime; a moral desert, beautiless, joyless, utterly unsatisfying to all the best and noblest instincts of their hearts. Do you realize that in a tenement-house district there is absolutely not one lovely thing on which the eyes can rest? Even the sky is often robbed of its fairness by the clouds of smoke and dust. The glories of sunrise and sunset are unknown. The sun crawls up from among the chimney-pots, and goes down behind brick walls and tin roofs. The streets are always filthy, the houses ugly, the shop-windows cheaply gaudy, or neglected and covered with dust; the blocks are wearily monotonous, the school-rooms are bare and uninteresting, the factories are filled with fluff, and dirt, and noise; the air is charged with foul odors from close courts, open drains, or the neighboring oil and varnish works; the river is foul with mud and ooze and the refuse of a great city; the district ends in heaps of rubbish and empty lots, waiting for a rise in the market. And the rooms are often worst of all. There is many a "home" where a boy or girl over fourteen years old would not think of passing an evening unless compelled to do so. Think of coming back after a hard day's work in a shop to find the only sitting-room half filled with wash-tubs, the baby crying, children squabbling on the floor, or perhaps tumbling about on the bed; the walls hung with the soiled clothes and dresses of the family; the whole place reeking with the smell of fat and garlic from the hot stove; the table "set" with coarse, broken china, strewn on a dirty board; a kerosene lamp, without a shade, smoking in the middle; a loaf of bread, in the brown paper in which it was wrapped at the bakery; and a coffee-pot of black, bitter coffee. That is the scene which welcomes many a girl or boy, just beginning to realize how differently other people live. Is it strange that they gulp down their sugarless coffee, and at the first chance slip out into the street beneath, glad, perhaps, if they escape without a harsh scolding or a blow? And what has the world outside their homes to offer them? An avenue lighted by electricity, with plenty of young people with whom to "carry on," without any interruption from father or mother; the bright, warm saloon, with every chance of pleasant companionship and obsequious attendance; or the gay theater or dance hall, where all the troubles of life can be forgotten for a few hours in excitement or sin. Is it strange that as we go about from house to house, every few weeks some mother tells us, with an affectation of indifference, but with a quiver in her voice, "Rosie isn't at home now; she's boarding. We don't just know where she is. She was a bad girl; she wouldn't work. Father licked her, and then she went away." Or, "Charlie done something wrong at the shop; he took some money from the boss, and we ain't seen him since." Is it strange that a young woman, attractive, intelligent, who has gone astray and found the misery of that, and now is trying to do right, and support a father and mother and little brother, should have said to me the other day: "There's nothing in the world that makes me happy; the only thing I can do is to keep working. I work at tailoring all day. Noontimes I work as soon as I've eaten my lunch. I bring my work home and sew until I fall asleep. That's the way I keep from going mad with my wretchedness."

I am quite aware that much of what I have written will seem overstated. It seems so to me, and yet I know that it is not. Every single fact has been verified, and can be verified in thousands of cases. And this is not more than half the truth. If any one is disposed to be skeptical, I can only ask him to make investigation on his own account. But let him be thorough. Let him not merely walk through the streets some breezy Monday morning; let him spend days and nights here; let him live, as we have done, in a tenement block; let him visit the people at all hours; let him, above all, spend a public holiday here; let him see the carnival of sin of a Fourth of July or a New Year's night. I do not say that he will even then understand the conditions of tenement-house existence; but I know that his incredulity will give place to a sad, bewildered realization of the horrors of a state of things where manhood is brutalized, womanhood dishonored, childhood poisoned at its very source.

That is the present witness of those who have looked unflinchingly at the facts. Two clergymen,[1] one of them the rector of one of the largest of our city churches, the other now a missionary bishop, formerly a hard-working priest among the city poor, have recently given public utterance to the statement that in many tenement-houses morality is practically impossible.

One question remains: Can anything be done to set things right? I can almost hear some one saying, "Oh, well, it is all very bad, no doubt; but it always has been, and I suppose it always must be." There is an answer to that. This is not a matter for sentiment, or pious condolence, but for justice. Thirty years ago Christian communities in many parts of this country were content that thousands of human beings should live in a condition of life where the marriage relation was unknown, and children grew up in utter ignorance and vice. But at last the conscience of the American people awoke to the wrong inflicted, and in its highest legislative assembly assured to the negro slaves of the South the rights of men. And have not the tenement-house people of our own race, our own blood, capable, many of them, of education and refinement quite equal to our own — have not they and their children a right to live pure and good lives? And if this is their right, then the enjoyment of it must be theirs sooner or later. If there is a God in heaven, and if righteousness and judgment are the habitation of his throne, it cannot be his will that one of these little ones should perish. Shall we work with him that his will be done, that even the weakest and poorest shall find the way open before him to purity and peace; or shall we longer withhold the poor from their desire, and turn away the stranger from his right, and plunder the heritage of the needy, and so be called to answer to the God of the poor in the day when he shall arise to shake terribly the earth? Already many hearts, among working people at any rate, are rising up to echo the call of a great English thinker:

> "Charitable persons suppose that the worst fault of the rich is to refuse the people meat; and the people cry for their meat, kept back by fraud, to the Lord of Multitudes. Alas! it is not meat of which the refusal is cruelest, or to which the claim is validest. The life is more than the meat. The rich not only refuse food to the poor; they refuse wisdom; they

[1] *Two clergymen:* The Reverend Dr. Rainsford, Rector of St. George's Church, New York City, and the Right Reverend W. D. Walker, S.T.D., bishop of Northern Dakota, formerly in charge of Calvary Chapel, New York City. [Huntington's note.]

refuse salvation. Ye sheep without shepherd, it is not the pasture that has been shut from you, but the Presence. Meat: perhaps your right to that may be pleadable; but other rights have to be pleaded first. Claim the crumbs from the table, if you will; but claim them as children, not as dogs. Claim your right to be fed; but claim more loudly your right to be holy, perfect, pure."[2]

Let us acknowledge that claim, and strive for the destruction of the tenement-house system, for the bringing in, even in the midst of the darkness of our great cities, of the kingdom of light, liberty, and love.

[2] *"Charitable persons . . . pure"*: John Ruskin, "Unto This Last," *Ad Valorem*. [Huntington's note.] Ruskin (1819–1900) was a prominent British essayist, aesthetician, and moralist.

JACOB A. RIIS

"The Problem of the Children"

Jacob A. Riis (1849–1914) was born in Denmark and emigrated to the United States in 1870. He worked a variety of odd jobs before obtaining newspaper work, which eventually led to a position as police reporter for the *New-York Tribune* and later the *Evening Sun*. His job as a crime reporter often took him into New York's tenement districts. The living and working conditions of the poor working class prompted Riis to contribute many newspaper articles describing their plight and to organize support for their relief. His efforts to garner public sympathy for New York's poor largely failed until he began using flash photography to depict the squalid interiors of their homes. He published several photographs in *How the Other Half Lives* (1890), which went through numerous printings. Riis continued to discuss the poor in subsequent works including *The Children of the Poor* (New York: Scribner's, 1892), from which the following selection has been excerpted (1–7).

The problem of the children is the problem of the State. As we mould the children of the toiling masses in our cities, so we shape the destiny of the State which they will rule in their turn, taking the reins from our hands. In proportion as we neglect or pass them by, the blame for bad government to come rests upon us. The cities long since held the balance of power; their dominion will be absolute soon unless the near future finds some way of scattering the population which the era of steam-power and industrial development has crowded together in the great centres of that energy. At the beginning of the century the urban population of the United States was 3.97 per cent. of the whole, or not quite one in twenty-five. To-day it is 29.12 per cent., or nearly one in three. In the lifetime of those who were babies in arms when the first gun was fired upon Fort Sumter it has all but doubled. A million and a quarter live to-day in the tenements of the American metropolis. Clearly, there is reason for the sharp attention given at last to the life and the doings of the other half, too long unconsidered. Philanthropy we call it sometimes with patronizing airs. Better call it self-defence.

In New York there is all the more reason because it is the open door through which pours in a practically unrestricted immigration, unfamiliar with and unattuned to our institutions; the dumping-ground where it rids itself of its burden of helplessness and incapacity, leaving the procession of the strong and the able free to move on. This sediment forms the body of our poor, the contingent that lives always from hand to mouth, with no provision and no means of providing for the morrow. In the first generation it pre-empts our slums;[1] in the second, its worst elements, reinforced by the influences that prevail there, develop the tough, who confronts society with the claim that the world owes him a living and that he will collect it in his own way. His plan is a practical application of the spirit of our free institutions as his opportunities have enabled him to grasp it.

Thus it comes about that here in New York to seek the children of the poor one must go among those who, if they did not themselves come over the sea, can rarely count back another generation born on American soil. Not that there is far to go. Any tenement district will furnish its own tribe, or medley of many tribes. Nor is it by any means certain that the children when found will own their alien descent.

[1] *pre-empts our slums:* It is, nevertheless, true that while immigration peoples our slums, it also keeps them from stagnation. The working of the strong instinct to better themselves, that brought the crowds here, forces layer after layer of this population up to make room for the new crowds coming in at the bottom, and thus a circulation is kept up that does more than any sanitary law to render the slums harmless. Even the useless sediment is kept from rotting by being constantly stirred. [Riis's note.]

Indeed, as a preliminary to gaining their confidence, to hint at such a thing would be a bad blunder. The ragged Avenue B boy, whose father at his age had barely heard, in his corner of the Fatherland, of America as a place where the streets were paved with nuggets of gold and roast pigeons[2] flew into mouths opening wide with wonder, would, it is safe to bet, be as prompt to resent the insinuation that he was a "Dutchman," as would the little "Mick" the Teuton's sore taunt. Even the son of the immigrant Jew in his virtual isolation strains impatiently at the fetters of race and faith, while the Italian takes abuse philosophically only when in the minority and bides his time until he too shall be able to prove his title by calling those who came after him names. However, to quarrel with the one or the other on that ground would be useless. It is the logic of the lad's evolution, the way of patriotism in the slums. His sincerity need not be questioned.

Many other things about him may be, and justly are, but not that. It is perfectly transparent. His badness is as spontaneous as his goodness, and for the moment all there is of the child. Whichever streak happens to prevail, it is in full possession; if the bad is on top more frequently than the other, it is his misfortune rather than his design. He is as ready to give his only cent to a hungrier boy than he if it is settled that he can "lick" him, and that he is therefore not a rival, as he is to join him in torturing an unoffending cat for the common cheer. The penny and the cat, the charity and the cruelty, are both pregnant facts in the life that surrounds him, and of which he is to be the coming exponent. In after years, when he is arrested by the officers of the Society for the Prevention of Cruelty to Animals for beating his horse, the episode adds but to his confusion of mind in which a single impression stands out clear and lasting, viz., that somehow he got the worst of it as usual. But for the punishment, the whole proceeding must seem ludicrous to him. As it is he submits without comprehending. *He* had to take the hard knocks always; why should not his horse?

In other words, the child is a creature of environment, of opportunity, as children are everywhere. And the environment here has been

[2] *nuggets of gold and roast pigeons:* Here, Riis echoes the hyperbolic imagery of early American promotion literature, which Benjamin Franklin had refuted in *Information to Those Who Would Remove to America* (1784:978): "In short America is the Land of Labour, and by no means what the English call *Lubberland*, and the French *Pays de Cocagne*, where the streets are said to be pav'd with half-peck Loaves, the Houses til'd with Pancakes, and where the Fowls fly about ready roasted, crying, *Come eat me!*"

bad, as it was and is in the lands across the sea that sent him to us. Our slums have fairly rivalled, and in some respects outdone, the older ones after which they patterned. Still, there is a difference, the difference between the old slum and the new. The hopelessness, the sullen submission of life in East London as we have seen it portrayed, has no counterpart here; neither has the child born in the gutter and predestined by the order of society, from which there is no appeal, to die there. We have our Lost Tenth to fill the trench in the Potter's Field; quite as many wrecks at the finish, perhaps, but the start seems fairer in the promise. Even on the slums the doctrine of liberty has set its stamp. To be sure, for the want of the schooling to decipher it properly, they spell it license there, and the slip makes trouble. The tough and his scheme of levying tribute are the result. But the police settle that with him, and when it comes to a choice, the tough is to be preferred to the born pauper any day. The one has the making of something in him, unpromising as he looks; seen in a certain light he may even be considered a hopeful symptom. The other is just so much dead loss. The tough is not born: he is made. The all-important point is the one at which the manufacture can be stopped.

So rapid and great are the changes in American cities, that no slum has yet had a chance here to grow old enough to distil its deadliest poison. New York has been no exception. But we cannot always go at so fast a pace. There is evidence enough in the crystallization of the varying elements of the population along certain lines, no longer as uncertain as they were, that we are slowing up already. Any observer of the poor in this city is familiar with the appearance among them of that most distressing and most dangerous symptom, the home-feeling for the slum that opposes all efforts at betterment with dull indifference. Pauperism seems to have grown faster of late than even the efforts put forth to check it. We have witnessed this past winter a dozen times the spectacle of beggars extorting money by threats or violence without the excuse which a season of exceptional distress or hardship might have furnished. Further, the raid in the last Legislature upon the structure of law built up in a generation to regulate and keep the tenements within safe limits, shows that fresh danger threatens in the alliance of the slum with politics. Only the strongest public sentiment, kept always up to the point of prompt action, avails to ward off this peril. But public sentiment soon wearies of such watch-duty, as instanced on this occasion, when several bills radically remodelling the tenement-house law and repealing some of its most beneficent provisions, had passed both houses and were in the hands of the Governor before a voice was raised against them, or anyone beside the politicians and their backers seemed even to have heard of them. And this hardly five years after a special commission of distinguished citizens had sat an entire winter under authority of the State considering the tenement-house problem, and as the result of its labors had secured as vital the enactment of the very law against which the raid seemed to be chiefly directed!

The tenement and the saloon, with the street that does not always divide them, form the environment that is to make or unmake the child. The influence of each of the three is bad. Together they have power to overcome the strongest resistance. But the child born under their evil spell has none such to offer. The testimony of all to whom has fallen the task of undoing as much of the harm done by them as may be, from the priest of the parish school to the chaplain of the penitentiary, agrees upon this point, that even the tough, with all his desperation, is weak rather than vicious. He promises well, he even means well; he is as downright sincere in his repentance as he was in his wrongdoing; but it doesn't prevent him from doing the very same evil deed over again the minute he is rid of restraint. He would rather be a saint than a sinner; but somehow he doesn't keep in the *rôle* of saint, while the police help perpetuate the memory of his wickedness. After all, he is not so very different from the rest of us. Perhaps that, with a remorseful review of the chances he has had, may help to make a fellow-feeling for him in us.

That is what he needs. The facts clearly indicate that from the environment little improvement in the child is to be expected. There has been progress in the way of building the tenements of late years, but they swarm with greater crowds than ever — good reason why they challenge the pernicious activity of the politician; and the old rookeries disappear slowly. In the relation of the saloon to the child there has been no visible improvement, and the street is still his refuge. It is, then, his opportunities outside that must be improved if relief is to come. We have the choice of hailing him man and brother or of being slugged and robbed by him. It ought not to be a hard choice, despite the tatters and the dirt, for which our past neglect is in great part to blame.

WILLIAM T. ELSING

From "Life in New York Tenement-Houses"

William T. Elsing, a clergyman from the Dewitt Memorial Church, devoted much time to New York City's poor. "Life in New York Tenement-Houses" first appeared in *Scribner's Monthly* 11 (June 1892): 697–721, the source of the following text, and was later collected in *The Poor in Great Cities: Their Problems and What Is Doing to Solve Them* (1895). Crane likely knew the *Scribner's* article, some passages from which anticipate *Maggie*. Compare Crane's "The girl, Maggie, blossomed in a mud puddle" with Elsing's "Some of the noblest young men I have ever known have worthless, drunken parents. Some of the most beautiful flowers grow in mud-ponds" (p. 139).

For nearly nine years I have spent much of my time in the homes of the working people, on the East Side, in the lower part of New York City. I have been with the people in their days of joy and hours of sorrow. I have been present at their marriage, baptismal, and funeral services. I have visited the sick and dying in cold, dark cellars in midwinter, and sat by the bedside of sufferers in midsummer in the low attic room, where the heat was so intense and the perspiration flowed so abundantly that it reminded me of a Turkish bath. I have been a frequent guest in the homes of the humble. I have become the confidant of many in days of trouble and anxiety.

I shall in this article tell simply what I have heard, seen, and know. I shall endeavor to avoid giving a one-sided statement. I have noticed that nearly all those who work among the poor of our great cities fall into the natural habit of drawing too dark a picture of the real state of things. The outside world has always been more inclined to listen to weird, startling, and thrilling statements than to the more ordinary and commonplace facts. If I were to crowd into the space of one magazine article all the remarkable things which I have heard and seen during the past nine years, I might give an absolutely truthful account and produce a sensation, and yet, after all, I should give a most misleading idea of the actual condition of the homes and the people with whom I have been so intimately associated. We must not crowd all the sad and gloomy experiences of a lifetime into a history which can be read in an hour.

What I have said applies especially to the homes of the people in the tenement-houses. An ordinary tenement-house contains five stories and a basement, four families usually occupying a floor. The halls in nearly all the houses are more or less dark, even during the brightest part of the day. In the winter, just before the gas is lighted, dungeon darkness reigns. When groping my way in the passages I usually imitate the steam crafts in a thick fog and give a danger-signal when I hear someone else approaching; but even when all is silent I proceed with caution, for more than once I have stumbled against a baby who was quietly sitting in the dark hall or on the stairs. In the old-style halls there is no way of getting light and air, except from the skylight in the roof, or from the glass transoms in the doors of the apartments. In the newer houses a good supply of air comes directly from the air-shafts at the side of the hall. The new houses are not much better lighted than the old ones. The air-shafts are too narrow to convey much light to the lower floors. In the older houses the sink is frequently found in the hall, where the four tenants living on the same floor get their water. These sinks in the dark halls are a source of great inconvenience. A person is liable to stumble against them, and they are frequently filthy and a menace to health. In the new tenements the sink is never placed in the hall. In addition to the owner and agent, in connection with every large tenement-house, there is a housekeeper. The housekeepers are usually strong and thrifty housewives who take care of the halls and stairs, light the gas, sweep the sidewalks, and show the rooms to new applicants, and frequently receive the rent until the agent or landlord calls for it. Sometimes the housekeeper deals directly with the landlord, who comes once or twice a month to look at his property and collect the rent. The housekeeper is frequently a widow, who gets free rent in exchange for her work, and by means of sewing or washing is able to provide food and clothing for her children. It pays the landlord to have one tenant rent free in order to have a clean house. If the house is small the housekeeper usually receives her rent at a reduced rate in exchange for her services. There is never any difficulty in getting a good housekeeper. The landlord or agent sees to it that the housekeeper does her duty and the housekeeper watches the tenants. If they soil the stairs and halls, she reminds them of the fact in no uncertain way. If a careless tenant gives unnecessary labor to the housekeeper that tenant will soon be compelled to seek other quarters. The result is that the stairs and halls in all the large tenement-houses are remarkably clean. I have visited a great number of them, and can confidently say that I have never seen the halls of a large tenement-house in as neglected and dirty a condition as the corridors of the New York Post-Office. But the moment you enter the rooms of the occupants you often step from cleanliness into filth. The influence of the housekeeper and the sight of the clean halls and stairs is to some the first lesson in cleanliness, and is not without its beneficial effects. There is a slow but constant improvement in this direction, and every year strangers from many lands are getting gradually acquainted with the use, value, and virtue of clean water.

The housekeeper is frequently wanting in the older and smaller houses, which were formerly occupied by one family, but now serve as homes for three or four. Every tenant is here expected to perform a portion of the housekeeper's duty without remuneration. These houses are sometimes extremely dirty, and the death-rate is higher than in the larger and better kept tenements.

Let us leave the hall and enter some of the homes in the larger houses. To many persons, living in a tenement-house is synonymous with living in the slums, yet nothing is further from the truth. It would be an easy matter for me to take a stranger into a dozen or more homes so poor, dirty, and wretched that he would not forget the sight for days, and he would be thoroughly convinced that a home cannot exist in a tenement-house; but I could take that same person to an equal number of homes in the same section of the city, and sometimes in the same house, which would turn him into a joyful optimist, and forever satisfy him that the state of things is not by any means as bad as it might be. To the casual observer the tenement-houses in many portions of New York present a remarkable degree of uniformity. The great brick buildings with their network of iron fire-escapes in front, their numerous clothes-lines running from every window in the rear, the well-worn stairs, the dark halls, the numerous odors, pleasant and otherwise, coming from a score of different kitchens presided over by housewives of various nationalities — these are all similar; but the moment you enter the rooms, however, you will find every variety of homes, many of them poor, neglected, wretched, and dirty; others clean, thrifty, and attractive; indeed, as great a variety as exists in the interior of homes in an ordinary town. There are homes where the floor is bare and dirty, the furniture broken and scanty, the table greasy, the bedlinen yellow, the air foul and heavy, the children pale, frowsy, and sticky, so that you squirm when the baby wants to kiss you; but there is also another and brighter side. There are at the same time thousands of cheerful, happy homes in the tenement-houses. The floor is frequently as clean and white as soap, water, and German muscle is able to make it. The tablecloth and bedlinen, although of coarse material, are snowy white. The stove has the brightness of a mirror, the cheap lace-curtains are the perfection of cleanliness, and the simple furniture shines with a recent polishing. There is nothing offensive about the well-washed faces of the children. A few favorite flowers are growing on the window-sill. The room contains a book-shelf with a few popular volumes. A bird-cage hangs from the ceiling; the little songster seems to feel that his music is appreciated in this tenement-kitchen, and pours forth more rich and tender notes than are ever heard in the silent chambers of the wealthy. In such homes the oft-recurring motto, "God Bless Our Home," is not an idle mockery.

A large number of tenement-houses in the lower portion of New York are only a little below the common up-town flat. It is often difficult to tell where the flat leaves off and the tenement begins. You get about as little air and sunshine in the one as in the other. The main difference lies in the number of rooms and the location. If some down-town tenement-houses stood up-town they would be called flats. The word *tenement* is becoming unpopular down-town, and many landlords have dubbed their great caravansaries by the more aristocratic name of "flat," and the term "rooms" has been changed to "apartments."

There are three distinct classes of homes in the tenement-houses; the cheapest and humblest of these is the attic home, which usually consists of one or two rooms, and is found only down-town. These are generally occupied by old persons. Occasionally three or four attic rooms are connected and rented to a family, but as small single rooms are sought after by lonely old people, the landlord often rents them separately. An old lady who has to earn her bread with the needle finds the attic at once the cheapest and best place for her needs. The rent of one or two unfurnished attic rooms ranges from $3 to $5 per month.

A large number of very poor people live in three rooms — a kitchen and two dark bedrooms. Where the family is large the kitchen lounge is opened and converted into a double bed at night. The rent for three rooms is generally from $8 to $12 per month.

The vast majority of respectable working people live in four rooms — a kitchen, two dark bedrooms, and a parlor. These parlors are generally provided with a bed-lounge, and are used as sleeping-rooms at night. The best room is always carpeted and often provided with upholstered chairs. The walls are generally decorated with family photographs and inexpensive pictures, and in some of them I have found a piano. These parlors compare very favorably with the best room in the house of the average farmer. The rent for four rooms is from $12 to $16 per month.

The rent is an ever-present and unceasing source of anxiety to a great many poor people. The family is sometimes obliged to go half clothed and live on the cheapest and coarsest food in order to provide the rent money. The monthly rent is a veritable sword of Damocles.[1] To a poor woman who dreads the coming of the landlord, the most enticing and attractive description of heaven which I have been able to give is a place where they pay no rent. The landlords are of necessity compelled to be peremptory and sometimes arbitrary in their demands. If a landlord were even a little too lenient his tenement property would certainly prove a losing investment. The apparently unreasonable harshness of many landlords is often justifiable, and the only means of securing them against loss. Generally where a good tenant is unable to pay the rent on account of sickness or lack of work the landlord is willing to extend the time a few weeks. I frequently find families who are two or three months in arrears. In the majority of cases where dispossess papers are served, the landlord does not know his tenant sufficiently well to trust him, or the tenant is unworthy of trust. Very few of those who are evicted are compelled to take to the street. In most cases sufficient money is collected from friends, neighbors, and charitable people to procure another place of shelter. Occasionally, however, all the worldly possessions of an unfortunate tenant are placed on the street. It is a pathetic sight to see a small heap of poor household stuff standing on the sidewalk guarded by the children, while the distressed mother is frantically rushing from one charitable organization to another in search of help.

A poor German woman came to me last year and informed me that her furniture was standing on the sidewalk, and she knew not what would become of her. She had with her a beautiful little girl. The child cried continually, but the mother's distress was too great for tears. She begged me in God's name to help her. I gave her but little encouragement, and dismissed her with a few kind words. She left without heap-

[1] *sword of Damocles:* Damocles, a courtier of the ruler of ancient Syracuse, was elevated to a position that allowed him to admire the crown's wealth and splendor. He accepted only to realize in terror that a sword hung above his head, suspended by a horsehair.

ing abuse on me or cursing the church for its neglect of the poor. A little later I went to the place where she informed me her furniture was and found all her earthly goods on the sidewalk. I inquired of some of her former neighbors about her character, and on being convinced that she was a worthy woman, rented two small rooms in a rear tenement. I found some young street-corner loafers, told them about the woman, and asked them to lend me a hand in getting the furniture moved. There is no man so bad that he will not do a good turn for another if you approach him properly. These young roughs went to work with a will, and when the poor woman returned from her last fruitless attempt to collect enough for a new home she found everything arranged. She was thankful and happy. I did not see her until two months later. Then she appeared in as great distress as before, and showed me a new dispossess paper. She informed me that she had failed to find work, everything had been against her, but she hoped to get on her feet if I would once more help her. I told her it was impossible for me to do anything more for her; so she thanked me for my former kindness and departed. That afternoon I heard of a lady in Orange, N. J., who wanted a house-servant and a little girl as waitress. I immediately thought of the German woman and promised if possible to send her out to Orange as soon as arrangements could be made. I was soon in the little rooms of the widow and her daughter and expected to be the bearer of joyful tidings. When I finished she looked sadly at the few scanty pieces of furniture and said:

"If I go to the country what shall I do with the stuff?"

"My good woman," I said, "the stuff is not worth fifty cents; give it to the boys to make a bonfire, and do what I tell you."

"But I have not money enough to leave the city."

I provided the fare, the boys had a glorious time around their fire, and that night, instead of sleeping in her comfortless room, the poor woman was on Orange Mountain. It would have been a losing investment for any landlord to have given an extension of time to that woman, and yet she was a thoroughly worthy person, as the sequel proved; her old misery and trouble were at an end. She found a good home and gave perfect satisfaction. . . .

The population of the tenement-houses in lower New York is continually changing. There is a constant graduation of the better element. As soon as the circumstances of the people improve they want better homes. A foreigner who took up his abode in a tenement-house fifteen or twenty years ago may be perfectly contented with his surroundings, but when his children grow up and earn good wages they are not satisfied with a tenement-house, and give the old people no peace until a new home is found. Sometimes a man who has led a bad life reforms and immediately seeks a better home for his wife and children. I know several men who were at one time low and degraded drunkards, who would have been satisfied with a pig-sty, who had torn the clothes from their children's backs, the blankets from their beds, and taken them to the pawn-shop to get money for drink; but through the good influences that were thrown around them, the wise counsel of friends, and the saving power of the gospel they became changed men. Their circumstances began to improve, the children were provided with clothes, one piece of furniture after another was brought into the empty rooms, until the place began to look like a home again. These men were charmed with the new life. Home became so dear a place that they are willing to travel an hour each morning and evening in order to make it still more attractive. They began to see the disadvantages of life in a tenement and found a new home on Long Island or in New Jersey. . . .

The hope of our great cities lies in the children of the poor. If we can influence them to become upright, honorable men and women, we shall not only save them, but produce the most powerful lever for lifting up those of the same class who are sinking. I know scores of children and young people who are far better than their parents. Some of the noblest young men I have ever known have worthless, drunken parents. Some of the most beautiful flowers grow in mud-ponds, and some of the truest and best young women in our city come from homes devoid of good influences; but in all such cases uplifting outside help has moulded their characters.

While the people in tenement-houses are compelled to sleep in rooms where the sunlight never enters, and suffer many discomforts from overcrowding, especially in summer, there are certain compensations which must not be overlooked. The poor in large cities who have steady work are, as a rule, better fed and clothed than the same class in rural districts. Fresh vegetables, raised in hot-houses, or sent from Southern markets, are sold throughout the winter at reasonable prices, and in the early spring strawberries and various other fruits are for sale on the streets in the tenement district long before they reach the country towns and villages. In the poorest quarter of the city you find the so-called "delicatessen" shops, where the choicest groceries, preserves, and canned meats are sold. The clothing, too, worn by the young people is stylish and sometimes expensive; anyone who walks through these districts will be astonished at the number of well-dressed young people. A young woman who earns from $6 to $8 a week will often be dressed in silk or satin, made according to the fashion. The teeth, finger-nails, and shoes are often the only signs of her poverty. When visiting a stylish young woman's plain mother, I have sometimes seen all the finery in which the daughter appeared at church on Sunday hanging on the wall of a bare, comfortless bedroom not much larger than a good-sized closet.

The tenement-house people are not all thriftless, as the records of the down-town savings-banks clearly prove. Seven hundred out of every thousand depositors in one of the banks on the Bowery live in tenement-houses, and if it were not for tenement-house depositors several of our down-town savings-banks would be compelled to give up business. An abundance of cruel and bitter poverty, however, can always be found. The "submerged tenth" is ever present.

A widow, for instance, with three or four young children who is obliged to earn her bread by sewing, is in a most pitiable and terrible position. Hundreds of such weary mothers continue their work far into the night, with smarting eyes, aching backs, and breaking hearts. There is nothing which makes a man who has any feeling for the suffering of his fellows so dissatisfied with our present social system as the sight of such a poor woman sewing shirts and overalls for twenty-nine cents a dozen. There are good people in all our large cities who live just above the starving point. The average earnings of the unskilled laborers with whom I am acquainted is not over $10 per week. When a man is obliged to spend one-fourth of this for rent, and feed and clothe his family on the remainder, it is impossible to lay by anything for a rainy day. When the father is out of work for a considerable time, or when sickness or death enter the home, distress, hunger, and an urgent landlord stare him in the face.

It is easy for those who have never felt it to overlook the constant strain of poverty and the irritation which it causes in families which in circumstances of ordinary comfort would be contented. In such cases particularly can great good be accomplished by a visit from some clear-sighted and sympathetic person.

Recently I was invited to act as referee between a husband and wife. There were three little children and a grandmother in the family. The man worked in a cigar-box factory; business was slack and he was employed only half time. His average weekly earnings were $5. They had a debt of $11 at a grocery-store and another of $35 at an undertaker's shop. I know the family; both husband and wife are honest, sober, and industrious people. The wife wanted to break up house-keeping; the husband was opposed to this plan, and they had agreed to abide by my decision. I examined each one separately. I began with the husband and said:

"When a physician prescribes a remedy he must first know the disease. I want you, therefore, to tell me plainly why your wife wants to break up the home. There may be good reasons why her plan should be adopted. If you two cannot possibly agree, and are fighting like cats and dogs, then I may be in favor of breaking up. Tell me just how the matter stands."

He informed me that he and his wife had always lived in perfect peace. They never had any trouble except poverty. The wife had become completely discouraged, and the only way she saw out of the difficulty was to put the children into an orphan asylum and go out as a house-servant until she could earn enough to clear off the debt, after which she hoped to get her home together again. The wife and grandmother gave me the same account. The perpetual strain of poverty was the only reason for breaking up the home. For the sake of the three little children I decided that the home must not be broken up and promised to see that the debt at the grocery-store was wiped out and the family clothing was taken out of the pawn-shop. The grandmother was so pleased with the decision that she determined to become a servant and begged me to find a place for her.

In our large cities there is too much isolation between the rich and the poor. The charitable societies are often the only link between them. If the mother of every well-to-do home in our large cities would regularly visit, once a month, a needy family, a vast amount of good would be accomplished among the worthy poor, and distress would be unknown. Human nature is too selfish for such a happy state of things ever to be realized, but it is possible to bring the givers and receivers of charity closer together than they are. If some of the wealthier ladies who now give a few dollars each year to the charitable societies would seek through these societies to come into direct personal contact with the recipients of their charity, they would experience a deeper happiness and fully realize the blessedness of giving. Business men are too much occupied to make a monthly visit to the tenement-houses, but if their wives and daughters would undertake this work a new day would dawn for many a poor, heartbroken mother who is now hopeless and longing for death to end her misery. We are frequently asked, "Is it safe for a lady to visit these great tenement-houses?" We answer unhesitatingly, perfectly safe. The young ladies connected with the

City Mission go unmolested into the darkest portions of New York. The first visit to a tenement-house might be made in the company of a city missionary, after which the most timid could go alone.

Nothing is easier than to make paupers out of the poor. Great discretion must be exercised, but the Charity Organization Society, the Society for Improving the Condition of the Poor, the City Mission, the Children's Aid Society, and other equally worthy institutions[2] are ever ready to give direction to individuals who desire to do personal work. A few persons have through the City Mission come into personal contact with the poor, and the results are most gratifying.

While in a small town the distress of the poor is easily made known through friends and neighbors or the clergyman, in our large cities the most deserving are often overlooked and suffer most intensely; and it is these cases which are reached by personal visitation. The worthy poor are generally the silent poor. Their sufferings must be extreme before they make their wants known. There are many poor, upright, God-fearing old people who struggle against fearful odds to keep body and soul together, and yet they drift daily toward the almshouse on Blackwell's Island, the last and most dreaded halting place on the way to Potter's Field. I have nothing to say against the administration of the almshouse or the treatment of its inmates, but I do not wonder that old men and women who have led a good moral life would rather die than be stranded on the island and take up their abode among the broken wrecks of humanity which fill that institution.

It is very unwise to give aid without a thorough investigation. Not long ago a Polish Jew asked me the way to a certain street. I directed him, and he said: "Dear sir, I am in great distress; my furniture is standing on the sidewalk in Essex Street, and my children are watching

[2] *Charity Organization Society . . . and other equally worthy institutions:* Of these various philanthropic organizations, the New York City Mission Society is the oldest, having been established in 1827 as the New York City Tract Society for the purpose of disseminating religious tracts. It expanded its role at midcentury and began operating mission stations. The Association for Improving the Condition of the Poor was formed in 1843 and oversaw upper-class male (and later female) volunteers who visited the needy and helped advance sanitary reform. The Children's Aid Society, formed by Charles Loring Brace in 1853, opened lodging houses for boys and girls, ran industrial schools, and provided reading rooms, gymnasiums, and other wholesome recreational facilities. Also, it found foster homes in the Midwest for thousands of New York City children. The Charity Organization Society, established in 1882 by Josephine Shaw Lowell, sought to evaluate applicants for charity and emphasized the importance of preventing poverty. Together the four received the United Charities Building on East Twenty-second Street as a gift from John M. Kennedy in 1891.

the stuff, while I am trying to collect a little money to get another place." He drew from his pocket a few coppers, and asked me to add my gift. I said: "I do not know you, and I am acquainted with a great many poor people whom I would like to help, but I have not the means; how, then, can you expect any help from me?" Two streams burst from his eyes. The big tears rained down his beard and coat. "It is hard," he said, and bowed his head, buried his face in a red handkerchief, wiped off the tears, and passed on. I crossed the street. The tears of that sad man touched me. I turned, ran after him, and said: "Where is the stuff?" "In Essex Street." "What have you?" "A table, bureau, bed, and looking-glass," he replied. "Have you nothing small that I can take with me and loan you money on?" He pointed to his well-worn greasy coat, and said: "I have this." "Show me the stuff," I said. We walked together, and I endeavored to carry on a conversation with the stranger in German, for he was ignorant of English, but suddenly he seemed to have lost all knowledge of the German tongue in which he had before addressed me, and was perfectly dumb. When we reached Ridge Street he finally spoke, and asked me to wait for him a moment while he went to see a friend. I said: "Look here, I want you to take me to the stuff immediately." He looked amazed and said: "What have I to do with you?" "A good deal," I replied; "you either take me to the stuff or I take you to the police station." "Do you think I am a liar?" I said: "You must take me to the stuff or you are a liar." "Come," he said, "I will take you to the stuff." It was wonderful to see how that old man, who had moved so slowly before, walked through the crowded streets. I had all I could do to keep up with him. We soon reached Essex Street. It was Friday afternoon and Essex Street was in all its glory — old clothes, decayed meat, pungent fish, and stale fruit abounded. The Ghetto in Rome and the Jewish quarters in London and Amsterdam are nothing compared with Essex Street. At one place it was almost impossible to get through the crowd, and I left the sidewalk and took the street. In a moment my new acquaintance disappeared, and I have not seen him since. I have no doubt this man and many others like him are making a good deal of money by playing on the sympathies of poor people.

I have made it a rule never to give a homeless man money, but when his breath does not smell of whiskey I give him my card containing the name and address of a lodging-house. The card must be used the same day it is given. As some of those who ask for a lodging never use the cards, my bill is always less than the number of cards given out. One night a man told me he was tired of his bad life and he wanted to become a better man. I spoke a few encouraging words to him and was about to dismiss him, when he told me he was sick and needed just five cents to get a dose of salts. I took him at his word and immediately sent for the drug and made him take it on the spot. It is needless to say that he never troubled me again.

There remain many cases where charity is of no avail. Where poverty is caused by crime, no relief can come except by breaking up the home. Not long since I was called to take charge of the funeral of a little child. I groped my way up the creaking, filthy stairs of a small, old-fashioned rear tenement. I knocked, but heard no response; I pushed the door open, but found no one in the room, yet this was the place — "Rear, top floor, left door." I made no mistake. I entered the room and found a dead baby wrapped in an old towel lying on a table. I learned from the neighbors that the father and mother had been out collecting money to bury the child and had both become beastly drunk. I returned to the dead child, read the burial service, and thanked God that the little one was out of its misery. A little later a man came and took the body to Potter's Field. The parents had buried (it would be more accurate to say starved to death) six children before they were two years old. Very little can be done for such people. Cumulative sentences ought to be imposed upon them each time they are arrested for drunkenness, so that prison-bars may prevent them from bringing the little sufferers into the world.

Shops, Saloons, Concert Halls

JULIAN RALPH

From "The Bowery"

Julian Ralph (1853–1903), born in New York City, entered the newspaper business at fifteen as a printer's apprentice for the Red Bank, New Jersey, *Standard*. He worked as a reporter for many different papers during the next several years, but eventually his lively wit and hard-nosed reporting gained him a position with Charles A. Dana's well-respected New York *Sun* in 1875. He reported for the *Sun* over the next twenty years. During that time, he also contributed a variety of articles to magazines. He traveled widely and wrote articles describing his adventures to such faraway destinations as Russia and to places closer to home such as the Bowery, the subject of the following selection. This piece has been excerpted from an essay that appeared in the *Century Magazine* 43 (December 1891): 227–37. In the mid-1890s, William Randolph Hearst hired him away from the *Sun* to serve as London correspondent for the *New York Journal*. When the war between Greece and Turkey broke out in the spring of 1897, Ralph went to the front to report the war for the *Journal*. The *Journal* hired another man to report the Greco-Turkish War as well: Stephen Crane.

It was the opinion of the most observant traveler I ever knew that no city in Christendom possesses a street comparable with the Bowery in New York City. His comment on the Bowery was that it is the only noble and important thoroughfare which is foreign to the city and country that possess it. I think it is the belief of nearly all traveled Americans that the Bowery is the most interesting thoroughfare in America. If there are any who are inclined to dispute the belief, it will repay them to consider the Bowery even more closely than did my friend who called it foreign to its country, for he supposed it to be a German street in America. It is largely German, but it is much else besides, and the more it is studied the more cosmopolitan it will seem, and the more peculiarities it will reveal.

In endeavoring to compare it with some other crowded, humming, Babylonish artery of petty commerce and jostling human surplusage,

Broadway Traffic, ca. 1892. Courtesy of the New-York Historical Society.

the mind turns to the Strand in London. But it does not rest there, for though the Strand is about as long as the Bowery, it is a lane by comparison, and though the Strand lives one life by day and another by night, as the Bowery does, it is as English as the rest of London, and it is mainly dignified, respectable, and well-to-do. It is comprehensible to any one who walks the length of it once; but the oftener you walk the

Bowery the more heterogeneous and contradictory you will find it. It is good to the pure in heart, criminal to the wicked, abandoned and disreputable to the outcast. It is the main boulevard of a population of nearly 300,000 East-Siders — their Strand for practical, matter-of-fact shopping by day, and for the pleasures of the theater and the concert-garden by night. But they maintain only two sides of it. Its half-dozen other characters rely for maintenance on strangers from every corner of the world — because to the immigrant and the poor new-comer it is the great show street of the town.

The Bowery is very old. It got its name from the first settlers of Manhattan, and dates with them. The word *bouwerij* is Dutch for farm, or country-seat, and our street derives its name from the fact that it ran through the bowery, or farm, of Peter Stuyvesant, Governor-in-chief of Amsterdam in the New Netherlands, and of the Dutch West India Islands. His estate reached from the highway to the East River, and the Stuyvesant mansion, just north of St. Mark's Church on Second Avenue, remains in a modern and enlarged form. His dust is hidden from us by a great stone that incloses a vault under the east wall of the present church, which is called "St. Mark's in the Bowery," though it was built in 1795, more than a century after the Dutch governor died.

In English colonial days the Bowery was the beginning, or the end, of the Boston Road, and during the Revolution, the present Atlantic Garden was the Bull's Head Tavern, or sojourning place and exchange of the New York drovers and butchers of that day. Next door, on the site now occupied by the famous old Bowery Theater, was the cattle-market, an inclosed lot for the herding and sale of cattle. There the British made it a custom to enjoy bear-baiting, that sport to which it was afterward so wittily said that the Puritans objected, not because it hurt the bears but because it amused the people.

Then came a period when the Bowery had grown to be not only a long and important street, but a respectable one. Tom Hamblin[1] was the manager of the old Bowery Theater at that time, and the first players of the country and of England performed there to notable audiences. They cannot have escaped severer criticism than their sons, the Booths and Wallacks of our day,[2] have been accustomed to, for a

[1] *Tom Hamblin:* Thomas Sowerby Hamblin (1800–1853) emigrated to America in 1825 and managed the Bowery Theatre from 1830 until his death.

[2] *Booths and Wallacks of our day:* The Booths were a family of actors that included Junius Brutus Booth (1796–1852) and his sons, Edwin Thomas Booth (1833–1893) and John Wilkes Booth (1838–1865) (Abraham Lincoln's assassin). The Wallacks were another famous family of actors.

preacher of that time made a solemn sensation by saying that when he passed that theater he saw the people jostling one another down the steps into a great black, yawning hole under the ground, and over their heads he read the awful, the ominous words, "The Pit."[3] In those days many rich and aristocratic families lived over on the East Side beyond the Bowery. The Quakers, now few and seldom heard of, were numerous and notable among them, and East Broadway — the heart of the Polish Hebrew quarter — was a splendid street. But the city grew, and with its growth came the development of the Volunteer Fire Department, and with that the Bowery changed again. Many of the finest young men of the town belonged to the fire-companies at first; sons of rich men and young mechanics pulled shoulder to shoulder at the ropes. But an era of ruffianism was at hand — an era that produced in New York, Philadelphia, and Baltimore such scenes and conditions as we can scarcely comprehend to-day. Gangs of fighting men infested various localities and terrorized the community. The rivalry and strife of the fire-companies in part attracted them and in part developed them. From striving to see which company could reach a fire earliest they came to striving to prevent each from getting to the fires at all. In some degree they were the cause of fires — when fate was kind, and conflagrations were too infrequent to please them. In this era was developed "the Bowery boy," the queerest product of America in his day.

The Bowery boy began with more good than evil in his composition. In the daytime he worked for his living; at night he aimed only to be a dandy and a fireman. He sang negro melodies very prettily, danced well, was a devoted patron of the theater, and worshiped good women. But with the growth of the city he came to have his own way to a greater extent than was good for him, and his type grew worse and worse, until the Bowery often became a bloody battle-ground between the police and the ruffians that the Bowery boys had become. In time he became a drinking, fighting, and gambling character, with a modicum of the high principles and stern morality in heroic directions that we afterward found in some of Bret Harte's Pacific Coast characters.[4] Desirous of punching somebody at all times, he especially liked

[3] *"The Pit":* The word literally refers to the place in an auditorium located on the floor, yet it figuratively denotes hell.

[4] *Bret Harte's Pacific Coast characters:* Bret Harte (1836–1902) set some of his best-loved short stories and tales in northern California.

to punch persons who were rude or cruel to the female sex. He was intensely patriotic if he happened to be American, and it was in his time that Americanism, or Know-nothingism,[5] was very rampant and bellicose. There are men alive to-day — old men, to be found at Washington Market[6] or behind fast horses on "the Road" — who are given to wailing over the degeneracy of the times, and to boasting that they knew the day when the greatest prize-fighters and thugs and punchers were all true Americans!

The Bowery boy was very proud and full of an affectation of rough airs that he considered exquisite. He dyed his mustache jet-black, oiled his hair profusely, and was much given to loud perfume. He wore a lustrous silk hat, a flannel shirt with a huge black-silk scarf under its collar, trousers that were very tight and needed no suspenders, a coat that he usually carried on his arm, well-polished boots (not shoes), and carried a cigar tilted heavenward above his nose, and spread his elbows apart so that nobody could pass him on a narrow pavement without jostling him. Of course if any one jostled him he was insulted, and when he was insulted he fought. In the days of his glory he scorned to use any weapon but his fists. His voice was modeled after that of the fire-trumpet, and he had a language all his own. He called to his sweetheart, "Here, gal," "Come, gal," and when he wanted any one to hold the nozzle of a hose he said, "You, dere, take der butt."

It is said that Thackeray[7] much enjoyed meeting a Bowery boy. The great novelist desired to go to Houston street. He was not certain whether he was right in pursuing the direction he had taken, so he stepped up to one of these East-Side Adonises and said: "Sir, can I go to Houston street this way?"

"Yes, I guess yer kin, sonny," said the boy — "if yer behave yerself."

If you walk down the Bowery to-day you will see traces of all these eras except the Dutch, and that remains in the queer title of the street, as I have said. Though no other street shows such a blending of discor-

[5] *Know-nothingism:* The Know-Nothing Party was an anti-Catholic, anti-immigrant political party that originated in secret societies and rose to and fell from power during the 1850s.

[6] *Washington Market:* The nation's largest food market during the early nineteenth century, Washington Market, vastly overcrowded and rife with corruption, was condemned before the Civil War but rebuilt as a retail market during the 1880s.

[7] *Thackeray:* The British novelist William Makepeace Thackeray (1811–1863) visited the United States on lecture tours in 1852–53 and 1855–56.

dant qualities, it is yet true that no artery in the town has yielded so slowly to the modernization that the rest of the city has undergone. It is true the elevated railway, of the original single-legged pattern, skirts each pavement, but it passes many and many an old-time New York dwelling the third story of which still consists of the old dormer windows piercing a tilted roof, which, with the slanting wooden cellar doors, were the characteristics of the best houses of the city fifty or sixty years ago. Farther down the street the railway passes two or three wooden houses of that earlier era when it was permitted to build with wood in down-town New York. It even passes over a mile-post bearing the legend, "1 mile from the City Hall." It rattles the windows in the old Bull's Head Tavern of Revolutionary times, and it keeps a-trembling more than one queer, crooked relic of the English days, like little Doyers street, which is also mainly wooden, and which, though only a couple of blocks long, turns and dodges in several directions like a thief eluding a policeman. It is not a nice street, and it looks as if it were doubling upon its own unsavory reputation.

The Bowery is something less than a mile in length. It reaches from Chatham Square to the little wedge in front of the Cooper Union at Eighth street which splits it in twain, sending one half up-town to be the great Third Avenue, and one half close beside it to be the Fourth Avenue. It has the width of both these wide avenues together. Its width varies, as becomes an ancient thoroughfare, but I think it averages more than one hundred feet from house-line to house-line, sixty-five feet being the roadway. If you are a stranger, and walk down the Bowery in the daytime without a guide, you will be apt to notice nothing more particular about it than that it is an enormous, crowded, noisy street of retail shops, lodging-houses, and museums. Any old New Yorker will show you some very old and respectable shops — notably a grocer's, a baker's, and a shop for the supply of firemen's goods — which were established there in the days of other generations. But these are not so interesting to a stranger as the many little stores that give a distinct character to the street. Except in the main street of Havre,[8] I never saw so many shops for the sale of jewelry as there are on the Bowery. Most of them display new, cheap, and flashy ornaments; half a dozen are what are called pawnbrokers' sales shops, or shops for the sale of unredeemed pledges; one is a mart for duplicated

[8] *Havre:* Le Havre, the great seaport located on France's north coast.

presents received by persons on their wedding-days, on anniversary occasions, or at Christmas.

The pawnbrokers' sales shops have held me before their windows many and many an hour since childhood, and to-day when I pause before one I feel a keener touch of the impulses of youth than anything else can bring back to me. There is much humbug in the Bowery, but there is no humbug in what these stores display. Pathos and tragedy are constantly exhibited and enacted on every block of that throbbing avenue, but it all seems to me as nothing beside the tragic and pathetic tales that are told by the goods in these store-fronts. The vanity of man is felt by every poor stranger who is knocked about and jostled by the crowds that throng the pavement; but for a sermon upon vanity I know no text in all New York like the contents of one of these windows.

The very manner in which the dealers have shoveled the goods out for exhibition is impressive. It is usually their rule to heap the bottoms of the windows a foot or two deep with the less showy and bulkier relics of misfortune, and then to display the more peculiar and tempting goods on swinging shelves hung close to the panes. Here you see medals presented for heroism in saving life or for bravery in battle, swords given to men for taking part in actions that are household words, badges of bejeweled gold bearing the arms of petted militia regiments, all showing that their owners were once confident of fortune and yet must have come to desperate passes. A medal of silver to the best scholar in a great sectarian school, one of gold to the champion clog-dancer of Australia, a golden-headed malacca cane won by ——— ———, the most popular police officer in ———, these call to the mind happy scenes that no one dreamed would have such forlorn sequences. But what of the scores of opera-glasses and bracelets engraved with such mottos as "To Laura on Christmas," or "Isabel," or "With J. M. F.'s love to Sadie"? Rings, bracelets, breastpins, jewels especially devised, and curios which no one would part with except from stern necessity, are in the heaps and on the shelves — literally in burden by the ton when you take them all together; and yet every article in the mass carries its sermon of happiness despoiled, of security that was only fancied, of vanity that toppled, or of applause that beckoned anguish.

Whether the taste for cheap jewelry is stronger with our adopted fellow-citizens than with ourselves I am not sure, but one sees the force of foreign inclination unmistakably in other features of the street. The frequency of signs painted with Hebrew characters in German words, even in the windows of the banks, is no more mistakable than the occasional "delicatessen" shops, as the Germans call those places which are nearly like our "fancy groceries." The number of places for the sale of musical instruments is so great as to indicate that the majority of their customers are from continental Europe, and in the still larger number of cheap photograph-galleries the same influence is apparent. To stop and examine the tintypes and *cartes-de-visite*[9] displayed by the photographers is to carry yourself out of America at once. Not only are the types of faces mainly Teutonic and Slavonic, but the sitters have shown a very foreign fondness for being pictured in fancy costumes and maskers' dresses. They pose as kings and queens, as huntsmen, as Swiss and Polish and Magyar peasants, the matrons and maidens in very short skirts and the men in feathered caps and velvet knee-breeches. Those other men and women who are plainly dressed have kept their hats and bonnets on more often than is customary elsewhere, and the babies appear to be victims of a strange rule which requires them to be photographed in nudity or the state closest to it. The source of the fancy costumes is seen in the many places for the hire of masquerade dresses that are in the Bowery and close beside it in the cross-streets, these places being always up one flight of stairs. The costumes are hired for use at masquerade-balls, and it is on the morning after such a ball, before the dresses are returned, that the dancers wear them once again in the photograph-galleries. . . .

Merely in passing I spoke of the "lodging-houses" as notable features of the Bowery. They are almost peculiar to it. There must be a score of them. Invariably they occupy the upper stories of the larger and newer buildings along the huge and swarming thoroughfare, and therefore passengers in the elevated cars get the clearest idea of their interiors. From the pavement all that is seen of them are their signs, which read about like this:

EAST SIDE HOUSE.
FOR GENTLEMEN ONLY.
Rooms, 15 cents.

or

AMERICA HOTEL.
LODGINGS FOR MEN ONLY.
Nice rooms, 25 cents.

[9] *cartes-de-viste:* Small photographic portraits mounted on a card and originally used as visiting cards.

Within recent years these have multiplied to such an extent as to bring about a keen competition, and he who runs may read the force of this in single lines that have been added to many of the signs. These addenda all indicate a general desire to do more than supply mere rooms as of old. "Baths free of charge," is the announcement of one landlord; "Reductions by the week," another offers; "A Cup of Good Coffee served Mornings to Each Lodger," says a third. As you look into each house from the Elevated Railway you invariably see a large assembly-room, bare-walled but clean, and set with tables and chairs. There is no hour of the day when there are not many men in each room, some merely lounging in the chairs, some reading papers, some playing dominoes, and nearly all smoking. In passing some of these lodgings a glimpse is had of bedrooms which rent for a quarter of a dollar a night with a cup of coffee gratis. They are mere closets made by running partitions up five feet apart from the floor to the ceiling. Each contains a cot, and sometimes a chair. There is no appurtenance for anything except sleeping, a common wash-room being elsewhere provided. The men one sees in these places are nearly all young, mainly at the threshold of manhood. It is a general impression that they are either criminals or hardened characters, and though I am certain this does them injustice, I have never been able to satisfy myself to what extent they are injured by the suspicion. That there are among them many petty thieves and parasites who live upon outcast women is certainly true, and I suspect it requires great strength of character for a poor, stranded victim of circumstances who drifts into one of these places to resist the overtures that come to him from such wretches. Yet I know that many a poor huckster and sober wage-earner who has only a bare foot-hold in the town is obliged to put up at these lodging-houses, and it stands to reason that in the course of every year thousands of decent, ambitious strangers who come to the great city to make a living or a fortune must perforce begin their new career in these honeycombs. Now and then such a man shoots himself in one of these places or throws himself out of the window upon the pavement below.

By the way, it would not be easy to make most readers believe how trifling a thing a suicide is in the Bowery. It is not because there are so very many, since death's harvest by that means does not exceed two hundred and forty a year throughout the whole city, but it is rather on account of the preoccupation of the people and the summary action of the authorities. The shot is heard by very few. Neighbors of long standing do not know one another, so that the persons in the house where the death occurs deal only with the authorities, and no one spreads the news along the block. An ambulance calls for the body, and then there is the greatest stir, for a knot of idlers always gathers to find out what called the ambulance. The little crowd collects, and hides what is brought out of the house. The average busy New Yorker feels no interest at all in the matter, for it is his life habit to avoid crowds. The ambulance drives away, and it is not until they read the papers next day that the people on the very block on which the tragedy occurred become aware that it took place.

Three notable Bowery institutions that attract attention in the daytime have not been mentioned. They are the drinking-places, the dime museums, and the eating-houses. It will seem like an exaggeration, but I carefully counted them before I put down the number of places in which liquor is sold on the ground floors, alone, of the buildings along the Bowery. There are eighty-two such places, or nearly six to every block. The street is fourteen blocks long, and there are sixty-five places where drink is sold on its east side and seventeen on its west side. As there are five blocks on the west side of the street on which no such places occur, the reader can imagine how thick the bars must be on other blocks. This total number includes four music-halls, as many restaurants and oyster-houses where bottled beverages are sold, two or three wine-houses, one wholesale liquor-store, and the bars connected with several theaters and variety-halls. Some of the saloons have glittering exteriors and costly fittings, but not one is of the so-called first class. In the main they are cheap places of a low class, the number of them being so great as to reduce the profits to a minimum. A few staid and respectable German places are in the number, and one orderly resort — the Atlantic Garden — boasts one of the most profitable bars in a city where there are single counters over which $500 is passed every day in exchange for drinks. Lager beer is of course the standard tipple of the Bowery, and it flows there in such torrents that I am not guilty of the slightest exaggeration in saying that early on Sunday morning, after a busy Saturday night, the very air that is breathed in the great avenue is weighted with the odor of soured beer.

The eating-houses are not nearly so numerous, though their comparison with the drinking-saloons is greater than the proportion of bread to sack which Falstaff deemed sufficient.[10] The lodging-houses

[10] *the proportion of bread to sack which Falstaff deemed sufficient:* In Shakespeare's *Merry Wives of Windsor*, act 3, scene 4, lines 3–4, Sir John Falstaff orders, "Go fetch me a quart of sack, put a toast in't." Sack was a potent, sweet white wine generally imported from either Spain or the Canary Islands.

support many restaurants, and as the Bowery is a principal artery, the transient trade in food is sufficient to maintain as many more. Again competition shows its paring hand, for in front of some of the eating-houses one sees announcements that "large portions" of roast beef, mutton, lamb, pork, and veal are offered at eight cents, with bread and potatoes thrown in. Ten cents is the standard price for such provision, and, since milk, coffee, and tea are usually sold at five cents, it is possible to purchase a solid and nutritious meal for a dime and a half. A moment's calculation shows, therefore, that a man may eat and lodge in the Bowery with a good bed and three meals a day for $4.90 a week, and with a fifteen-cent bed and eight-cent dishes for $2.73 a week.

It sometimes seems to me that there is no avenue of profit or of commerce that is so illuminated by genius as the Bowery museum business. If ingenuity be a form of genius, there cannot be any doubt that I am right. A few visits to these resorts will satisfy the more intelligent citizens, and the visits will naturally be paid in early youth. But the populace as a whole is not characterized by the greater degrees of intelligence, and it is surprising to note how skilfully the managers of these places keep astir the ready curiosity of the mob. As much color and oil as have distinguished the galleries of the Louvre have been spent upon the huge canvases that all but cover the museum buildings. Sometimes the garish signs and pictures completely conceal the façades of the building and block up the windows, it having been found that many of the wonders on exhibition suffer less by gaslight than by the blaze of day. A museum is fairly started when it has a mass of gorgeous paintings, a tout, or crier, at the door, a ticket-taker in the lobby, and a band of three musicians limping, squeaking, and pounding just within the inclosure. I have known little of the interiors within recent years, but I see the signs frequently, and I have observed the progress that has been accomplished in the science of museum management since my boyhood days. The "fattest woman on earth" was sufficient in that era, but now she is represented twice as fat as of old, and yet dancing like a fay. There is most ingenious "faking" (the museum term for humbug) as of old, but there is also much reality — real "heroes" of trips over Niagara in barrels, of the bridge-jumping mania, of criminal life, and of distorted natural history. The more pretentious of these museums are so conducted that the only advantage that is ever taken of a stranger lies in the presumption that he will believe what he hears and credit what he sees. Yet in at least two of the six museums which illuminate the Bowery a fool or a too trustful stranger will be certain to be robbed. The tricks by which such persons are despoiled of their money are as old as sin itself, yet age does not wither nor custom stale a single one. An example of the dark ways of the robbers who lurk in these dens is this: Within one of the lower class of museums the visitor will notice a door through which he is invited to pass in order to have his cranium examined by a phrenologist, and to receive a present of a chart setting forth his proclivities and possibilities. Within is a room, a chair, and the alleged phrenologist. The visitor notices that the walls are bare; at least he perceives nothing to interest him as he glances around him. But just as the phrenological inspection is finished, a click is heard, a piece of a partition falls down upon a set of hinges, and the victim reads, "Professor Blinkum's charge is $2." If the victim is wise he will pay the fee; it will be cheaper than the drubbing and perhaps the actual robbery by violence to which he must otherwise submit.

The museums are brilliant at night, and it is then that the Bowery becomes newly and doubly interesting. It is probably the most brilliantly lighted thoroughfare on this planet. The money spent in lighting it is prodigious; the illumination is prodigal; the effect is dazzling. But the method adopted for this lighting is cheap and vulgar, and emphasizes the popular meaning which the word "Bowery" has taken on. The English word "brummagem"[11] fails to convey half the definition of the term "Bowery." The words lean in the same direction, but to be Bowery is to be twice what is meant when we say a thing is brummagem. Whatever has the Bowery stamp is not merely an imitation, but it is a loud and offensive falsity. In New York, when the people see a great glass stud, cut to look like a diamond worth $10,000, and worn on the shirt of a store clerk, they call it a Bowery jewel, and they say of the man that he looks very Bowery. The extremes of fashion are caricatured and intensified in the Bowery, where the cut of men's trousers, the size of plaid patterns, the shape and style of the shoes, the gorgeousness of the waistcoats worn by the mock dandies — not to speak of the swagger and swing of the East Side belles — often surpass endurance if not belief. A Bowery dude is constitutionally unable to put on his hat unless he may balance it on one ear. It suits the street, therefore, to boast the most brilliant illumination of the coarsest and most dazzling sort.

I counted its surplus lights the other night, — the mere electric arc-lights which dangle before the stores and resorts, — and I found that

[11] *brummagem:* Cheap imitation goods.

they numbered 263. On the west side there were 189, and on the east side 74, or, altogether, about 19 to each block. The arc-light is that variety of electric lamp which is produced between two thick carbon-pencils inclosed in a great cocoanut-shaped shell of glass. Let the reader who is familiar with this added burden upon human existence, this ingenious instrument of torture, fancy, if he can, the hissing and sputtering, the lightning-like starts and jumps, the alternating flashes and depressions that the glare of the Bowery undergoes. A tour of this street by night is a never-to-be-forgotten experience, but in the main the street is like a great electric lantern. It is the most brilliant eye in the Argus head of New York, and it is the eye that never sleeps; for when the rest of the town is dim, and its bustle is all but hushed, the eye of the Bowery looks out into the night with a gleaming stare that only the rising of the sun is able to intimidate.

The great wholesale houses have closed, but the people of a vast network of streets walled with high tenements have come home from work, have supped, and are out on the Bowery for the night's shopping, amusement, or exercise. The sidewalks are almost packed with people bathed in the brilliant light of such a number and variety of shops as are not to be found in any other equal area in the city. But the outcasts of society are in the throng; the tenth of the town that lives by night is astir. Poor creatures, indeed, are these Bowery miscreants — the product of that same tenement region where, a careful missionary says, one hundred thousand persons have moved in and fourteen churches have moved out within the past ten years. The criminals found in the Bowery are of the stunted, half-starved type of which the tenement house is the matrix. Undersized, wizen-faced, aged while yet of tender years, little-eyed, cunning, shabbily dressed and constantly hunted, they are rather like human rats than men and women. Their haunts are in the cellars, the rum-shops, and in the disorderly places on upper floors — for it is a peculiar fact that not only does the Bowery contain liquor-stores side by side in places, but it contains rows of buildings in which every floor is given over to disreputable uses. I shall not dwell upon that phase of the Bowery life except to answer the question that is asked of every citizen by every stranger who is curious to visit that quarter — "Is it safe?" It is. Better than that, it is worth while. It is not well for a lady to walk out alone in any part of the city at night. Yet a woman without an escort, walking briskly along, is less likely to be affronted on the Bowery than on Fifth Avenue, by day or by night. There is one rule for escaping annoyance in New York city. It is the same for women as for men. That is to walk straight along without stopping or staring. It is the gawk, the gaby, the idler, and the over-curious meddler who invites insult and annoyance.

By half-past nine o'clock the shopping-places have closed, and the fourfold procession of shoppers has come to an end. The last family group, headed by the husband, with the wife a step behind him, and her babies trailing after her each clutching the other's clothing, has been swallowed up by the darkness of the side streets. The Bowery now belongs to the seekers of recreation and of vice. They are moving in and out of the museums, the gin-shops, the concert-halls, and the theaters. They have the choice of ninety-nine such places. Seven of these are theaters, six are museums, and four are music-halls.

The English theaters (or American theaters in which English is spoken) are what are called "gallery-houses"; that is to say, the gallery forms the most important if not the largest part of each. To enter certain ones costs only ten cents, and fifty cents secures an orchestra chair. In two, which are handosme theaters, the best plays and nearly the best companies are seen. They are operated as the theaters of small cities are, being considered as part of the provincial circuits to which New York successes are sent after their runs in first-class up-town houses. But the other English theaters are for the exhibition of variety-shows, or music-hall performances. What has always interested me most about them is the fact that they attract the newsboys and street Arabs with irresistible magnetism. The average New York newsboy, when he counts the cost of a day's living, includes ten cents for "de tee-a-ter" as regularly as he figures upon the amount for lodgings and for his three meals of "beef and beans." As there are thousands of these boys, the number that have earned the price of a gallery-seat is very great each night, and in consequence the strife for an early choice of seats is vigorous. The result is that the ragged little shavers form a line long before the theater doors are opened, and this line grows, and lengthens, and tails along the sidewalk until it makes what would be a notable picture for a Mrs. Stanley to fix upon her canvas. There are fights now and then in the line, and a babel of cries and whistles and shouts goes out from it. When the doors are opened the rush up the theater stairs is like a mountain freshet reversed. Like stampeding cattle the boys fling themselves down the aisles and over the seats until there is not a vacant place left. Then they take their coats off and fold them in their laps, and the air fills with the aroma and crackle of peanuts. Monitors, with long ratans and uncommonly bad tempers, endeavor to keep the little savages in some sort of order, and it is to these guardians that reference is made in the frequently repeated cry of

"Cheese it! de post!" There is no time here for a study of that queer sentence. "Cheese it" is the warning cry of the New York street-boy, and though many have guessed at it, I have never known any one who was able to give its derivation. "The post" is the monitor, but why he is called a post in a Bowery theater, and nowhere else, some one else must explain. . . .

In parting with the subject, let me add that the survival of the ancient "true American" spirit (always suspicious of danger to the Republic and always belligerent) still leads some good citizens to harbor deep suspicions of all that the Bowery typifies. They tremble lest foreigners, in numbers sufficiently great to maintain Old World customs, should endanger the existence of our own institutions. I do not read any danger in any feature that makes up the Bowery except in its vices, and they are human rather than peculiar to any nationality. The "true Americans" of the first half of the century were themselves the offspring of foreigners, and so, by no greater removes, are many of those who now carry forward the old patriotism. This is, in some degree, true of all of us except the red men, but it is especially true of New Yorkers. This city has always been an open door to foreign immigrants, and lately it has been their principal gateway. A few always linger here at the threshold of the New World, and, being thrown together again, establish so-called colonies or foreign quarters. Therefore we have the Bowery as it is. It does not offer any new problem or confront us with an unfamiliar condition. For more than two centuries the city's population has contained a very considerable admixture of persons foreign to those who have ruled it, and at times some of the new blood has been far less desirable than any considerable element which we are now taking into the national system.

DAVID GRAHAM PHILLIPS

"The Bowery at Night"

David Graham Phillips (1867–1911) was born and raised in Indiana. After graduating from the College of New Jersey (later Princeton) in 1887, he worked as a journalist, first in Cincinnati and then in New York City. This essay appeared in *Harper's Weekly* 35 (September 19, 1891): 710, shortly after Phillips had moved to New York, so it provides a good example of an outsider's impressions of the Bowery. Some years later, Phillips

would become a highly successful novelist. His muckraking novels attacked corruption in government and in the financial world. His later novels, several published posthumously, treat women's issues. His most important is a story of a prostitute, *Susan Lenox: Her Fall and Rise* (1917).

When I was a school-boy in a Western town I had the habit — I and many of my schoolmates — of reading those romances of the adjective and the exclamation point behind a geography tilted high to conceal the novel from the teacher. As my favorite writers divided their attention pretty evenly between the far West and the New York Bowery, I got, or imagined that I got, a clear idea of life in both those places.

To me the Bowery was a wonderful place — fascinating, full of romance, yet terrible and most dangerous. It seemed to me that its streets were inhabited by beings of superior wit and intelligence, renowned detectives of many disguises, gentlemanly but desperately wicked criminals, honest keen young men who were always getting into difficulty, but always escaping in some creditable heroic way.

And I have found that this idea in a modified form prevails generally to a great extent, in the minds of New York people even. The Bowery is supposed to be the seat of certain high classes of criminals; of certain kinds of merchants shrewder than any other kind and less scrupulous; of a Bowery boy and a Bowery girl.

So wide-spread is this impression without New York that the Bowery is inspected daily in summer by crowds of tourists. Some of them walk its streets with eyes looking right and left for signs of danger, with coats tightly buttoned, and money hidden in deep-lying pockets. Others — and they, for the most part, have their families with them — observe it open-mouthed from the safety of a hired hack. By a queer twist in human nature few of these people are undeceived. Confident of finding their ideal realized, they are not disappointed, and returning home, tell of the wonders they have seen.

As a matter of fact, the Bowery is no such place as out-of-town New York thinks it to be. It is a remarkable place — interesting, fascinating almost, but neither superhuman nor dangerous. Time was when the latter could not be said; but the times have changed. Five Points has given way to Paradise Park; Armory Hall is torn down; old landmarks of crime and vice have vanished; and the Bowery boy and the Bowery girl are traditions. So the wicked romancer must put his stories in the past tense — his Bowery stories as his Indian and road-agent tales — Bowery Joe and Deadwood Dan.[1]

I confess that my first sensation in examining the Bowery was strong disappointment; the same sense of having been deceived that accompanied the facts about Santa Claus, or, better still, the first acquaintance I had with a detective.

Yet at first view the Bowery impresses you as a place which ought to come up to the romance. As you come up Park Row, itself a curious place, you find yourself getting into a deep shadow, like the entrance to a cave. The air smells close and musty. The sunlight has taken the freshness with it. There are elevated tracks hanging low, and so closely covering the street that rain seldom falls upon it. This is Chatham Square, the entrance to the Bowery. Leaving the deep twilight of Division Street to the right, you go straight up into the broad yet dark highway of east-side life, crowded both in street and sidewalks, noisy with a multitude of sounds, some of which are to be heard nowhere else, shadowed by the tracks of the elevated, which hang low over the entire width of the street from Chatham Square to Grand, and in these ten or fifteen blocks is contained the true Bowery. Here are jumbled together a queer and varied lot of enterprises.

In and around Chatham Square the chief business is the lodging-house. On the outer walls are hung great signs, bearing pretentious names, The Windsor, The Grand Windsor, The Atlantic, The Pacific, The Grand. Some of these places, most of them in fact, have a front as inviting as the names; there is gaudy paint, shining brasswork, an air of cleanness, luxury even. This splendor is strangely out of keeping with the price-lists hung beneath the signs and over the sidewalks, generally on cloth, through which a light shines at night, that he who runs may read. The prices range from fifteen to fifty cents. Clean sheets are offered at some places as an especial inducement. All this appears from the street.

If you enter you will see how hollow is the outward show, how exorbitant the price for the accommodation offered. There is a general room where lodgers may sit in the daytime. Then there is a clerk's box, with a narrow way in front of its window, through which access is had

[1] *Bowery Joe and Deadwood Dan:* The Bowery detective was a popular character type in the nickel and dime novels of the period. See, for example, William Henry Manning's *Bowery Bob, Detective* (New York: Beadle and Adams, 1891). Dandy Dan of Deadwood was a character created by Luis Senarens for a series of nickel westerns during the early 1890s.

to the sleeping-rooms — either a single room with rows of beds, or divided into little box-like places, the partitions not reaching to the ceiling. And what squalor, what odors, what bare and gaunt horror is here!

On these narrow dirty beds what misery finds oblivion or partial oblivion! For sleep here is sometimes heavy and sotted, deeply, wretchedly breathing; again, it is broken by curses and moans, by wild startings and clutchings. If you go in at night and the wide-shouldered watchman lifts his light that you may look along the rows of beds, you will see in the dim, hot, foul air faces and forms to remember and to dream about. Matted of hair and beard, the sorrows, the bitterness of life risen to the face, and glowering there without the sullen concealment of the waking expression. And by each bedside the rags and tatters in which life is lived.

Here are beggars and vagrants: broken-down criminals; lonely old men, deserted of friends and family; murderers — all that which has been utterly wrecked in the storm. And in the late afternoon these lodgers come out to stand or slouch up and down in the Bowery. They go to the hospital, to the police station, to the Potter's Field, rarely to happier days.

Here live the Bowery thieves — unwashed, badly clad, sullen of aspect, moving in gangs, with hands in the bursted-out pockets of the trousers. They have the courage of despair at times — in hard winters especially. They have also a timidity which makes them look guilty on the approach of a policeman.

After the lodging-houses you will notice the dime museums. And here again is that pretentious exterior — the gay paint, the big signs, all the promise of good things within. There are pictures of curiosities that would appeal to the most blasé museum-goer. Inside it is a sad swindle. "Ladies and gentlemen, this is an image of the horned lady now living in Asia. This is an image of the two-headed calf now exhibiting in England." It is all fraud, all fake; and though you may have gone in expecting a cheat, you will have so far surpassed your expectations that you will look sneakingly about as you come out. The shooting-galleries with the gaudy swinging targets — lions, tigers, elephants — are equally a delusion, and you will never get the quarters that reward hitting the bull's-eye five times. The photograph galleries, where tintypes are to be had at phenomenally small prices, will give you likenesses to make you wonder at your own possibilities of homeliness.

Then there are the shops. The Bowery is a great trading-place. The lower east side does most of its shopping there — buys its hats and shoes and shirts and gowns and furniture. There are queerer shops in other streets not far from the Bowery; but the Bowery is the headquarters for swindling in merchandise, as in amusement.

Bargain signs hang over every door. Every one has just assigned, has just burned out, is leaving business. Everything is going at a sacrifice. All things are at bargain price; yet nothing is a bargain. There is everywhere a dreary monotony of cheapness — the cheapness that shines in the windows and falls to pieces in the arms of the purchaser on his way home. A dollar will buy more here than anywhere else in the city — and also less.

Among these shops are the pawnbrokers — pretenders and swindlers, as the others. For they are not pawnbrokers, for all their three-ball signs and their conspicuous directions as to how to reach the private entrances. They, like the auction houses, sell bad jewelry, fire-gilt watches, nickel-plated chains, gold-washed rings. They pretend that these things are unredeemed pawns. In reality not a pawn is in all the display.

In fact, the whole Bowery, in all its amusements, in all its business enterprises, is based upon the grand primal principle of the philosophy of humbug — that you can get something for nothing. That is the wherefore of all this tawdriness, all this vain show, this paint and gilding and glitter. These fakirs cater to the belief of low intelligence in its own shrewdness and cunning.

A stranger to New York walked up one side of the Bowery and down the other. When he emerged from its gloom, he said, "I never before knew how ugly the human race is." The Bowery is ugly, and the men and women who move about in it share its homeliness. Faces lower as the air. Real laughter — vim of health and spirits — is as rare as bursts of sunshine. The people are clearly on the edge of life, fighting anxiously, harassedly, for a foothold, and seemingly never quite gaining it. Few frames are stalwart; few shoulders do not stoop. Now and then the face of a young girl passes before you. There may be prettiness in it, but it is the sickly beauty of a potted flower in a sixth-story tenement back window. And you see clearly that even the trace of fairness that comes with all youth in all places will fade, vanish utterly, in the stifling atmosphere of its environment. And for this one touch of freshness, how much that is jaded and hollow-chested and scrawny-necked you will see! Generations of toil — hard, grinding, desperate toil — have produced these faces, these forms. Here, too, are the shadows of ancestors' sins that live behind uninviting faces. There is also the track of fresh vice, of avarice, of drunkenness, of sleepless nights followed by days of labor.

You need not go home with these people to find out their wretchedness. They have brought away the damp and poisonous exhalations of narrow halls and dirty rooms in their clothing. They have brought away the miseries of poverty in their faces. So each nationality in its own way explains to you why it is here and how it is struggling since it arrived.

All this is quite plain, to be sure. All human faces are interesting. But among the richer people the emotions are hidden by long schooling or softened by the comfort of the routine of life. Among these people there is no training to keep the face smooth and calm. The edge of vice and poverty is not turned. Poor food, poor lodging, dissipation, cut and tear and scar. So the mind and heart are bare upon the face.

And then you begin to see that the Bowery is the true cosmopolitan thoroughfare of New York. There are few nations not represented, few languages that do not fall upon the ear in the course of an afternoon. It is as though the workmen had returned to Babel. And there are glimpses of all costumes — a bit of gay color from South Italy here, the wooden shoes and yarn stockings of a German peasant girl there, a fez from Turkey or Arabia, a coat from China, a fur cap from Russia. No costume will be in its native completeness. The effect of each is spoiled by a touch or a good deal of the cheap ugliness so generously displayed in the shop windows.

Much that has been said of the Bowery by day applies to the night as well. But there are some differences. The electric lights, the gas jets, lift the shadow of the elevated tracks, and change the street to its betterment at first glance. Some of the ugly things are covered up. The places of amusement, now become the centre of interest, are brilliant. The crowd is looking for pleasure, for relief from squalor at home, from toil at shop and factory. The faces are lighted up somewhat, but there is still little sign of enjoyment that is not coarse and brutal and marked all over with the promise of swift and sure retribution. But there is a certain air of enjoyment, a glamour of pleasure. There is music in this garden; a woman is singing. There are lights in that second-story hall, and sounds of a waltz measure timed by shuffling feet. A drunken man — a score of drunken men in a block — reels along, muttering and chuckling. Girls — such girls as Kittie Lynch has made famous at Harrigan's[2] — are abroad, exchanging jests not without wit with young men who swagger.

Sometimes you will find things that are amusing. But before you are done laughing you will see something that will make the laugh die — something to disgust or to excite pity or indignation. There is a comedy element in this life. But the comedy is for the actors, not for spectators. The tragedy is too near the surface, peering from rents in garments or blanching under the paint of the cheeks. In this pleasure, or semblance of pleasure, you can have no part, any more than you could live in the smells and sounds and sights of a rickety towering tenement.

In this gathering of the nations in the Bowery lies the reason for the change that has come over it. The native element has been driven out. The little part of it which remains seems almost submerged by the grand in-rush. And these people, driven from many lands and from the islands of the sea by the press of hunger or by misfortune alone, separate each to a distant quarter. Seeking business or pleasure, they meet and mingle in the Bowery.

And of necessity they touch and are touched by the American element — a low American element, yet American. They get our customs, our manners, our methods, in a certain depraved form. Thus they reach the first step in their fight upward.

The Bowery is no longer a distinct quarter, inhabited by a distinct type of American life. It has become a social clearing-house.

STEPHEN CRANE

"Where 'De Gang' Hears the Band Play"

Stephen Crane published this fictional sketch anonymously in the *New York Herald* (July 5, 1891): 21, the source of the following text. Preparing a biography of Crane during the late 1940s and early 1950s, Melvin H.

[2] *Harrigan's:* Ned Harrigan (1845–1911) earned his reputation as a stage comedian and playwright with his numerous plays depicting life on the Lower East Side. Harrigan depicted a variety of ethnic types sympathetically and ably captured their various dialects. Harrigan's Theatre was located on West Thirty-fifth Street off Sixth Avenue.

Schoberlin discovered the sketch and attributed it to Crane. Schoberlin died before he could publish his findings, however, and this piece is not included in the Virginia Edition of Crane's writings. It was ignored until Thomas A. Gullason edited it, along with several other early Crane articles, in "The 'Lost' Newspaper Writings of Stephen Crane" ***Syracuse University Library Associates Courier*** **21 (Spring 1986): 57–87.**

Hard featured is the "tough youth." Hard mannered is the tough girl. She abounds on the east side. Down around Tompkins square[1] she and her striped jersey are particularly prevalent. There are *musicales* in Tompkins square these hot summer nights — band concerts they are called — and great is the rejoicing thereabouts each season at the advent of the band.

This particular tough girl's name was Maggie. Her intimates call her "Mag." And Mag goes.

After supper last Wednesday in the apartments of her parents in a Stanton street tenement house Mag announced to her brother: —

"Say, Jimmy, I'm going to the band play to-night an' I want de watch."

"Oh, you do, do you! Well, now, I'm just goin' out meself to-night an' I need de ticker mor'n I need a dollar. So you don't get it, see?"

"Don't be fresh, Jimmy, you know it's me own watch, an' you said you'd give it me back last week. Give it me now or you don't get it again."

"Again? Rats! I don't need to get it again. I got it now."

A cloud of suspicion settled in the narrow strip between Mag's bang and her eyebrows.

"Say, Jim, give it to me straight. Have you soaked that watch? If you have, I'll tell dad."

"Soakin' nothin'. You'r always thinking people are soakin' things. If the band loses the air to-night you'll think they've hocked it. An', say, if I catch you doin' the walk to-night with that dude mash I'll spoil his face, see? You'r getting too lifted, anyway, since ye got in de factory."

Mag resorted to the feminine reply of slamming the door, and ten minutes later was lounging down the street with both hands stuck in

[1] *Tompkins square:* Tompkins Square Park, located in the Lower East Side of Manhattan between Seventh and Tenth Streets.

the pockets of her black jacket, a flat brimmed glazed hat on the back of her head and some pliable substance in her mouth which she assiduously chewed upon. "Mag" was out for a good time, not a bad time in the sense of viciousness, mind you, but simply to lounge around that dreary, house hemmed, overcrowded, electric lighted square and to listen to the brassy melodies of the band.

Hundreds of other girls were there in the same way, and yet Local Propriety did not think of holding up its hands, scarcely of washing them.

It was Leiboldt's band that furnished the attraction for the streams of infantile, young and middle aged humanity that poured into the square from the big tenements roundabout, meandered aimlessly along the paths and overflowed the benches. Babies with thin faces, that had breathed the air of stuffy little kitchens all their lives, were there to get a mouthful of what passed current for ozone.

The music was like a tonic to them. They danced shadow dances on the grass plats where the hissing electric lights cast fantastic figures for them.

"Them kids are havin' all the fun," observed Mag to her "ladifriend," who had met her.

"Did you hear the heat laid out Bella's baby 'safternoon? Umha, so Ed told me. Hard on Bella, ain't it? That's four she's lost, you know. She's takin' on awful, Ed says."

"Didn't she have no doctor?"

"Ayap, young fellow from de hospital. Didn't do no good. Let's go round the square and touch some of the gang for soda. I'm broke; you got any coin?"

"Nope, only a half, an' I'm a savin' up hard now."

"What you goin' to get?"

"One of them yachting waists, with a white cap. I think they real neat an' stylish lookin', don't you?"

"Do fur your figger all right. I'm goin' to have one of those long coat basques, with lace around the bottom of the skirt, made o' that blue sateen I got. Going to get it made next week so I can have it for the excursion. Goin', ain't you?"

"Was, but Jimmy ain't in de 'sociation now; says they ain't nathin' but speelers[2] in it, and when I told him I was goin' swore he'd hock me shoes and give me away to the old man about Fred. Nice way to treat a

[2] *speelers:* Swindlers.

sister, ain't it, after me puttin' up me wages fur him to rush the growler with ever since he lost his job. Jimmy's getting awful tough and sometimes I feel real mortified about him. He don't seem to care nathin' about society nowadays. He's left off keepin' company with May too."

"What's he going to do, anyway? Goin' back to his trade in the fall when they open up?"

"I dunno. He talks a good deal about primaries and such things. Guess he'll be a politician, if he don't get some steady work this summer."

It is eight o'clock and the music has begun. They are playing "The Star Spangled Banner," and that melodious expression of a nation's pride floats out over a queer and cosmopolitan audience. Hun and Hebrew, German and Gentile, Gaul and Celt, they all applaud it, though it is safe to say that at least one half the listeners don't know what the air is.

It is music, however, and music of any kind is balm to the work-worn souls of the teeming east side. Old men are there puffing stubby pipes and listening attentively.

Two grizzled sons of Erin, whose raiment bore plenteous marks of mother earth and mortar, were sitting solemnly together and discussing the quality of the music.

"To me moind this band do play the classical tunes pretty well, Tim," said one, "but O'im thinking ye'll agree wid me that fur rale music the Sixty-ninth regiment band do beat out anything there be in this country."

"These min do play smooth, but they don't hav' the shnap to um. Oi'll say they do play smooth."

"Your right; they do play smooth."

"Yis, they do."

"They do."

"Yis."

"O'i do say they play smooth."

"So they may, but O'i know — ."

And the argument then grew heated.

Dark browed Italians slouched about in little groups chattering like magpies and then relapsing into sullen silence. Great admirers are they of Leader Leiboldt and his players. So, too, are the Italian women. Now and then you see one made picturesque by a gay colored native headdress or a bright hued gown. Queer colors there are in some of those costumes, shades that if introduced into a Broadway window would immediately be fashionable from very novelty.

But outnumbering all the rest are the Germans, the blue eyed, good natured Germans, who stand around and hum the airs to themselves.

And mingling with them all, giving a spice to every nationality, is the "tough girl" and her "tough brother."

They are of no nation.

They are just "eastsiders."

"Jimmy" gets there with the proceeds of his sister's pawned watch, and generously buys beer for "the gang." Everybody buys beer for that matter, it seems; not by the glassful; the financial resources of the east side could not stand such a pressure: but by the pailful. "Growler" is the only word known for it.

No matter what the size of the receptacle may be seven cents is the price. The hot nights have come and how the "growlers" are rushed!

The bearers hurry away in a stream, either to a convenient truck, there to lie down, gulp the beer and listen to the music, or to the roof of some neighboring tenement, where a little air comes stirring from the river, laden with suggestions of Hunter's Point.[3] Highly flavored, but enjoyable nevertheless.

"Say, Jimmy, what was ye sayin' to me girl last?"

"Oh, I was just a stringin' her about you; just guff. She said you hadn't asked her to de picnic, and I told her ye had an important engagement with a cop and couldn't get away to see her."

"Ye just cut yer string; ye hear, young feller. I'll take care of me girl meself. See?"

"All right, but, just the same, you don't go to de picnic 'less ye giv' me de case fur de ticket first, see? I got stuck one already."

"Who did you?"

"Billy."

"What's he doin' now, anyway?"

"Nothin'. He's married."

"Come, now, you fellows, move on there. Move on, I tell you, or I'll fan ye." This from a policeman.

And they moved.

Only to stop on the next corner, join raspy voices with the band while it played "Annie Rooney" and loudly demand an encore.

The music programme, however, was a good one.

[3] *Hunter's Point:* A neighborhood across the East River in northwestern Queens.

JAMES D. McCABE

"The Beer Gardens"

James D. McCabe (1842–1883) was born in Richmond, Virginia and educated at Virginia Military Institute. During the Civil War, he wrote plays with war themes and edited the *Magnolia Weekly*. After the war, he wrote war poems that achieved considerable popularity in the South. McCabe wrote several book-length works as well, his most important being *Lights and Shadows of New York Life* (Philadelphia: National Publishing Company, 1872), from which the following selection is taken (550–54). McCabe describes the Atlantic Garden, a place Crane often visited with his friend and roommate Frederick Lawrence. In *Maggie*, Pete takes Maggie to a beer garden, and although Crane does not name the place, his description suggests the Atlantic Garden.

In some respects, New York is as much German as American. A large part of it is a genuine reproduction of the Fatherland as regards to manners, customs, people, and language spoken. In the thickly settled sections east of the Bowery the Germans predominate, and one might live there for a year without even hearing an English word spoken. The Germans of New York are a very steady, hard-working people, and withal very sociable. During the day they confine themselves closely to business, and at night they insist upon enjoying themselves. The huge Stadt Theatre[1] draws several thousand within its walls whenever its doors are opened, and concerts and festivals of various kinds attract others. But the most popular of all places with this class of citizens is the beer-garden. Here one can sit and smoke, and drink beer by the gallon, listen to music, move about, meet his friends, and enjoy himself in his own way — all at a moderate cost.

From one end of the Bowery to the other, beer-gardens abound, and their brilliantly illuminated signs and transparencies form one of the most remarkable features of that curious street. Not all of them are reputable. In some there is a species of theatrical performance which is often broadly indecent. These are patronized by but few Germans, although they are mainly carried on by men of that nationality. The

[1] *Stadt Theatre:* The New York Stadt Theatre, located at 43 Bowery, first opened in 1864. From 1878 it was also known as the Windsor.

The Atlantic Garden, which Crane often visited. It was probably a model for the beer hall Pete and Maggie visit. From James D. McCabe, *Lights and Shadows of New York Life* (Philadelphia: National Publishing Company, 1872).

Rough and servant girl elements predominate in the audiences, and there is an unmistakably Irish stamp on most of the faces present.

The true beer-garden finds its highest development in the monster Atlantic Garden, which is located in the Bowery, next door to the Old Bowery Theatre.[2] It is an immense room, with a lofty curved ceiling, handosmely frescoed, and lighted by numerous chandeliers and by brackets along the walls. It is lighted during the day from the roof. At one side is an open space planted with trees and flowers, the only mark of a garden visible. A large gallery rises above the floor at each end. That at the eastern or upper end is used as a restaurant for those who desire regular meals. The lower gallery is, like the rest of the place, for beer-drinkers only. Under the latter gallery is a shooting hall, which is usually filled with marksmen trying their skill. On the right hand side of the room is a huge orchestrion or monster music-box, and by its side is a raised platform, occupied by the orchestra employed at the place. The floor is sanded, and is lined with plain tables, six

[2] *Old Bowery Theatre:* The Bowery Theatre was located at 48 Bowery between Bayard and Canal Streets, next door to the Atlantic Garden.

feet by two in size, to each of which is a couple of benches. The only ornaments of the immense hall are the frescoes and the chandeliers. Everything else is plain and substantial. Between the hall and the Bowery is the bar room, with its lunch counters. The fare provided at the latter is strictly German, but the former retails drinks of every description.

During the day the Atlantic does a good business through its bar and restaurant, many persons taking their meals here regularly. As night comes on, the great hall begins to fill up, and by eight o'clock the place is in its glory. From three to four thousand people, mainly Germans, may be seen here at one time, eating, drinking, smoking. Strong liquors are not sold, the drinks being beer and the lighter Rhine wines. The German capacity for holding beer is immense. An amount sufficient to burst an American makes him only comfortable and good humored. The consumption of the article here nightly is tremendous, but there is no drunkenness. The audience is well behaved, and the noise is simply the hearty merriment of a large crowd. There is no disorder, no indecency. The place is thoroughly respectable, and the audience are interested in keeping it so. They come here with their families, spend a social, pleasant evening, meet their friends, hear the news, enjoy the music and the beer, and go home refreshed and happy. The Germans are very proud of this resort, and they would not tolerate the introduction of any feature that would make it an unfit place for their wives and daughters. It is a decided advantage to the people who frequent this place, whatever the temperance advocates may say, that men have here a resort where they can enjoy themselves with their families, instead of seeking their pleasure away from the society of their wives and children.

The buzz and the hum of the conversation, and the laughter, are overpowering, and you wander through the vast crowd with your ears deafened by the sound. Suddenly the leader of the orchestra raps sharply on his desk, and there is a profound silence all over the hall. In an instant the orchestra breaks forth into some wonderful German melody, or some deep-voiced, strong-lunged singer sends his rich notes rolling through the hall. The auditors have suddenly lost their merriment, and are now listening pensively to the music, which is good. They sip their beer absently, and are thinking no doubt of the far-off Fatherland, for you see their features grow softer and their eyes glisten. Then, when it is all over, they burst into an enthusiastic encore, or resume their suspended conversations.

On the night of the reception of the news of Napoleon's capitulation at Sedan,[3] the Atlantic Garden was a sight worth seeing. The orchestra was doubled, and the music and the songs were all patriotic. The hall was packed with excited people, and the huge building fairly rocked with the cheers which went up from it. The "German's Fatherland" and Luther's Hymn were sung by five thousand voices, hoarse or shrill with excitment. Oceans of beer were drunk, men and women shook hands and embraced, and the excitement was kept up until long after midnight. Yet nobody was drunk, save with the excitement of the moment.

The Central Park Garden, at the corner of Seventh avenue and Fifty-ninth street, is more of an American institution than the Atlantic. It consists of a handsome hall surrounded on three sides by a gallery, and opening at the back upon grounds of moderate size, tastefully laid out, and adorned with rustic stalls and arbors for the use of guests. At the Atlantic the admission is free. Here one pays fifty cents for the privilege of entering the grounds and building. During the summer months nightly concerts, with Saturday matinées, are given here by Theodore Thomas and his famous orchestra[4] — the finest organization of its kind in America. The music is of a high order, and is rendered in a masterly manner. Many lovers of music come to New York in the summer simply to hear these concerts.

The place is the fashionable resort of the city in the summer. The audience is equal to anything to be seen in the city. One can meet here all the celebrities who happen to be in town, and as every one is free to do as he pleases, there is no restraint to hamper one's enjoyment. You may sit and smoke and drink, or stroll through the place the whole evening, merely greeting your acquaintances with a nod, or you may join them, and chat to your heart's content. Refreshments and liquors of all kinds are sold to guests; but the prices are high. The Central Park Garden, or, as it is called by strangers, "Thomas's Garden," is the most thoroughly enjoyable place in the city in the summer.

[3] *Napoleon's capitulation at Sedan:* In 1870, France declared war on Prussia. France suffered several defeats, leading to early September when Napoleon III capitulated at Sedan and was taken prisoner for the remainder of the war, which lasted until the following year.

[4] *Theodore Thomas and his famous orchestra:* Theodore Thomas (1835–1905) briefly conducted the New York Academy of Music and conducted the Brooklyn Philharmonic Society from 1862 to 1891, the New York Philharmonic Society in the late 1870s and through the 1880s, and his own ensemble, the Thomas Orchestra. The performance of the Thomas Orchestra in Central Park helped popularize German classical music in the United States.

GEORGE FREDERIC PARSONS

From "The Saloon in Society"

George Frederic Parsons (1840–1893), born in Brighton, Sussex, England, spent several years at sea, ending up on North America's West Coast, where he worked as a journalist in British Columbia and California, becoming editor of the Sacramento *Record* in 1869. While in Sacramento, he wrote *The Life and Adventures of James W. Marshall: The Discoverer of Gold in California* (1870). He moved to the East Coast in 1883 to take a position with the *New-York Tribune*, for which he primarily worked as a reviewer. In New York, he also contributed to the day's leading periodicals. The following text is excerpted from "The Saloon in Society" *Atlantic Monthly* 59 (January 1887): 86–98. Here, Parsons ably articulates the viewpoint of temperance advocates and also voices the idea of paternalism, the notion that it was civilized society's responsibility to serve as guardian of the childlike unfortunates.

"Use," says an old proverb, "is second nature." The truth of the observation is shown in the assimilating power of custom. There is nothing, however inherently vicious or dangerous, to which we cannot become so accustomed as to lose sight of its evil qualities. There is no habitual apprehension among the operatives in powder-mills and manufactories of the higher explosives, though statistics show that all such works are certain to be destroyed, sooner or later. The most unhealthy trades are as well supplied with labor as the most healthy. Custom and conservatism are sometimes so strong that men whose lives were being shortened by breathing a vitiated atmosphere, or by inhaling fine particles of stone or metal, have resisted every effort to improve their surroundings. Exposure to danger begets indifference to it. Exposure to vice begets indifference to it. The ancient Greeks had inherited the practice of infanticide from savage ancestors. They became so inured to it that, when the custom of exposing children gradually superseded child-murder, there were not wanting moralists to deplore the change as denoting the rise of what moderns would call a "sickly sentimentalism." Nor was there altogether lacking an element of plausibility in these complaints, for humanity was so little developed that the fate of exposed children, sad though it was, elicited no practical sympathy. When a Greek mother thus abandoned her infant, knowing that in the first place it was very likely to perish from the exposure, and that, should it survive, it would be doomed to slavery if a boy, and worse if a girl, the consequence of her act did not deter her. It was social crime disguised by custom.

A distinguished ethnologist, addressing a section at the late meeting of the British Association, expressed the opinion that mankind has made comparatively little physical and mental progress in the last five thousand years or so. Of course not many persons will agree with this view, and yet occasionally facts are encountered which seem to justify a doubt as to whether, despite the march of what we call civilization, the race has grown much since the dawn of the historical period. If a general disposition to exalt ourselves over those who preceded us were a proof of progress, the case might be regarded as settled. But if we are prepared to be candid with ourselves, it may not prove difficult to show that even the barbarism of ancient Greece is not so far behind us as we should like to be able to feel that it is. We do not kill our children. We do not any more expose them habitually, after the antique mode. We cherish them in their infancy, rear them to adolescence, and then send them forth to pit their immature vitality against the most elaborate, skillfully devised, and comprehensive machinery for ruining human beings, both mind and body, that the world has ever seen.

We are so accustomed to the saloon that its relations to and effects upon society are apt to seem merely part and parcel of the existing order of things; and those of us who are conservative from inheritance or education naturally look upon it as an institution which has its proper functions in the general scheme, and which could not be eliminated without creating a painful void, to say the least. The object of this paper is to induce those who read it to clear their minds of prepossessions and the fixed impressions derived from habitude, and to regard the saloon as it is, realizing the true nature of its influence, and perceiving the actual character of the functions it discharges. The chief backing of the saloon undoubtedly comes from our inherited aptitudes. We are all sprung, no matter what European race we claim as ancestors, from men who drank heavily a few generations ago. The predominance of the animal life has indeed not yet ceased, and, while the intellectual side of man is struggling to assert itself, the old Adam has hitherto been too strong for it. Progress in temperance there certainly has been, but when we look abroad and see what the saloon still means in our community life, how large a feature it is in social problems, how deeply it colors and affects not only the character and tone of our politics, but those of our morals, of our intellectual status, of our domestic existence, it will have to be admitted that hitherto popular sensibility on this question has been sluggish and dull, and that, for a nation commonly credited with capacity for self-government, we have a great deal to learn and a great many crying evils to get rid of.

Let us begin at the foundation, and trace the saloon through society. The first fact to be noted is that the institution is planted most thickly where the poorest people live. It has been alleged by some philanthropists and penalogists that poverty is in many instances the cause of drinking, and not the effect; that men drink to forget their sorrows and remember their misery no more. No doubt there is some truth in this. Indigence, squalor, destitution, innutrition, produce morbid conditions of body and mind, probably reinforce inherited tendencies to alcoholism, and move to that desperation which seeks relief in the oblivion either of death or drunkenness. But, however true such considerations may be, they cannot affect the obvious fact that the influence of the saloon among the poor is wholly mischievous. Even with those whose case is almost desperate the effect of resorting to this means of forgetfulness is pernicious. The man who drowns memory at the saloon does not facilitate his subsequent proceedings. He is poorer, in worse health, less able to confront difficulties, on the morrow. But one tendency in him is strengthened, and that is the tendency to repeat the debauch. The poor man who, being a husband and father, frequents the saloon runs the risk of betraying his most sacred trusts. His home may be uncomfortable, his meals may be unsavory, but the saloon cannot improve his surroundings. On the contrary, it absorbs his scanty earnings, thus depriving those dependent on him of the pittance of food and clothing he alone is able to supply. The drinking husband, too, often makes a drinking wife. For the women among the poor have most need of all the moral support their husbands, fathers, brothers, can bring them. Being physically weaker, they are more liable to faint under the heat and burden of the day; and when they see the man who should be their staff and prop cravenly giving way to despair and selfishly seeking oblivion for himself, they are apt to succumb to the pressure and to follow the evil example.

Then the fate of the children is generally sealed. The saloon has orphaned them. It will be henceforward their school and training ground. It will bring them up to be petty thieves, while they are developing the muscle that is to qualify them for ruffianism; and when they are old enough they will fall naturally into some gang, if they live in a city, and become material for professional criminals, loafers, and wrecks. But suppose that the poor frequenter of the saloon stops short

Jacob A. Riis, *A Growler Gang in Session* (1890). Courtesy of the Museum of the City of New York. Beer was easy to obtain, even for boys, provided they had money. Two members of this gang were arrested for robbery shortly after the photo was taken.

of drunkenness. Suppose that he confines his expenditure there to what is called "a few social glasses." In that case the evil he suffers will be less, but he will still suffer evil. He will squander a part of his earnings, and that he cannot afford. He will neglect his family, and that he cannot afford. He will fall into bad company, and that he cannot afford. He will most probably also become one of the chronic accusers of society, who, while wasting their own substance in riotous living, foolishly and inconsequently complain of the poverty whose continuance their own thriftlessness and self-indulgence are mainly accountable for. The Knights of Labor[1] refuse to admit a saloon-keeper or a liquor-dealer to their ranks. The prohibition looks as if it meant something, at first sight, but it has no real significance. The Knights of Labor and all the labor organizations must take to themselves a full share of the reproach that labor in the United States is every year spending upon drink which it does not need, and which hurts it, fully six hundred millions of dollars: a sum so large that if even fifty per cent. of it were saved a fund could be established in two years which, under wise management, would render destitution among the poor of this country impossible forever after.

Let us remember, however, that if the workingmen do not show the highest wisdom here, they are quite as advanced as the society to which they belong. We are so accustomed to have the worship of God and of the devil carried on side by side, so used to see the saloon standing cheek-by-jowl with the church, that we find it very difficult to perceive things as they really are. The relationship of the saloon to crime is notorious. It is not indeed true, as sometimes alleged, that professional criminals, such as burglars, sneak-thieves, pick-pockets, confidence men, etc., are prone to drink. These classes are strictly temperate, as a rule, and as a matter of business. But it is true that three fourths of all crimes of violence, such as assaults, assaults to murder, homicides and murders, are committed under the influence of drink. Our police justices are called perpetually to deal with a mass of brutal crime which is almost wholly the sequence of drink; and in all our great cities the taxpayer has to provide constant sustenance for the considerable number of chronic drunkards who have been reduced to hopeless degradation by the saloon. All this we know as we know that the sun rises and sets, and no doubt many, if not most, of us question its continuance as little as we do that of the solar phenomena. It is a new idea that such misery and sin and lawlessness are not unavoidable, but that they are evidences of the backwardness of the civilization on which we pride ourselves, and testify not less strongly to the insensitiveness of the general conscience.

[1] *Knights of Labor:* A labor union formed in 1869 among skilled garment workers in Philadelphia, the union expanded greatly over the next two decades and extended membership to most workers and some professional classes as well, though not to lawyers, bankers, or liquor dealers.

The saloon of the poor man is usually marked by vulgar decoration or careless squalor. It offers him more warmth and better shelter, perhaps, than he can get at home. It appeals to his most selfish feelings. It tickles his self-indulgence. He finds there, too, an excitement, a companionship, which are seductive. It is no wonder the poor man yields. The classes above him in posessions yield too, in their turn, and with not half his excuse. But what is the sociability encountered at the saloon? It is a sham, like all the alleged pleasure obtainable from the institution. Saloon friendship is a Dead Sea apple.[2] The men who strike hands with one another as comrades and boon companions will stand by while one of their number goes headlong to destruction; will observe with stolid passivity the victim's advance from one fatal symptom to another; will register his successive downward steps, marking in cold blood his loss of one holdfast after another; and when ruin, complete and irreversible, has concluded the drama, they will turn away and pursue their own affairs. The truth is that the spirit of the saloon is incompatible with the germination or growth of real friendship, as it is with any good thing. Indeed, its influences and tendencies are so palpably and wholly evil that any nation voluntarily adopting such an institution, and accepting it as part of the fixed order of things, could hardly escape Manicheism[3] but for that curious aptitude of the average human mind to separate religious beliefs from every-day experiences.

The lowering influences of the saloon react directly and with energy upon the poorest classes. The abuse of drink does not necessarily or immediately involve personal degradation or personal privation among such as possess some property. But with those for whom sustained sobriety can procure only a narrow sufficiency, intemperance means swift descent into discomfort and suffering. The poor man cannot drink without falling behind in everything. The saloon not only deprives him of reason and the full use of his faculties, it drives him to the pawnshop with his few possessions; it strips him, his wife, and his children of the clothing they need; it bares the walls of his poor rooms; it makes all his material surroundings meaner and shabbier, at the same time that it implants a distaste for the steady industry which is

[2] *Dead Sea apple:* Also known as Apple of Sodom, the legendary fruit was beautiful outside yet empty within. The term refers to anything hollow or specious.

[3] *Manicheism:* A dualistic theology holding that good and evil must coexist and that Satan and God are coeternal.

the one means of redemption. Into brains diseased and inflamed by drink socialist doctrines fall with fructifying power. The man whose own weakness and self-indulgence have brought him to indigence and misery is prone to shift the blame for his condition on the shoulders of society. A spurious self-respect is engendered by nursing the fallacy that, if justice were done, the street-corner loafer would fare equally well with the capitalist.

Yet these victims of a national vice must not be judged harshly. Their temptations are manifold. Their education is narrow. The world has been a cold stepmother to them, and they cannot be found fault with for not exhibiting an intelligence and a discrimination too commonly sought in vain among much more fortunate people. To a considerable extent they are helpless, just as the people may be so regarded in whom the passion for gambling is fostered and stimulated by governmental lottery schemes. When Great Britain went to war with China[4] to force the opium trade upon the Middle Kingdom, her neighbors were shocked, and with reason. Those, however, who perceived clearly enough the immorality of England's policy on that occasion have failed, for the most part, to see in the national support of the saloon a betrayal of the masses bearing an ugly resemblance to that involved in the opium war. It is the general acquiescence in the evil which makes it so hard to deal with. Nothing is easier than to point out the effects of the saloon upon various classes. Nothing is easier than to prove that human selfishness and frailty have much to do with the persistence of the evil. But nothing is so difficult as to secure comprehension of the vital truth that beyond the foibles of individuals and classes lies the national supineness or tacit approval of the saloon, as a determining cause.

There are among the poor everywhere many distinctions. There is the considerable element of crude, unformed humanity, which has not advanced beyond the capacity for manual labor of the simplest kind. This element has many of the defects and limitations which belong to childhood. It is easily led and misled. It picks up, as we all do, vices quicker than virtues. It cannot be said that it is repressed by the squalor in which it lives, for it has never known any better conditions,

[4] *war with China:* The first Opium War in China broke out in 1839 after the Chinese government attempted to stop foreign merchants from importing opium into China. The conflict ended with the Treaty of Nanking, which ceded Hong Kong to Great Britain, but the treaty said nothing about the opium trade, and hostilities broke out again in 1856. This Anglo-Chinese War was settled by the treaties of Tientsin (1858), a provision of which legalized the opium trade.

and it is painfully difficult to educate it out of the filthy habits which it has brought from the state of savagery whence it so lately emerged. This class seeks and finds in the saloon the excitement which the red Indian obtains in the same way. The brutality of these coarse natures is stimulated by drink, and they fight among themselves, or go home and beat their wives and children. The saloon operates with them as a conservative influence; it does much, that is to say, to keep them in their pristine condition of semi-savagery. But it has another and even more sinister function. The United States are absorbing vast numbers of the least civilized and enlightened peasantry of Continental Europe, people who know no English, have the vaguest idea of our form of government, and belong to the most conservative class extant; who bring with them obstinate prejudices, superstitions, delusions, and antipathies; who for generations have been taught to look up from their lowly positions and hate the classes above them. The saloon unites these unfit citizens, gives them a focus, and helps to develop the socialistic or anarchist sentiments which they imbibed in Europe. The saloon, which poisons the minds of the poor with the same drugged liquors which destroy their bodies, prepares them for hostility to the law by encouraging them to believe that every whiskey-muddled loafer is somehow entitled, without work, to a competence at the hands of society, and that the men whose brains and industry have put them above want or earned them abundance are somehow the enemies of the poor. No doubt there is in this enmity a basis of genuine feeling, though the feeling is not estimable. To the indolent and self-indulgent man, whose own laziness keeps him poor and in want, the sight of prosperity attained by energetic work must be irritating, because it involves self-reproach; and it is one of the vices of human nature to hate that which humiliates it by emphasizing its own inferiority.

All such meanness, however, all such impotent malevolence, all the envy, hatred, and uncharitableness which waste in transforming into evil impulses the energy which should find vent in industry, are fostered by the saloon. It fixes the squalor and misery of the poor, just as alcohol fixes the tissues plunged in it. Its alchemy converts the potentialities of successful labor into the seditious bitterness of the dissipated socialist. It makes the poverty which, sweetened by mutual forbearance and the hope of better things, might often be tolerable, maddening and unendurable, by adding to it the intensified selfishness of intoxication, the chronic shame and despair of semi-convalescence. And it has still more serious effects. Who that has lived in cities has not marked a peculiar growth of such centres? — a class of children who have never been young; boys and girls whose sharp features denote the experience of sordid age, whose anæmic bodies and fleshless limbs bespeak habitual innutrition, and whose sinister and hard expression shows how fervid is the hate they already bear to a world which has acquainted them only with suffering and sin. Everywhere this class of city-bred children is on the increase. Everywhere anxious philanthropists, sociologists, penalogists, are discussing its treatment, and wondering what shall be done with it.

The organization of charity, the organization of criminal administration, are alike busy with the problem, and systems of corrective and educational institutions arise in all our large cities, founded for this express purpose. Meanwhile, the supply of material does not diminish, as how should it? The saloon is at work night and day to maintain it. Not content with debauching the adult generation, it plants its deadly seeds in the systems of unborn children; its projects the curse from generation to generation. The city-bred children, whose premature knowledge of evil, precocious savagery, and anti-social proclivities may well cause apprehension for the future, are made what they are more by the saloon than by any other agency. It influenced their lives before they saw the light. It gave them weakness or disease on both sides. It handicapped them in the race of life from the start. It is doing this steadily, continually, and society nevertheless continues to tolerate such an agency; continues to act precisely as though its elimination were unthinkable; continues to evolve its houses of correction, industrial schools, asylums, probationary institutes, and shifts of all kinds in profusion, keeping its eyes shut to the perennial procreative energies of the saloon, whose unremitting exercise renders these remedial measures as futile as the effort to fill the bottomless jars of the Danaides.[5]

What would be thought of the medical administration which, in the time of a great epidemic, concerned itself chiefly in providing additional hospital accommodations, and paid no attention to the origin and mode of prevention of the prevailing disease? In the case of the saloon in society the facts are continually in evidence. Our police courts are mainly occupied with the petty offenses which spring directly or indirectly from drink. Through them drift the myriad wrecks which strew the path of progress. In them is exhibited, every day and all day, the extent, depth, paralyzing influence, of the saloon. It is bad enough in politics, but its social effects, especially among the

[5] *bottomless jars of the Danaides:* For murdering their husbands, the Danaides, the daughters of Danaus, king of Argos, were punished in Hades and compelled to fill with water a vessel so full of holes that the water ran out as soon as poured into it.

poor, are as those of a pestilence. The cruder element of the community is brutalized and retarded in its growth by this influence. Another element, that of the physically or intellectually feeble (always considerable, and increasing with the growth of competitive pressure), is condemned to a wretched fate by the same instrumentality. The people who have not the energy of mind or body to form clear and practical purposes, or to put them in operation if formed, are the easiest victims of the saloon. As a rule they are sensitive, often morbidly so. They brood over their weakness and their failure. Naturally prone to depression, they become jaundiced and desponding. From that state of mind to the craving for any kind of stimulant the transition is natural and swift, and the saloon does the rest. There are thousands of families doomed to indigence, disappointment, misery, through life, that might have lived at least in decent poverty and with self-respect, but to-day are plunged in hopeless ruin by drink, and are sinking out of sight in the quicksand. . . .

"Tempting Poor Women: They Buy Beer and Have It Charged as 'Potatoes'"

Giving Maggie's mother a drinking problem, Stephen Crane incorporated into his fiction another serious social issue facing nineteenth-century urban America: female alcoholism. Since saloons were more than willing to sell alcohol to women and children, housewives could get beer and liquor if they had the money. Availability, in other words, was a matter of domestic economy. Since most poor, married women had no personal income, their husbands supplied them with their household funds. Those men who held their wives accountable for every penny would hardly let them purchase growlers of beer for themselves. However, as the following investigative report shows, local grocers had a way to overcome the difficulty. It appeared in the *New-York Tribune* March 30, 1890: 15.

Liquor-Shops Thinly Disguised as Grocery Stores — A Flagrant Abuse of the Excise System

The farcical administration, or rather maladministration, of the excise laws in the City of New-York has so long been a standing joke that reference to the subject has lost the power to elicit a smile. The glaring monstrosities of this branch of the city's government have been so thoroughly ventilated and held up to the gaze of the community that the whole matter may be regarded as being completely threshed out and further discussion as futile until the arrival of the long delayed but surely approaching day when the tentacles of the whiskey octopus shall have been torn from the throat of the State. There is, however, one evil growing out of and linked with the unrestrained liquor traffic in the city that seems to have escaped general notice. Yet this evil is of such magnitude that it is passing strange that more extended notice of it has not been taken by those who are laboring with the object of having proper restrictions placed upon the selling of liquor. The question referred to is the prevalence of the places known as "grocery bars."

Scattered pretty nearly all over the city and particularly in the poorer districts are small grocery stores which are merely liquor shops under a thin disguise. The casual observer might pass a hundred of these places without having any suspicion aroused as to their true character. To the uninitiated they are grocery stores pure and simple, but they afford a fine exemplification of the truism that things are not always what they seem. Viewed from the exterior there is nothing in sight that does not properly pertain to a provision dealer's trade, but if the store is entered and a tour made to the rear an entirely different condition of affairs reveals itself. Regularly fitted up bars, equipped with all the appendages of a liquor store, are found, and business conducted on exactly the same principles as in the ordinary drinking places. Once in the bar, which in most cases is separated from the store proper by a screen door, there is not the slightest secrecy observed. The visitor walks up to the counter, orders his drink and is served just as he would be in any liquor shop which he had entered by the "family entrance" on Sunday.

If there was nothing else open to censure in the present practically untrammelled excise system, the existence of the grocery grogshops would in itself present to the minds of thinking people prima facie evidence of the necessity for a discriminatory liquor license law. The evil results brought about by the grocery bar can hardly be overestimated. It forms in itself a constant menace to the happiness and prosperity of the working classes, and sows the seeds of discord among humble homes that otherwise would be peaceful and contented.

A Tenement-House Episode

A typical case may be easily pictured. A man goes off to his work in the morning leaving his wife and little ones at home. Some neighbor calls to visit the wife, and after the usual tenement-house gossip has been exhausted the suggestion is made that a can of beer or a little whiskey would be highly desirable by way of lubricating the vocal cords. So the stuff is sent for, a child acting as the messenger. Of course the grocery bar is the place patronized. After the liquor is drunk in nine cases out of ten there is a desire for a further supply, but perhaps the supply of ready money has been exhausted. It is imperative that another "growler" or "flask" should be obtained and the question resolves itself into one of ways and means. It would never do to obtain liquor on credit and have it charged in the little passbook, as the husband might object when he came to look over the same passbook at the end of the week preparatory to paying his week's grocery bill. But why should the wife not go over to the grocery, and obtaining the liquor have the amount requisite to pay for it charged in the book as potatoes or butter or flour? This is done, and done repeatedly until when the husband comes home and finds his wife intoxicated. In many such cases the criminal court records tell the sequel.

Herein lies the evil of the grocery bar. It is a standing temptation to poor women of slightly bibulous tendencies to drink. Thousands of women who would be ashamed to enter a liquor shop and call for liquor have always before their eyes the open screen door of the grocery bar where they can slink in and drink while ostensibly purchasing household necessaries. The craving for drink is thus easily established, and all kinds of subterfuges are resorted to in order to make the breadwinner of the family believe that he is paying for provisions and not for the whiskey or beer consumed by his wife. Little children sent on errands to bring liquor are told to have it charged as groceries, and are thus brought up in an atmosphere of lying and deception. The storekeeper is only too glad to aid in the deceit, the profits on the liquors he sells being immeasurably larger than on grocery goods.

Now for the legal aspect of the question. Section 11, Chapter 28 of the Excise Law of the State of New-York, 1857, sets forth: "In all licenses that may be granted (excepting to inns, taverns or hotelkeepers) to sell strong or spirituous liquors or wines in quantities of less than five gallons, there shall be inserted an express declaration that such license shall not be deemed to authorize the sale of any strong or spirituous liquor or wine to be drank in the house or shop of the person receiving such license, or in any outhouse, yard or garden appertaining thereto, or connected therewith."

How the Law Is Violated

Yet, in open defiance of these provisions, which relate expressly to grocery stores, bars running in full blast in these places are to be found all over Manhattan Island and on almost every block in the downtown wards. That the excise inspectors can be unaware of this state of things seems hardly credible. These officials are notoriously near-sighted, of course, and it is suggested that their dimness of vision can be materially increased for a consideration.

The liquor-shop keepers naturally enough feel considerably aggrieved at the manner in which their trade is caused to suffer by the opposition of the grocery bars. They do not, however, dare to complain lest they should incur the wrath of the Excise Board, with which they form a community of interest, so to speak. A prominent member of the Central Liquor Dealers' Association speaking of the matter said: "Don't print my name, for by doing so the vials of wrath of the Excise people would be poured upon my head. But don't you think it is rather unfair that these grocery bars should be permitted to run in opposition to the regularly licensed shops? Take my own case, for instance. Here am I running a reputable saloon, obeying the laws and paying a license of $200 yearly. Not a stone's throw from here is a grocery which only pays $25 a year, and yet has a regular bar trade. The other fellow sells the worst kind of stuff, on which he makes a larger profit than I can on my liquors, and he only has to pay about one-tenth of the license fee that I do while his expenses are not one-tenth as large as mine. I think it would be only fair to have all persons running bars placed on an equal footing, and I should be glad to see a law passed which, even if it did increase fees, would insure this."

Before the passage of the High License law in Pennsylvania there were hundreds of grocery bars in the larger cities working the same bad effects as those in New-York now do, but to-day they are entirely exterminated. When the day comes for New-York to have a properly regulated liquor traffic the same results will undoubtedly follow, but under the present obnoxious system, particularly in New-York City, the grocery rumseller will probably continue to flourish in his nefarious trade secure from legal interference.

J. W. BUEL

From Metropolitan Life Unveiled; or the Mysteries and Miseries of America's Great Cities

J. W. Buel (1849–1920), born in Golconda, Illinois, travelled and wrote extensively. In addition to numerous travel narratives, he wrote adventure stories set in places through which he had travelled. Buel's finest overseas travel narrative is *Russian Nihilism and Exile Life in Siberia* (1883), a work that anticipates George Kennan's *Siberia and the Exile System* (1891). Buel's best domestic travel narrative is *Metropolitan Life Unveiled; or the Mysteries and Miseries of America's Great Cities* (St. Louis: Historical Publishing, 1882), from which the following selection has been taken (45–49, 52–61). Here he describes what were euphemistically known as concert halls but that more precisely were drinking establishments where women were employed for the purpose of coaxing drinks from male patrons. Buel's witty, urbane persona distances him from the sordid places he describes, that is, until he gets to the "Black and Tan," a concert hall where black men associated with white women and black women with white men.

Of the large number of concert saloons and *maisons de joie*[1] which once lined Water street, only two remain as reminders of the past and its gross iniquities. There is one kept by a faded old slattern at No. $337\frac{1}{2}$, which we will enter for a moment's observation. The deserted appearance is marked as we step upon the threshold, though our reception has some elements of warmth in it. There is a bevy of eight highly painted and seasoned girls, in a theatrical make-up of tights and tonsorial[2] stockings, reeling through a mechanical dance, as badly lacking grace as the fiddle, bass-viol and harp, attuned by three wild looking gin-guzzlers in a raised corner, are of harmony. As we advance toward the bar in the rear way, many flattering remarks are passed upon us by the besotted females, which give unmistakable expectation that we have come either to treat, or on more reprehensible business;

[1] *maisons de joie:* Houses of pleasure.

[2] *tonsorial:* Literally the word refers to a barber and his work and was often used humorously. Here, it appears to be used figuratively to suggest another figurative phrase, cutting a rug.

our appearance, however, is too respectable to suggest the latter, so we "stand treat," and shy off to avoid the caresses which are proffered, after which, learning from the madame that business is distressingly dull since the missions were established, we take our departure.

Going north two squares we turn on James street, and passing through a file of wretched men, women and children, one half drunk and the other half crazy, we will stop at No. 96 and take a peep at the interior of another concert hall very unlike the one just visited. There is a furious noise inside, made by clanging cymbals, explosions from a bass-drum, and the tooting of horns. A combined smell of garlic, pretzels, bologna and Limburger leaves no possible room for doubt respecting nationality and patrons, though suspended from a rope, which encircles the room, is a display of national bunting comprehensively cosmopolitan.

Though many of the surroundings are eminently Teutonic, yet a glance suffices to show that the gay revellers may claim a varied nativity. Sailors, however, predominate, and we cannot suppress a generous smile while watching the mazy varsouviennes, rigadoons, waltzes and the saturnalia and wassail.[3] There are more than a score of brightly dressed bawds, each giving a cunning display of bust and limbs while whirling through the room in lascivious suggestiveness. The male participants, in partial intoxication, exhibit no sensitiveness, and every girl who has a partner regards whatever charms she may possess as legitimate property for critical and prying inspection, showing as little sense of modesty as a gelding. This hall bears the patriotic title, "Flag of our Union," and the display of variegated gonfalon[4] makes the name peculiarly appropriate.

In order to depart without exciting undue attention or insult we must minister to the remorseless cravings of the girls who have infested us since our entrance; this is best accomplished by calling for two dollars' worth of stale beer, or ten wine-glasses full of that demoralizing liquid at the very reasonable rate of twenty cents each. This acts like a ticket of honorable exit, and leaving the place, which has a thousand prototypes, we stray around to Chatham Square, and drop into the Sultan Divan. There are some attractive features in this concert resort and beer garden, for its domiciliaries are rather handsome,

[3] *varsouviennes, rigadoons, waltzes and the saturnalia and wassail:* The first three are dances, while the last two suggest boisterous carousing in general.

[4] *variegated gonfalon:* Banners or flags, frequently ending in several streamers and sometimes used in political or religious processions.

and are arrayed in silk crinoline cut decidedly *decollette,* with skirts to match. Immediately beneath a pretty corset, half revealing, may be seen enough to excite a wish for larger audience, but the girls affect a coyness that assimilates surprisingly well with their business, which is illustrated by the anxiety a stranger almost invariably manifests. These girls can serve you with wine or beer in a manner a novitiate would declare perfectly divine; at the same time they appear so young and innocent, with just a perceptible disposition to venture within speaking distance of impropriety! Sometimes they may be prevailed upon to sing a song — a nice, every-day, warranted to wash kind of a ditty, that may have a very indistinct insinuation, which leads a dull fellow to think, how decidedly clever she is!

The waiters at the Divan are peculiar in one respect, to thoroughly appreciate which it is necessary to try the experiment. A good-looking, charming young man, with plenty of money and allective[5] graces, is privileged to indulge in any conversation he may desire with any of these girls; may, indeed, make assignations, to which request an appointment is readily granted. When the proceedings have gone thus far the young man will always spend a goodly sum for wine or other drinks, passing the waiting time thus in patronizing the bar; for remember, none of these girls can leave their places until after midnight. When the wished-for hour arrives, the young man is told to go outside and abide the coming of his selected companion. He obeys with alacrity, and stands there, too, until he realizes his deception and that he has been badly sold. The girl, however, has made a handsome profit on the drinks she has sold him, for all barmaids receive a commission of twenty per cent. on their sales. This cunning scheme is one that the Sultan Divan girls have practiced since the place was established, and they have long since found that it pays much better than criminal receptivity.

Not far from the Divan is the Atlantic Garden, the largest place of the kind in New York. Its specialty is a band of female minstrels, all of whom dress with most captivating taste and play their silver instruments with real excellence. The proprietor of this enterprise came to New York several years ago *sans* money, credit and everything. He began life in America as a beer slinger and gradually worked up until he established the Garden which has yielded him such large revenue that he is now one of the Metropolitan millionaires. The place is quite

[5] *allective:* Enticing.

orderly and is entirely free from all the naughtiness usually indulged in at beer gardens. . . .

If we estimate the importance of our visit to New York by the *outre* and degrading sights we shall witness, a visit to Harry Hill's theatre,[6] on the corner of Houston and Crosby streets, must not be neglected. Therefore, in pursuance of our determination, we leave the Bowery and climb a pair of stairs which have their ascent from the interior of Hill's saloon. At the moment our eyes fall upon Harry's aggregation of song and minstrelsy, purveying to a mixed assemblage of bilks, blokes and *canaille,*[7] our gorge almost rises at the fumes and depravity before us.

Harry Hill sees that we are strangers, and with much courtesy conducts us to a seat which commands an excellent view of the house. On the stage is a double quartette of men and women, the former being in the disguise of low-down Ethiopians, while the latter are quite evil enough appearing without any aid of costumer or diabolizing painter. In fact, the girls must surely have been born with unnatural homeliness, and grown uglier every day of their lives thereafter. One of the more weazened, shattered voiced, debilitated travesties on women starts a song which is to the audience a palatable sandwich between the rhodomontades of alleged negro jokes. Turning our eyes from the stage, however, we will roll our vision to the wine room on the left, and in there we shall see something, which if not more refining is at least less monotonous. There are several young girls, fairly well dressed, apparently extracting comfort from the laps in which they loll, sipping wine between obscene conundrums, and laughing immoderately at suggestive stories. Modesty forbids that I should tell the rest, for prying hands too oft indulge in dalliances not to be described; but while refraining from expression we do all the more thinking, though our conclusions must be a comforter to none save ourselves.

Harry Hill has acquired a large fortune exhibiting the human menagerie of animal natures which he manages, and his name is familiar to nearly all Americans. Though his patrons belong to the most infamous classes, he suffers no one to be robbed in his establishment. He was formerly a professional prize fighter, and has a large acquaintance among the thugs and thieves, who regard him with such feelings

[6] *Harry Hill's theatre:* Harry Hill's famous dance hall, located at 26 East Houston Street, had the reputation of being the most respectable disreputable house in New York City.

[7] *bilks, blokes and canaille:* Riffraff.

of friendship that none of them ever transgress the rules of his house. Besides this, Harry is familiar with all the tricks of Metropolitan thieves, and his eyes are ever on the alert. Should he discover any attempt at pocket-picking or swindling, the thief would suffer a dreadful punishment at the pugilist's hands. His influence, however, ceases the moment his patrons leave his premises, and the hall is therefore used for plotting hundreds of robberies which occur within a stone's throw of, and upon men just leaving, the theatre.

Theo. Allen's concert and dance house is located in a basement on Bleecker street. Allen acquired much fame by the title given him several years ago when missionary work first began on a redoubtable basis among the foul and impious denizens of Five Points.[8] He kept a den at No. 301 Water street, which was so festering with infamies that the full effect of missionary work was turned upon him, but it at first only served to give him the distinguishing cognomen, "Wickedest man in New York." This name proved to him a capital prize in fortune's lottery, for his place at once assumed a popularity never before nor since enjoyed by any saloon-keeper. He received the mission laborers with becoming generosity, gave them liberty to conduct sacred meetings in his bar-room, and even compelled his besotted scarlet women to attend the services. At length Allen professed repentance and went to lecturing publicly, but he kept the saloon until he found it advantageous to sell out to the missions, when, having netted more than $100,000, he cracked his fingers at the world and changed from low to high living. The result of this was approaching poverty in a few years, when he again opened a house of prostitution, and now conducts a flourishing business, where we will visit him.

The room into which we have descended is an evidence of human endurance and vitality, for it is so pregnant with noisome vapors that to those who study life theoretically, — like the anatomist who acquires his physiological knowledge by clinics on a manikin, — suffocation would result after about fifteen minutes confinement in such pestilential and pyæmia[9] producing exhalations. Nevertheless, this

[8] *Five Points:* A neighborhood named for the intersection of five streets: Mulberry Street, Anthony (now Worth) Street, Cross (now Park) Street, Orange (now Baxter) Street, and Little Water Street (now defunct). It had a reputation as the foulest, most depraved neighborhood in New York City. Travelling through Salonika, Greece, in 1856, Herman Melville wrote in his journal: "Aspect of streets like those of Five Points Rotten houses. Smell of rotten wood" (Journals 55). New York City acquired and condemned most of its tenements between 1887 and 1894.

[9] *pyæmia:* A condition of blood poisoning.

inferno is always crowded after nightfall with a bachanalian throng of debased humanity. The women, who are compelled to live in such a poisonous atmosphere and drink for the profit of the bar, are like wraiths of consumption and fever; their eyes are lustreless, their limbs shrunken, and despite the veneering of high coloring cosmetics, their skin is harsh and dead as parchment. Yet there are always men, young and old, who find enjoyment in the private society of these sapless hulks of womanhood, and who spend their time and substance whirling through mechanical waltzes with arms thrown tightly around their almost bloodless victims of direst vice.

Near the corner of Bleecker and South Fifth Avenue is a dive still lower than any we have yet visited, and since it is in the neighborhood of Allen's, let us drop in, taking the chances for our safe exit. This place has a double name; some of its patrons stick to the title, "Black and Tan Concert Hall," while a majority insist that the founder gave it the allective euphemism, "Chemise and Drawers."

This concert hall is also in a basement, and its patrons are chiefly descendants of Ham,[10] or, rather, Americans of off color, though there is always a large per centage of white persons, male and female, mixed promiscuously with the pungent effluvium of the room. All fresh air being entirely excluded, and a red hot stove kept roaring in consonance with the singing and simmering water pan on top, there is some resemblance noticeable between the den and a Turkish bath sweating room. The Augean Stables,[11] which for some thousands of years have held their reputation unsullied for obnoxious exhalations and excrements, which only Hercules could endure, must lower their pennant in recognition of this modern compeer.

But despite the befouling odors and associations, we will essay an investigation of this putrescent *pot-pourri*. Passing along an ample hallway, in which the bar-room is located, along the counter of which is a closely packed row of drinkers, we halt in the rear, which expands suddenly into a ball-room. Here are negroes with white women for partners, and white men in the arms of oleaginous black wenches, all twirling in the lascivious waltz, to the music of fife, fiddle, banjo and

[10] *descendants of Ham:* Tradition associates Ham, the second son of Noah, with Egyptian culture and, more broadly, anything African. Descendants of Ham refers to anyone of African descent.

[11] *Augean Stables:* Among his heroic labors, the mythical Greek hero Hercules had to clean the stables of Augeas, who kept an immense number of oxen and goats yet had never cleaned their stables.

piano. Streams of highly scented perspiration are seen tracing down the cheeks of the colored dancers until the exhalations can almost be felt as well as smelt; indeed the spiritual part of the occasion rivals a convocation of excited pole-cats, but the participants are unconscious of everything save the fun in hand, and even our weakened stomachs are partly forgotten in amazement at the loud sounding osculations of a buck negro practicing on a white girl's cheeks. When the dance is concluded the enraptured couples march up to the bar, and sometimes to a strictly retired part of the house, but they generally appear again and keep up their wild orgies until dawn.

This place is one of the filthiest in the city of New York, and this very fact and reputation renders its proprietor a most profitable service. Visitors to Gotham who seek the rare, though degrading, sights seldom fail to call at the "Chemise and Drawers," and as all callers are expected to treat as a compensation for admittance, the bar gathers a goodly patronage from the large number of these chance customers.

Before concluding our tour of the concert halls of the Metropolis, we must visit one of the higher grade, where our sight may be regaled, and the unpleasant effects of the places we have already examined may be in part effaced.

The most high-toned concert and beer-garden is perhaps the Cremorne, located on Thirty-second street near Broadway, though the Haymarket, which is only two blocks below, and Buckingham Palace, on Twenty-seventh street, are also very fine places and have an enormous patronage.

The Cremorne Garden is magnificently furnished, the walls being set with immense looking-glasses, and heroic sized statues of several of the Grecian gods being placed at intervals in the grand room. On a gallery, which traverses the entire hall-width, half-hidden by a profusion of vegetation, is the band of sixteen pieces, discoursing elegant music. There are seventy-five tables for beer and wine service in the room, and at each table sits a girl charmingly dressed, who invites customers. Let us go over to this first table on the right, for I discover a very beautiful girl stationed there all alone. Look! she is beckoning to me and as there is at least an outward appearance of her being a dear charmer we can certainly find some amusement around that table.

Taking seats — there are four chairs at each table — I call for some beer, because it is cheaper than wine, and forthwith the girl raps for a colored waiter who, taking a check from the pretty bar-maid, disappears, but soon returns with our order. By way of engaging the girl in conversation I say:

"You seem to be quite young, and I declare you are very pretty; how long have you been here?"

"Oh, I only started in this business a week since. How old do you think I am?"

"About eighteen I should think" — though I knew she was at least twenty-three.

"You are a pretty good guesser, only missed it one year, for I am just seventeen."

"Say, my dear girl, you have a ravishingly lovely arm, so perfect in mould, and skin like the Georgian women."[12]

"Do you really think so? I appreciate the compliment."

"I'll tell you what to do, go out West and your fortune will be made; for a pretty face never wants for admirers in that section. You could marry a millionaire gold-miner or cattle dealer in fifteen minutes after reaching Denver."

"Is that so? I'd like to go, but I have a poor old mother here who needs my support, and I can't leave her."

"Now, my dear, although I wear a sombrero, and have Western accent, don't try to play that poor-mother racket on me, for it won't wash any more than poke-berry colored calico. Say, tell me where you live and if this is your only means of support, for I see you wear the Fifth Avenue regulation clothes."

"If you'll buy a three dollar bottle of wine and go with me to the private wine-room, I'll tell you all you want to know; the proprietor won't let us leave our tables unless our customers buy a bottle of wine. Come, now, won't you?"

"Well, I guess I won't, not to-night, for your story is like that of all other bar-maids; you've been too confiding, dropped the casket of your chastity and broken the jewel which few like you try to mend. My conversation has been for the benefit of my companions, who are visiting New York, and not for my edification."

It was thus we parted.

There are hundreds of other concert halls and beer gardens in New York, but those we have visited, as described, are samples of the whole.

[12] *skin like the Georgian women:* The adjective Georgian refers to the time of the Georges, kings of Great Britain, which roughly coincides with the eighteenth century, when excessively white skin was a symbol of beauty and refinement.

2

The Plight of the Working Woman

In the collar-and-cuff factory Maggie takes a stool and a sewing machine in a room with twenty young women "of various shades of yellow discontent." Placing her within a large room among much machinery and many women who scarcely differ from one another, Stephen Crane made Maggie's situation that of all young female factory workers in the late nineteenth century who struggled to maintain their health, dignity, and self-respect amidst oppressive working conditions. Unlike other contemporary authors, however, Crane does not dwell on working conditions in the factory. While the reformers often described the number of hours women worked, the vast number of shirts they made, and the scant pay they received, Crane provides few particulars. The numerous magazine and newspaper articles treating the plight of the working woman that partially inspired Crane to write his book also obviated the need for him to write similarly. Instead, he could take a more aesthetic approach and use, for example, color imagery to depict female factory workers.

While working women in the United States gained public attention during the last third of the nineteenth century, in Great Britain, wage-earning women received notice some decades earlier. British seamstresses found their muse in Thomas Hood, whose 1843 "The Song of the Shirt" quickly became one of the time's best-remembered poems. Generally inspired by the pitiful working conditions among the London poor, "The Song of the Shirt" was specifically inspired by the story of a poor woman who worked steadily yet still could not earn enough to support her family.

Though some American women worked outside the home as schoolteachers and as domestics in other people's households during the early nineteenth century and more began working for wages with the growth of the textile industry during the 1820s and 1830s, the total percentage of wage-earning American women remained small until after the Civil War when positions as factory workers, retail salesclerks, and secretaries became widely available. In the last two decades of the century, the factories attracted many female workers. Though women helped produce a variety of goods, the largest industry employing female workers was the shirt industry. Women not employed in the factories often did needlework piecemeal in their homes, work that differed little from that done by Thomas Hood's London seamstresses a half century before.

With so many American women working under poor conditions in shirt factories or making shirts at home for scarcely enough money to survive, it is hardly surprising that "The Song of the Shirt" was remembered and often repeated in late-nineteenth-century America. Several of the following essays recall the poem. In *How the Other Half Lives,* Jacob Riis alludes to it, quotes lines from it, and calls for a poet to "tell of the sad and toil-worn lives of New York's working-women" (202). Katherine Pearson Woods, too, remembers the poem in her discussion of the workingwoman. The shirt finisher interviewed by the *New-York Tribune* expresses her familiarity with the poem and mentions that she had heard it recited at a club meeting. D. W. Griffith would later make a film illustrating the plight of the working woman called *The Song of the Shirt.*

The first group of documents in this chapter describe women's life and work from the perspective of others. Jacob Riis's essay, a chapter from *How the Other Half Lives,* scrutinizes the poor pay and working conditions of female factory workers and those who did piecemeal work independently. Examining a variety of employment opportunities for women, Katherine Pearson Woods challenges society's prevailing tendency, especially as articulated in the quotation from British essayist John Ruskin that opens her piece, to elevate the idea of womanhood beyond all practicality. Both Riis and Woods also discuss labor movements that worked for improving women's working conditions. The Working Women's Protective Union mentioned by Woods was the first of many such organizations devoted to defending the rights of wage-earning women.

John Sloan, *Night Windows* (1910). Sloan contrasts the nighttime activity of two different women; one dutifully goes about her housework while another strikes a seductive pose which catches the attention of the boy sitting on the opposite rooftop. Courtesy of the Hood Museum of Art, Dartmouth College.

Jane Cunningham Croly advocated women's right to work in her newspaper articles and exemplified it in her personal and professional life. A working journalist, Croly challenged long-standing middle-class tradition by continuing to work after she had married and started a family. Her Senate testimony provides a good indication of her attitude toward women and work and contains her interview with a woman who worked making cigars. The frank opinions of the "cigar girl" provide much additional insight into the working woman's attitude toward her coworkers and her employer.

Other articles in this section provide different perspectives. Abraham Hummel and William F. Howe perpetuate the stereotype of the promiscuous working woman. The last article in this section is, perhaps, closest in spirit to *Maggie*. In "The Woes of the New York Working-Girl," Edgar Fawcett treats several aspects of the poor, young working woman's life that anticipate Crane's novel: her shoddy tenement, her irresponsible alcoholic parents, the poor pay and working conditions of the factory, the public's insensitivity to the working woman, and the clergy's reluctance to save her.

The second group of articles purportedly describe the conditions of the working woman in her own voice. All come from the women's page of the *New-York Tribune*. Most are adapted from interviews of women from a variety of trades. Ostensibly the words of working women, these articles nevertheless show the reporter's hand shaping the material for inclusion for the *Tribune*'s generally upscale readership during a time when the *Tribune* was making way for a women's page as one of the paper's regular features. Unintentionally, perhaps, these various firsthand accounts react to one another. In one, the "shop-girl" — a female salesclerk — speaks. She works in a fairly affluent store, makes ten dollars a week (twice what many other female salesclerks made), speaks condescendingly of other working women, and expresses much concern about her personal appearance, recalling her fear that the store management would make her give up her bangs. The diminutive shirt finisher, on the other hand, resents store clerks and the department stores that used a woman's height as a criteria for employment. She deplores the fact that short women were often discriminated against and asserts that she could do a good job as a salesclerk, for she was not "one of the stuck-up sort who think they are put behind a counter only to show off their bangs and their bangles" (see p. 253). (The clever parallel construction suggests that the reporter rewrote the shirt finisher's words.) Taken as a whole, these interviews bring alive the personal appearance, outlook, and lifestyle of the working woman in late-nineteenth-century America.

Working Women from the Perspective of Others

THOMAS HOOD

"The Song of the Shirt"

Thomas Hood (1799–1845) devoted his professional life to miscellaneous writing and editorial work. He wrote in a variety of genres, from poetry to prose sketches to humorous essays. Generally inspired by the pitiful working conditions of London's working women, "The Song of the Shirt" was specifically inspired by a report in the London *Times* about a poor woman who worked steadily yet could not earn a decent enough wage to support her family, and who was arrested for pawning some pieces of material entrusted to her in order to feed her starving children. "The Song of the Shirt," Hood's best known and most popular work, first appeared in *Punch* 5 (December 16, 1843): 260, the source of the present text. The poem was reprinted widely in the United States where it became a theme for the working woman.

With fingers weary and worn,
With eyelids heavy and red,
A Woman sat, in unwomanly rags,
Plying her needle and thread —
Stitch! stitch! stitch! 5
In poverty, hunger, and dirt,
And still with a voice of dolorous pitch
She sang the "Song of the Shirt!"

"Work! work! work!
While the cock is crowing aloof! 10
And work — work — work,
Till the stars shine through the roof!
It's O! to be a slave
Along with the barbarous Turk,
Where woman has never a soul to save, 15
If this is Christian work!

"Work — work — work
Till the brain begins to swim;
Work — work — work

Till the eyes are heavy and dim! 20
Seam, and gusset, and band,
Band, and gusset, and seam,
Till over the buttons I fall asleep,
And sew them on in a dream!

"O! Men with Sisters dear! 25
O! Men! with Mothers and Wives!
It is not linen you're wearing out,
But human creatures' lives!
Stitch — stitch — stitch,
In poverty, hunger, and dirt, 30
Sewing at once, with a double thread,
A Shroud as well as a Shirt.

"But why do I talk of Death?
That Phantom of grisly bone,
I hardly fear his terrible shape, 35
It seems so like my own —
It seems so like my own,
Because of the fasts I keep,
Oh! God! that bread should be so dear,
And flesh and blood so cheap! 40

"Work — work — work!
My labour never flags;
And what are its wages? A bed of straw,
A crust of bread — and rags.
That shatter'd roof — and this naked floor — 45
A table — a broken chair —
And a wall so blank, my shadow I thank
For sometimes falling there!

"Work — work — work!
From weary chime to chime,
Work — work — work — 50
As prisoners work for crime!
Band, and gusset, and seam,
Seam, and gusset, and band,
Till the heart is sick, and the brain benumb'd, 55
As well as the weary hand.

"Work — work — work,
In the dull December light,

And work — work — work,
When the weather is warm and bright — 60
While underneath the eaves
The brooding swallows cling
As if to show me their sunny backs
And twit me with the spring.

"Oh! but to breathe the breath 65
Of the cowslip and primrose sweet —
With the sky above my head,
And the grass beneath my feet,
For only one short hour
To feel as I used to feel,
Before I knew the woes of want 70
And the walk that costs a meal!

"Oh but for one short hour!
A respite however brief!
No blessed leisure for Love or Hope, 75
But only time for Grief!
A little weeping would ease my heart,
But in their briny bed
My tears must stop, for every drop
Hinders needle and thread!" 80

With fingers weary and worn,
With eyelids heavy and red,
A Woman sate in unwomanly rags,
Plying her needle and thread —
Stitch! stitch! stitch! 85
In poverty, hunger, and dirt,
And still with a voice of dolorous pitch,
Would that its tone could reach the Rich!
She sang this "Song of the Shirt!"

JACOB A. RIIS

"The Working Girls of New York"

The first book by Jacob A. Riis (1849–1914), *How the Other Half Lives: Studies among the Tenements of New York* (New York: Scribner's, 1890), significantly influenced the way Stephen Crane, as well as the rest of middle-class America, saw the urban poor. For Riis's biography, see the headnote to "The Problem of the Children" (p. 128), and for Riis's influence on Crane, see the introduction (p. 8). The following selection comes from *How the Other Half Lives,* 234–42.

Of the harvest of tares,[1] sown in iniquity and reaped in wrath, the police returns tell the story. The pen that wrote the "Song of the Shirt" is needed to tell of the sad and toil-worn lives of New York's working-women. The cry echoes by night and by day through its tenements:

> Oh, God! that bread should be so dear,
> And flesh and blood so cheap!

Six months have not passed since at a great public meeting in this city, the Working Women's Society[2] reported: "It is a known fact that men's wages cannot fall below a limit upon which they can exist, but woman's wages have no limit, since the paths of shame are always open to her. It is simply impossible for any woman to live without assistance on the low salary a saleswoman earns, without depriving herself of real necessities. . . . It is inevitable that they must in many instances resort to evil." It was only a few brief weeks before that verdict was uttered, that the community was shocked by the story of a gentle and refined woman who, left in direst poverty to earn her own living alone among strangers, threw herself from her attic window, preferring death to dishonor. "I would have done any honest work, even to scrubbing," she wrote, drenched and starving, after a vain

[1] *tares:* A type of weed. Cf. Lord Byron, *Childe Harold's Pilgrimage,* canto 4, book 120: "Alas! our young affections run to waste, / Or water but the desert; whence arise / But weeds of dark luxuriance, tares of haste / Rank at the core, though tempting to the eyes."

[2] *Working Women's Society:* An organization established in 1888 to protect the interests of working women.

search for work in a driving storm. She had tramped the streets for weeks on her weary errand, and the only living wages that were offered her were the wages of sin. The ink was not dry upon her letter before a woman in an East Side tenement wrote down her reason for self-murder: "Weakness, sleeplessness, and yet obliged to work. My strength fails me. Sing at my coffin: 'Where does the soul find a home and rest?'" Her story may be found as one of two typical "cases of despair" in one little church community, in the *City Mission Society's Monthly* for last February. It is a story that has many parallels in the experience of every missionary, every police reporter and every family doctor whose practice is among the poor.

It is estimated that at least one hundred and fifty thousand women and girls earn their own living in New York; but there is reason to believe that this estimate falls far short of the truth when sufficient account is taken of the large number who are not wholly dependent upon their own labor, while contributing by it to the family's earnings. These alone constitute a large class of the women wage-earners, and it is characteristic of the situation that the very fact that some need not starve on their wages condemns the rest to that fate. The pay they are willing to accept all have to take. What the "everlasting law of supply and demand," that serves as such a convenient gag for public indignation, has to do with it, one learns from observation all along the road of inquiry into these real woman's wrongs. To take the case of the saleswomen for illustration: The investigation of the Working Women's Society disclosed the fact that wages averaging from $2 to $4.50 a week were reduced by excessive fines, "the employers placing a value upon time lost that is not given to services rendered." A little girl, who received two dollars a week, made cash-sales amounting to $167 in a single day, while the receipts of a fifteen-dollar male clerk in the same department footed up only $125; yet for some trivial mistake the girl was fined sixty cents out of her two dollars. The practice prevailed in some stores of dividing the fines between the superintendent and the time-keeper at the end of the year. In one instance they amounted to $3,000, and "the superintendent was heard to charge the time-keeper with not being strict enough in his duties." One of the causes for fine in a certain large store was sitting down. The law requiring seats for saleswomen, generally ignored, was obeyed faithfully in this establishment. The seats were there, but the girls were fined when found using them.

Cash-girls receiving $1.75 a week for work that at certain seasons lengthened their day to sixteen hours were sometimes required to pay for their aprons. A common cause for discharge from stores in which, on account of the oppressive heat and lack of ventilation, "girls fainted day after day and came out looking like corpses," was too long service. No other fault was found with the discharged saleswomen than that they had been long enough in the employ of the firm to justly expect an increase of salary. The reason was even given with brutal frankness, in some instances.

These facts give a slight idea of the hardships and the poor pay of a business that notoriously absorbs child-labor. The girls are sent to the store before they have fairly entered their teens, because the money they can earn there is needed for the support of the family. If the boys will not work, if the street tempts them from home, among the girls at least there must be no drones. To keep their places they are told to lie about their age and to say that they are over fourteen. The precaution is usually superfluous. The Women's Investigating Committee[3] found the majority of the children employed in the stores to be under age, but heard only in a single instance of the truant officers calling. In that case they came once a year and sent the youngest children home; but in a month's time they were all back in their places, and were not again disturbed. When it comes to the factories, where hard bodily labor is added to long hours, stifling rooms, and starvation wages, matters are even worse. The Legislature has passed laws to prevent the employment of children, as it has forbidden saloon-keepers to sell them beer, and it has provided means of enforcing its mandate, so efficient, that the very number of factories in New York is *guessed* at as in the neighborhood of twelve thousand. Up till this summer, a single inspector was charged with the duty of keeping the run of them all, and of seeing to it that the law was respected by the owners.

Sixty cents is put as the average day's earnings of the 150,000, but into this computation enters the stylish "cashier's" two dollars a day, as well as the thirty cents of the poor little girl who pulls threads in an East Side factory, and, if anything, the average is probably too high. Such as it is, however, it represents board, rent, clothing, and "pleasure" to this army of workers. Here is the case of a woman employed in the manufacturing department of a Broadway house. It stands for a hundred like her own. She averages three dollars a week. Pays $1.50 for her room; for breakfast she has a cup of coffee; lunch she cannot

[3] *Women's Investigating Committee:* An organization created to probe illegal labor practices including the employment of underage children.

afford. One meal a day is her allowance. This woman is young, she is pretty. She has "the world before her." Is it anything less than a miracle if she is guilty of nothing worse than the "early and improvident marriage," against which moralists exclaim as one of the prolific causes of the distress of the poor? Almost any door might seem to offer welcome escape from such slavery as this. "I feel so much healthier since I got three square meals a day," said a lodger in one of the Girls' Homes. Two young sewing-girls came in seeking domestic service, so that they might get enough to eat. They had been only half-fed for some time, and starvation had driven them to the one door at which the pride of the American-born girl will not permit her to knock, though poverty be the price of her independence.

The tenement and the competition of public institutions and farmers' wives and daughters, have done the tyrant shirt to death, but they have not bettered the lot of the needle-women. The sweater of the East Side has appropriated the flannel shirt. He turns them out to-day at forty-five cents a dozen, paying his Jewish workers from twenty to thirty-five cents. One of these testified before the State Board of Arbitration, during the shirtmakers' strike,[4] that she worked eleven hours in the shop and four at home, and had never in the best of times made over six dollars a week. Another stated that she worked from 4 o'clock in the morning to 11 at night. These girls had to find their own thread and pay for their own machines out of their wages. The white shirt has gone to the public and private institutions that shelter large numbers of young girls, and to the country. There are not half as many shirtmakers in New York to-day as only a few years ago, and some of the largest firms have closed their city shops. The same is true of the manufacturers of underwear. One large Broadway firm has nearly all its work done by farmers' girls in Maine, who think themselves well off if they can earn two or three dollars a week to pay for a Sunday silk, or the wedding outfit, little dreaming of the part they are playing in starving their city sisters. Literally, they sew "with double thread, a shroud as well as a shirt." Their pin-money sets the rate of wages for thousands of poor sewing-girls in New York. The average earnings of the worker on underwear to-day do not exceed the three dollars which her competitor among the Eastern hills is willing to accept as the price of

[4] *shirtmakers' strike:* In March 1890 unionized and unorganized shirtmakers struck for a ten-hour day and weekly wages.

her play. The shirtmaker's pay is better only because the very finest custom work is all there is left for her to do.

Calico wrappers at a dollar and a half a dozen — the very expert sewers able to make from eight to ten, the common run five or six — neckties at from 25 to 75 cents a dozen, with a dozen as a good day's work, are specimens of women's wages. And yet people persist in wondering at the poor quality of work done in the tenements! Italian cheap labor has come of late also to possess this poor field, with the sweater in its train. There is scarce a branch of woman's work outside of the home in which wages, long since at low-water mark, have not fallen to the point of actual starvation. A case was brought to my notice recently by a woman doctor, whose heart as well as her life-work is with the poor, of a widow with two little children she found at work in an East Side attic, making paper-bags. Her father, she told the doctor, had made good wages at it; but she received only five cents for six hundred of the little three-cornered bags, and her fingers had to be very swift and handle the paste-brush very deftly to bring her earnings up to twenty-five and thirty cents a day. She paid four dollars a month for her room. The rest went to buy food for herself and the children. The physician's purse, rather than her skill, had healing for their complaint.

I have aimed to set down a few dry facts merely. They carry their own comment. Back of the shop with its weary, grinding toil — the home in the tenement, of which it was said in a report of the State Labor Bureau: "Decency and womanly reserve cannot be maintained there — what wonder so many fall away from virtue?" Of the outlook, what? Last Christmas Eve my business took me to an obscure street among the West Side tenements. An old woman had just fallen on the doorstep, stricken with paralysis. The doctor said she would never again move her right hand or foot. The whole side was dead. By her bedside, in their cheerless room, sat the patient's aged sister, a hopeless cripple, in dumb despair. Forty years ago the sisters had come, five in number then, with their mother, from the North of Ireland to make their home and earn a living among strangers. They were lace embroiderers and found work easily at good wages. All the rest had died as the years went by. The two remained and, firmly resolved to lead an honest life, worked on though wages fell and fell as age and toil stiffened their once nimble fingers and dimmed their sight. Then one of them dropped out, her hands palsied and her courage gone. Still the other toiled on, resting neither by night nor by day, that the sister might not want. Now that she too had been stricken, as she was going to the store for the work that was to keep them through the holidays, the battle was over at last. There was before them starvation, or the poor-house. And the proud spirits of the sisters, helpless now, quailed at the outlook.

These were old, with life behind them. For them nothing was left but to sit in the shadow and wait. But of the thousands, who are travelling the road they trod to the end, with the hot blood of youth in their veins, with the love of life and of the beautiful world to which not even sixty cents a day can shut their eyes — who is to blame if their feet find the paths of shame that are "always open to them?" The very paths that have effaced the saving "limit," and to which it is declared to be "inevitable that they must in many instances resort." Let the moralist answer. Let the wise economist apply his rule of supply and demand, and let the answer be heard in this city of a thousand charities where justice goes begging.

To the everlasting credit of New York's working-girl let it be said that, rough though her road be, all but hopeless her battle with life, only in the rarest instances does she go astray. As a class she is brave, virtuous, and true. New York's army of profligate women is not, as in some foreign cities, recruited from her ranks. She is as plucky as she is proud. That "American girls never whimper" became a proverb long ago, and she accepts her lot uncomplainingly, doing the best she can and holding her cherished independence cheap at the cost of a meal, or of half her daily ration, if need be. The home in the tenement and the traditions of her childhood have neither trained her to luxury nor predisposed her in favor of domestic labor in preference to the shop. So, to the world she presents a cheerful, uncomplaining front that sometimes deceives it. Her courage will not be without its reward. Slowly, as the conviction is thrust upon society that woman's work must enter more and more into its planning, a better day is dawning. The organization of working girls' clubs, unions, and societies with a community of interests, despite the obstacles to such a movement, bears testimony to it, as to the devotion of the unselfish women who have made their poorer sister's cause their own, and will yet wring from an unfair world the justice too long denied her.

KATHARINE PEARSON WOODS

"Queens of the Shop, the Workroom, and the Tenement"

Katharine Pearson Woods (1853–1923) was born in Wheeling, Virginia (now West Virginia), but spent much of her youth in Maryland. After her father died a Confederate soldier in 1863, she moved into the rectory of her grandfather, the Reverend James D. McCabe. There her uncle, James D. McCabe, Jr., author of *Lights and Shadows of New York Life,* encouraged her literary interests. Deeply religious, Woods entered the Episcopalian sisterhood. Though frail health forced her to withdraw, the experience fostered her desire for social equality, an impulse that would guide her subsequent literary efforts. Upon returning to Baltimore, she secretly wrote and anonymously published her first novel and her most important book, *Metzernott, Shoemaker* (1889), a sensational work that vigorously asserted the importance of equality between management and labor. Woods wrote other socially conscious fiction and historical-religious novels, but after *Metzernott* her other important book is a biography, *The True Story of Captain John Smith* (1901), an impassioned defense of Smith that ably refuted his numerous detractors. The following essay is reprinted from *Cosmopolitan* (November 10, 1890): 99–105.

> "Queens you must always be; queens to your lovers, queens to your husbands and your sons; queens of higher mystery to the world beyond, which bows itself, and will forever bow, before the myrtle crown and stainless sceptre of womanhood."
>
> — Ruskin

> "As the unwise, inequitable and defective features of our present economic conditions inevitably tend to reduce all who live by their own labor to debasing poverty and dependence, and as the suffering and degradation resulting from this system bear most heavily upon women who support themselves by their own labor. . . . We have formed the Working Women's Society, believing that relief and rescue for those women now oppressed and wronged cannot come without their united effort and mutual association."
>
> — Preamble to the Constitution of the Working Women's Society

To enumerate the different trades by which women in New York are endeavoring — not to live — that for many of them is as utterly unattainable a goal as the end of the rainbow — but simply to postpone as long as possible their appearance at the morgue or the cemetery — to attempt to do this would be useless. Briefly they may be divided into certain broad classes, such as medicine, literature, education, manufactures and domestic service. Under medicine we include the lady doctor and the unskilled hired nurse; under literature we shade down from the editor or fashionable lioness, through typewriters, stenographers and compositors to the book stitchers and folders and the gold-leaf girl; while manufactures covers everything from silk weaving to buttonhole making. Now in all these trades or professions it remains emphatically true that there is "room at the top." The woman of exceptional ability, who knows her niche in life and climbs upward to it with unflinching courage and unswerving will, usually attains it, though often at the price of treading under her more feeble sisters. The editor of a popular paper or magazine does not often quarrel with her salary; the fashionable milliner or dressmaker can command her own price; the lady professor has her own work and her own reward.

But queens?

Which is correct, Ruskin or the Working Women's Society?

To the credit of the noble profession of letters let it be spoken, it knows no distinction of sex. "There is neither Jew nor Greek, bond nor free, male nor female," when one comes within the sound of a printing press, chiefly because what is wanted is work of a certain kind and grade; and also, in the lower ranks of the profession, because of the intelligence and strong organization of the Typographical Union, which admits women upon exactly the same footing as men. Compositors receive on an average twelve dollars a week; their work is piece work entirely, their hours are comparatively short, and the wages in almost every instance sure.

Stenographers and typewriters have often a hard struggle to secure a foothold; they have unions, but they are rather social clubs than trades-unions; their wages run from six or eight dollars a week up to twelve and even eighteen; their success usually depends upon their own business ability; and they receive in all but the rarest instances all that their employers agreed to pay them.

Education is considered the peculiar business of women; perhaps for that very reason it is one of the worst-paid businesses in the world; the salaries of men who engage in it are double those of the women, who do better work and more of it.

Into the servant-girl question we shall not go at present; it would in itself require a volume; and there remains therefore the one department of manufactures.

Among these there are four trades which are not injurious — that is a weak word — but murderous to women. These are artificial flower making, cigar or cigarette making, working on ostrich feathers, and sewing in all its forms. I may also mention the girls who work in soap factories, and whose business it is to wrap the separate cakes, while hot, in paper. The caustic soda used in the manufacture first turns their nails yellow, then eats away the ends of their fingers. There seems no way to help this, as the deftness of touch required would be of course impossible if the workers wore gloves. It is indeed only possible to any given set of workers for a very short time; but there are always plenty to take their places when they drop out, and though one wonders sometimes what becomes of them there does not seem to be any answer. A machine which should wrap the soap and save their fingers would also throw the majority of them out of employment, and they would probably bitterly oppose its introduction.

The arsenic used in making artificial flowers is, in about two years, almost invariably fatal to the workers, who exhibit all the symptoms of arsenical poisoning — sores on the face and hands, swelling of the limbs, finally nausea and convulsions. Arsenic is, however, about the cheapest dye that can be used.

Workers in tobacco suffer from nicotine poisoning, which kills in a less repulsive manner but no less surely; and the feather workers suffer also from poisonous dyes used in the manufacture; the slightest prick of a finger with the needle allows the dye to mingle with the blood.

The mention of the needle, that ancient emblem of womanhood, brings us to sewing women of all grades: cloak makers, shirt makers, everything makers. At first glance this trade seems healthful enough, and so indeed, in itself, it is. And it is so pleasant, so thoroughly womanly, to sew; there are so many bright fancies stitched into the work or evolved by the whir of the sewing machine.

It seems inhumanly cruel, therefore, to make this special trade the means of the most grinding oppression that can be or is practised upon woman.

But why should one trouble to write about this class of workers, or indeed, any class? What good does it do?

"Yes," said one woman with whom I spoke, "there was a lady around here about three years ago asking them same questions, but it didn't help nobody."

"No, I suppose not," I said.

"Then why do they ask them?" she returned, with absolute justice. This woman was out of work, but better off than some, inasmuch as she had neither husband nor children to support. She has worked hard all her life and is now past middle age, thin and worn, with a face of quiet hopelessness and long, thin, pathetic hands.

She is a very fair specimen of the American working woman, the development of the girl who comes to the city full of hope and energy to "get work." She has been told that industry and economy are the highroad to wealth, but she does not aim at wealth, only to lay by a little against a rainy day. So she hires a furnished room and does her own cooking — Heaven save the mark! — a cup of very strong tea and baker's bread! Upon this, with sometimes a "relish," she makes two meals a day, and she works twelve, fourteen, sixteen, eighteen hours. Consequently when youth leaves her, which it does very speedily, health goes with it; she has no reserve force of vitality to draw upon, for overwork and underfeeding have exhausted that as she went along; she drops out of the ranks and goes — where? God knows; may He help her!

The woman of whom I have spoken is or was a cloak maker. "I make the cloak," she said, "all but the machine stitching and pressing; yes, ma'am, buttonholes and all. If I'm kep' busy all the time, and no delays, I can make six dollars a week, but there's a many delays. The boss he says, 'Now, I'll give you a dollar and forty cents or a dollar and fifty cents for that jacket,' he says, 'or that plush coat,' and that doesn't sound bad. But when I baste it together and send to be stitched, the stitcher's work is ahead of mine, and I must wait half an hour or an hour to get it back again, for I've no other coat to work on between whiles. Then when I've done it all there's maybe no more work ready, and I wait — I've waited as much as three days — to get some more, and then been told there was no more for me. And the forelady, she can be very ugly when she likes, if she has a spite on you she gives you work you don't like, and if you name it to her, 'You can go,' she says. It's them Eyetalians that spoils everything," she went on; "they come over here and they'll work for next to no wages at all; an Eyetalian can live on ten cents a day, and no American can't do that, and they can run the machine faster than a woman."

"Them Eyetalians" and Polish Jews seem to be the bane of the clothing trade from the worker's side. In the department of ladies' cloaks as of men's clothing they reign supreme, and male foreigners are taking the places of American women because they work cheaper or, by reason of their greater muscular strength, more rapidly. There are 1200 women tailors in New York working on men's clothing. These work from 5 or 5:30 A.M. until 7 and 8 o'clock P.M. The male worker receives eighteen dollars a week, and is expected to stitch up from twelve to fourteen coats a day; the woman finishes the same number and receives six dollars a week. That limit of six dollars is one which it seems almost impossible to overpass. She who can count upon it is considered fairly well off; nine dollars for the very few who attain it is absolute wealth.

Dressmaking is also a favorite industry with Italians. Almost any morning upon Broadway one may see one or two Italian women, bowed, miserable and filthy, each of whom carries upon her head a bundle about ten feet long, four or five broad, and of the same thickness. My own first impression regarding this sight was, "What a big bundle of rags!" But they are costly rags. She has received them from a fashionable clothing house, and she is carrying them home to the tenement where she resides. Here, amid filth and vermin inconceivable they are made into robes of the latest style, returned to the factory to be draped, and then may be seen behind the plate-glass windows of up-town stores. Some idea of the risk run by this method of manufacture may perhaps be gained from the fact that foremen and "foreladies" who come in contact with these workers bring home living remembrancers to their up-town boarding houses.

Shirt-making has had a bad name as an industry since Hood wrote his Song of the Shirt; nor does the invention of the sewing machine appear to have benefited the worker. In this trade the average earnings are about four dollars a week; some make even less, others more. About five years ago, I am told, the average wages were about eight dollars; but within five months there were three reductions. The first the workers — at least those in one particular factory — took without rebelling, at the second they murmured, at the third they struck. "We were not organized," one of them said to me, "but we struck all the same, and organized afterward. Well, they held out for awhile then they gave us one-half; the other half we got in August without asking." "And yet wages have steadily gone down," I said. "Because they broke up our organization," was the reply. "The next August they closed their factory on purpose, and the girls being thrown out of work drifted off in various directions. The employers did it to break up our organization." "What can women get who make shirts that retail at fifty cents?" I said. "Oh! those are made in reformatories," was the reply. "The House of the Good Shepherd, the Westchester Protectory and others do this sort of work so cheap that business firms can't compete with them. Why, when we were on strike that time the House of the Good Shepherd worked straight along. The others all stopped, but that held straight on. They claimed it didn't interfere with us."

Let us be just. Perhaps it did not; perhaps the House of the Good Shepherd was then working on a special line of goods which did not compete with the strikers. Let us make every excuse possible. But oh! false shepherds unworthy called good, who foul the waters with your feet so that the flock cannot drink thereof; who take from the streets girls who have been driven there by poverty, and use them as instruments to beat down wages, to tread down their struggling sisters into the mire from which they have been temporarily lifted. Only temporarily; for of what avail is it if you wash and clothe a girl and fill her mind with new thoughts and purer hopes; if you accustom her to greater comfort in the way of shelter, food and clothing than she ever dreamed was possible; or if you create in her new wants of flowers, books, and pictures? All these things are good, if you do them; but how shall they profit her, if with them you teach her a trade that she cannot live by? which you have taken pains to ruin for her. Find her a place in a factory and leave her. At the first cut in wages — even sooner perhaps — she will remember that she already knows a trade far more lucrative. We try to be just; but does it not seem as though the saintly ministrants in these reformatories were as anxious to lay up treasures in heaven as are worldlings to do so upon earth, and so took pains to secure a perennial supply of the raw material?

All counters of cheap underwear are supplied from reformatories. Not long ago Mrs. L. M. Barry, well known as a Knight of Labor and defender of woman, found such preternatural bargains at Wanamaker's in Philadelphia that she determined to find out how he came by them. She obtained employment from him as a machine hand, and soon found out from the wages paid that the cheap goods were not of home manufacture. Further inquiry satisfied her that they, as usual, came from reformatories. Now, there is no reason for prejudice against prison or reformatory work as such, for in respect of cleanliness and good sanitary conditions it is preferable to much made outside. That to which the unions object is the low rate at which the work is contracted for, which injures those within the prison equally with those without.

Shopgirls, or salesladies, as they prefer to be called! Here the great evils are excessive hours, working overtime without extra pay, unwholesome sanitary conditions and excessive fines. Just here it may not be amiss to speak of the Working Women's Protective Union,[1] No. 19 Clinton Place, whose special mission it is to collect wages which the worker cannot collect for herself. It has been in operation for twenty-seven years, and has collected in that time thousands of dollars' worth of wages due without one cent of cost to the person wronged. But fines are beyond the reach of even this Union; from them there seems no redress — though upon what principle a woman who receives seven dollars a week is fined thirty cents for ten minutes' tardiness I confess myself unable to see. Seven dollars is by no means the usual wages per week, which range from two to eighteen dollars, the latter to a girl with a good figure who can show off cloaks in the cloak department. In one store the fines in one year amounted to $3000, which was divided between the superintendent and the timekeeper, and the former was heard to charge the latter with lack of strictness. So much for the slave-drivers! The owners also have their pick at the bones of the slave; for in many houses employees are expected to take from two to three weeks holiday in the dull season at their own expense. This on a salary of, say, three dollars a week!

Is it possible to live pure, upright lives under such conditions? Thank God! it is possible, as is attested by the thousands who maintain their integrity in spite of all hindrances; but it is more than hard. It has been well said that, while men's wages cannot fall below the starvation line women's can, since the paths of shame are always open to her. This is a terrible factor in our political economy.

Why write of these things? Where is the remedy? God help us if we cannot find one! For the souls of the coming generation lie in the hands of these women; and we shall never be the people we should and might be until we have learned that it is the first and most important business of a nation to protect its women, not by any puling sentimentality of queenship, chivalry or angelhood, but by making it possible for them to earn an honest living.

For this, the only method is union among women, the best hope is in the women themselves. For men, hard as they have been to women workers, are now being driven by the pressure of their competition, by the effect which women's low rate of wages has had upon theirs, to see

[1] *Working Women's Protective Union:* Established in 1863 under the auspices of *New York Sun* editor Moses Beach, the Working Woman's Protective Union (WWPU) advocated shorter hours and higher wages, but devoted most of its efforts to providing legal services for women victimized by unscrupulous employers.

that their own interest demands her enfranchisement and elevation. The unions are opening to her, she has long been "free of the guild" among the Knights of Labor, whose preamble sets forth among the things to be accomplished by organization: "Equal wages for equal work, without regard to sex." The newly formed clothing unions are ready to welcome her; but woman shrinks back from organization, Heaven knows why! It is perhaps because in organization one finds the truest freedom, and woman has been a slave too long to know what freedom means. Then, too, we are so hard upon each other, we women; it is so difficult to make us trust one another, to bind us together, to create in us a feeling of real sisterhood. And our weakest point is just where our strongest should be; it is in those women workers who have found or made a standing-place for themselves and who by no means wish to be classed as working women. What could not the educated workers of New York do for their struggling sisters — teachers, writers, stenographers, and such like? It would have been amusing to a student of human nature had it not been so infinitely sad, to watch the look of scorn which rose to the surface at the question, "Can you give me any points about your business? I am studying the working women of New York —" "I know nothing about working women," came the quick, short answer.

Some of the things that might be done are shown to us by the two societies already quoted. The Working Women's Society aims to organize women, to teach them the strength and self-respect that organization brings.

Among its remaining aims as set forth in the preamble are to enforce existing laws for the protection of women and children in factories, to investigate and protest against all violations of these laws, and to promote further legislation on this subject, to found a labor bureau, and to secure for both sexes equal pay for equal work.

On May 6, 1890, a mass meeting was held at Chickering hall under the auspices of this society and over 100 clergymen. "A Report on the condition of Women and Children in the New York retail stores" was read, which ought to have caused the very stones to cry out. A preamble and resolutions were adopted, and it was attempted to start a consumer's league, the members of which should pledge themselves to buy at only such stores as should be included in a white list, to be prepared by a committee. To this white list — the obverse of a boycott — there could be no possible objection, provided a sufficient number of stores could be found where employees are treated fairly well; but will it be possible to find consumers enough to found the league?

Wealthy women of New York, attention! This is your business. Will you give up your bargain counter — for it is the wealthy who seek bargain counters — for the sake of your suffering, starving sisters?

The work of the Protective Union, as already explained, is very different, but equally needful. It would seem that small as the wages are, it is a mere matter of course that the workers should receive them when they are due; but whether this be so or not the books of the Union abundantly testify. Some methods of defrauding an employee it has almost broken up, such, for example, as taking girls on trial without wages to learn a business, and when they asked to be paid, turning them off and taking on a new set. The Union has taught the workers to demand a written contract, the keeping of which it stands ready to enforce. Against other wrongs it is powerless, but this of violation of contract it sets straight with all its might; its scope is limited, but it does well all that it attempts without money and without price. No officer is allowed to receive any salary; the lawyer has given his services gratis for twenty-three years; each case is carefully and impartially investigated, and if the money is due payment is enforced if there is any property to levy upon. If not, the offender may be imprisoned for fifteen days if a man; if a woman there is no redress — a bit of chivalry on the part of the law which appears, after the facts we have been considering, exceedingly ill-timed, when taken in connection with the fact that your most arrant and barefaced defrauder of her working people is your high-class, fashionable dressmaker.

A small attempt on the part of the workers to help themselves is the Coöperative Shirtmakers, 770 Third avenue. It was a little pathetic to hear from them that they have been together five years, "longer than most coöperative things hold together." They are thoroughly bright, intelligent women, large-hearted and large-minded, with full sympathy and sisterly love for their sex. Not all of their members work together; of those who do, no one receives more than her regular wages; the profits, if any, are divided between a sinking fund to increase the business and a benefit fund for sick members.

I have not tried to exhaust this subject, in fact it is inexhaustible; only to say such things as may perhaps open the eyes of some one person to the lives that are being lived through around us. And yet, what good will it do? But God help us all unless we change this state of things, and that right speedily!

JANE CUNNINGHAM CROLY

"Senate Testimony from 'Jennie June'"

Jane Cunningham Croly (1829–1901) was born in England and, during her adolescence, came to the United States with her family where they settled at Poughkeepsie, New York. After her father's death in 1854, she went to New York City, where she began writing for the *New-York Tribune* and soon was contributing a regular ladies' column to the *Sunday Times and Noah's Weekly Messenger.* By the decade's end, she had begun sending correspondence, written under the pen name "Jennie June," to newspapers from Baltimore to New Orleans, and therefore is considered the first woman journalist to syndicate her writing. In 1856, she married fellow journalist David Goodman Croly. Together they had four children (three surviving infancy), but neither her marriage nor her burgeoning family stopped her from pursuing her writing career, and, over the next several decades, she contributed to many different newspapers and magazines and wrote several books. For middle-class women, she provided advice with such works as *Jennie June's American Cookery Book* (1866) and *For Better or Worse* (1875). She also published an advice book for the working woman, *Thrown on Her Own Resources* (1891). Her most important work is the *History of the Woman's Club Movement in America* (1898), and she herself was an important figure in the development of women's clubs in the United States.

Croly's reputation as an advocate for the woman's right to work made her an ideal choice to speak before the Senate Committee on Education and Labor in 1883. In the following excerpt, taken from the *Report of the Committee of the Senate upon the Relations between Labor and Capital, and Testimony Taken by the Committee,* vol. 2 (Washington, D.C.: 1885), 605–12, Croly gives her main ideas toward the working woman and provides the Senate with a case study of her own that supplies further insights into the world of the working woman.

Difficulties of Working Women

Q. Please tell us something of the difficulties that women encounter in the manner of their lives and in the expense of living. — A. The difficulty of the working girls in this city is because there is no relation between the money that they can earn and the supply of accommodation; that and one other; the principal hardship which they suffer is from the fact that they are not able to leave their miserable homes. It is as I said; the working girls of this city are not in a normal condition; they are not the natural outgrowth of the needs of the country; they are forced into the labor arena by the short-comings of some man, by the drunkenness or laziness of the father or husband. You see in that case it is not a normal condition. They have their work at home; they have the requirements of a household very often when married, and the girls have nearly always to turn the larger part of their wages into the home to sustain the mother and the rest of the family. That is the case with nine-tenths of the poor working girls of this city. They cannot accept even gratuitous homes that could be made for them because their little means are required to sustain the family at home.

Working Conditions in Europe and America Contrasted

It is not as it is in other countries, where labor for pay is the normal condition of the whole family, where the children are put to work as soon as they can begin to earn money, where they take some little branch of their father's trade and work at it. Here, if a man is a hard, industrious working man he is ambitious for his family. He wishes his family to be better than he was himself. He wants to educate them better. He sends them to school; above all, he sends his daughters to school as soon and as long as he can; but when he cannot do that any longer, when the girls are forced out, the conditions are doubly hard for them; it is the strain at home and the strain abroad, added to poor food and no comforts; no supplies necessary for strength and not even enough for warmth. These are the conditions that make it so hard to bear. I have myself seen whole families die off. I have seen the five daughters of one family die off, one after another, simply for the want of food and proper accommodation, whose little means of $5 a week were taken to their mother for the support of the family, and very often these were absorbed by the father in drunken freaks. I mention that as one instance. We very often think that those are isolated cases, but they are not isolated in this country or in this city. The normal condition of the working women, of course, is doubly hard because they have the strain upon them, as I have stated, both in their work and in their home. They have to buy, at high prices, poor articles, and have to pay very high prices for such accommodation as they get, for meats and everything that they need, for which their money is expended, while all the time the pressure from the outside is to reduce their wages and to make the circumstances so hard for them that they can hardly live.

Wages Dependent on Skill

The wages of girls in this city are, of course, graded by their skill; and that is why I place so much importance upon the training of girls in these trades. It is a curious fact that to-day, while girls are starving, while they are making miserable pittances by making shirts and overalls and doing slave work for $3 a week, yet if they were trained to labor they could earn from $10 to $15 a week. There are plenty of places that are open to women that they cannot fill. I myself have had a great many opportunities to put women in places, but I could not find the trained women.

Training Wanted by Women

It is training that we want for our girls and women in order to make them able to support themselves. I disagree with Mrs. Blake in her comparison between the laborer at $1.25 a day and the sewing woman at 50 cents a day. The sewing woman earns 50 cents per day because she cannot sew; the washerwoman earns a $1.25 per day because she *can* wash. Wherever a person can do anything, can *do* it in a proper sense, they can always earn a living by it; they can always get a certain amount for it; but, I do not approve of encouraging women or girls to think that because they can only earn $3 a week it becomes, therefore, only a question between starvation and vice. One can live on $3 a week a great deal better than one can die morally, whether for more or less.

Q. What is the cost of renting comfortable rooms such as are ordinarily rented or suitable for people of small means? — A. Within a few months I have had interviews with a great many working girls, because I have been writing a number of articles in regard to the way we live here in New York, in regard to the methods of living among the working women.

BY THE CHAIRMAN:

Q. Where are those articles published, please? — A. I have one of them here, and I will, if you will permit me, read you what a working girl herself told me.

STORY OF A WORKING GIRL

An intelligent cigar girl, recently on strike, said that of all the girls she knew, who had grown up with her and worked with her, the worst off were those who had married, for they had, almost invariably, to support the family, and suffer abuse and ill-treatment besides. This young woman is perhaps twenty-five or twenty-six years of age; she is by no means ill-looking, and is one of the most sensible and thoughtful of her class. She has always, as she says, had a comfortable home. Her father was an honest, industrious man; her mother a kindly, home-loving Englishwoman, who cooked, and washed, and ironed, and took the best care of her family. When the husband died she took in sewing and washing, and kept the children together, with their help, till a son and daughter married, and then the mother and only remaining daughter lived together, the daughter supporting the mother, and the mother taking care of the daughter, doing the little housework and all of the sewing, and occasionally helping by doing something in the way of sewing or taking care of children for a neighbor. "I never knew what it was to hear an unkind word," said the girl, when she related to me this little history, "and mother was never tired of doing anything she could to give me pleasure. I would far rather have had to support her than not all my days, and since she has been gone I still live alone and keep house, for I can't bear to part with her bits of things, and I feel more independent like in my own room, where I can come and go just as I please."

Cost of Her Support

"How much does it cost you to support yourself in your room?" I asked. "Well," she replied, "I have two little rooms in a rear house, but it's clean, and one of them has a window as looks out where it is quite fresh and pleasant. For these I pay $6 a month, and my living, I reckon, costs me, what with coal and wood and light, about $4 a week." Your washing, then, I remarked, would bring your weekly expenses, without extra fare, clothing, and the like up to $6 per week. "I do my washing and ironing myself," she said, "on Saturday afternoons, but you may count it $6 all the same, for it comes to about that."

Her Earnings

"How much are you able to earn, then, at cigar-making," I inquired. "Well, cigar-making used to be a pretty good business for first hands," she replied, with a pleasant look, at the thought of those old times. "I've seen the time when I could make $12 and $13 a week at cigar-making; but you can't do it now, and that was the reason why we struck; it was not for a rise; it was to have the superintendent back as was good to us, and give the girls some chance as well as the bosses."

Difficulties of Her Work

"How did the change of superintendents affect your interest," was asked. "Well, you see, we had been getting 80 cents a thousand, and a smart worker can make 2,000 cigars a day, counting a day from 7 o'clock in the morning till half past five in the afternoon, with half an hour for dinner, which is a regular ten-hours day. But sometimes, when we are very busy, like now, we'd have to work till 9 o'clock at night, and then we'd make 2,500, may be, and earn $2 in a day; but that would only be for a little while, and we could not do it for long. Besides, we never get the full count; if there is the least little flaw the inspector throws it out, and these do not count at all; but they always go through another sorting, and more than half of those that have been thrown out and are not counted in the girl's work are counted in as good all the same. Our superintendent and his inspector were very good to us; they didn't throw out unless there *was* a flaw; and they looked out that the tobacco was in good condition for rolling, neither too dry or too damp, for this makes a great difference in the doing of the work; so they made up their minds that this superintendent was too good and too much of a gentleman, and they discharged him and put a regular rough in his place, a man who bullied the girls and never spoke a civil word in giving an order, but just swore at them; and when he found how much the best hands could earn he cut the prices down from 80 to 65 cents a thousand and threw out at such a rate that the loss of pay went up from $1 to $2 a week; you see $1 a week out of each girl's wages, where 500 girls are employed, mounts up to $500, and is an item, but it cuts the wages of the girls down to a little over half what they used to be, and a good many of the best girls have given up the business. I think I'll give it up myself, for I can't earn enough to keep myself in a home of my own, and I don't like boarding."

The Manners of Rough Working Girls

Subsequently I asked her how it had happened that she had not married since her mother's death. Her reply was characteristic. She said, "Men was too sassy." She wouldn't mind so much supporting a man for company, but she didn't want to be abused for it. A good deal of the slang among the rougher girls, who work with men, is the same as that commonly used by the rudest men; but it would give a by-stander, who might be shocked by it, a very false impression, if it were supposed that they were moral delinquents, and not abundantly able to "take care of themselves." They pride themselves on being "tough," able to "give as good as they get," but it is the universal testimony of men themselves who work in these shops that it is not these apparently rude and aggressive girls who fall; it is more frequently the gentle and timid, who crave affection and companionship, and must take them in any form in which they can get them. The manner of the majority of the girls is derived from their often dreadful associates and surroundings. Even correctness of language would be looked upon as "airs," and punished accordingly. Yet that appreciation, and even ambition for a different life exists in common with rough habits, and almost brutal surroundings, there is sufficient evidence. In one of the worst neighborhoods in the lower part of this city there live in two rooms three girls. Dens of wickedness are on either side of them and over the way, and the girls themselves would perhaps be taken by a superficial observer for specimens of the lowest order. For they can swagger, and even swear, and "talk back," but if rude words proceed beyond their limit they would not perhaps hesitate to hit back, and hit "straight," for no man who knows them in or out of the shop but respects them for all their free talk, and is well aware that they "don't stand no non-sense," and that they "know how to take care of themselves." The girls work in a type-foundry, and earn from $5 to $7, and $8 per week. Their father died in a drunken fight; their mother, as soon as they were old enough to do regular work, they relieved from outside labor, keeping her at home, "like a lady" and maintaining also a brother at school. Upon this brother their hopes and ambitions rest, and they are devoting all their young lives to him. Their mother died, but they retain their home, such as it is, and, inside, it is home to them, after all, and they have educated their brother at a high preparatory school and now at a medical college, for he is to be a professional man and a gentleman, and he is to be made so by their labor and sacrifices. Will he remember and not be ashamed of the sources from which he derived his inspiration and opportunities?

Heroism of Working Girls

It is almost a hopeless task to present in one brief paper even a glimpse of the heroic lives which are to be found behind the dreary and common-place exteriors of many working girls' homes. Doubtless these are, in some of their details, exceptional; but the girlhood which grows up to young womanhood in the dark and cheerless places, the youth which knows nothing of beauty and brightness, the age which has no loving, cared-for past or restful future, these are not exceptional. Meeting with individual cases which present features of great severity, we are apt to think that these are unusual, until more intimate knowledge proves them so common as to have become commonplace and incapable of exciting sympathy. A quiet girl, employed in a foundry down town, attracted attention simply because of her reserve, her extremely gentle bearing, and freedom from the usual defects of loud talking and free manners. She proved to be the daughter of a man who came to New York after a fire which had destroyed his home and business, ruining him financially and, as it proved, mentally. The family struggled along for awhile, the daughter being the chief dependence. Then one by one two younger children died and were soon followed by the mother, who was worn out by grief and want of nourishing food. The repeated shocks prostrated the father, who had been able to find no steady employment, and he became an imbecile, dependent entirely upon his daughter, not only for support, but for care and protection. This daughter has, through all the years of her youth and young womanhood, since she was fifteen years old, worked in a shop and supported the family on her earnings of $7 per week. At the same time she has spent her evenings and early mornings in doing the work of the poor room they occupy, her nights, many of them, in nursing and tending the sick, and days in one eternal round of hard mechanical work. She is now more nearly broken down than at any previous time in her life, for her father is a constant anxiety as well as burden. She must have a place they can call home, because that is the only way they can live, and she cannot relinquish her steady hold upon the hard routine of her life for one day, one hour, or one minute. Talk to this girl about two weeks at a "seaside home!" One of the terrors of her life, though she can hardly drag herself to her shop, is that she may be deprived of her work.

Physical Strength an Element in Work

The silk manufacture of this country is of comparatively recent growth, and the papers frequently contain congratulatory paragraphs concerning its rapid development and future prospects. Already it has built up the fortunes of manufacturing firms engaged in its production and enlisted the labor of thousands of operators, the majority of whom are young girls. The work requires constant standing at a loom and incessant activity in manipulating threads; the standing behind a counter with the counter to lean upon, and the body exercised easily and in a diversity of ways, is nothing to it, yet a crusade was organized in favor of seats for the shop girls, while no one dreams of making a protest for the benefit of the loom-tenders, the silk weavers, carpet weavers, stocking weavers, woolen-cloth weavers, and others. Why? Because we do not see it, and do not realize it, and also because bread is the first necessity of life, and bread must be had by whatever hard and cruel means it is obtained.

People, particularly girls and women, who are naturally inclined to accept whatever falls to their lot, can "get used" to almost anything, but sometimes the flesh breaks down, even though the spirit be willing. The pay for weaving a yard of silk is from 7 to 10 cents, and an expert operator can weave from 10 to 12 yards per day; less skillful hands do not average so much, and wages range therefore from $4 to $8 per week. A young Russian girl, well brought up, tenderly reared, with property in her own right, left her own country and her possessions because, upon the death of her parents, she was tyrannically controlled by a brother and an aunt, and because she had become an enthusiast of republican ideas. She was willing, in her own words, "to go anywhere, and to do anything to be 'free,' and to say the words that were in her own mind." She arrived in New York with a party of thirty of her countrymen and countrywomen. Their dreams of a new Atlantis and of a model community in the New World were somewhat rudely dispelled, and a number of them returned to their native land. But the remainder resolved to stay for a few years at least, and see what there was in this land of promise which would be worth garnering and taking back to their own countrymen and countrywomen. Among the remainder was the young girl alluded to. They had made common cause and common purse. She obtained work in a silk factory, earning $4.50 per week, and living in a little attic under the eaves, and with only a skylight, that she might divide with those who were not earning anything. The work, to one unaccustomed to manual labor, was almost impossible. She suffered tortures from the heavy, incessant movement, the noise of the looms, and the standing from morning till night, with half an hour of interval for rest and dinner. At the end of four weeks she broke down, but rallied and went to work again. She can better sustain the difficulties of her position now, but she is not of the kind who will ever quite get used to them, her sensitive organization and more delicate bringing up precluding such a possibility. The one strong reason she gave for endeavoring to overcome the hardships of this species of mechanical labor was that it was "clean," which she remarked was so "all-important" to her.

Permanent, not Temporary, Means of Advancement Wanted

I could go on multiplying instances through a volume, for the class is so numerous, and they are re-enforced from sources so wide and diversified, as well as by the ever-increasing pressure of the labor-seeking millions who throng to these shores, that it must forever increase instead of growing less. What is needed, therefore, for their benefit is not temporary help so much as permanent means of advancement, higher ideals for the majority, associative effort for strength, protection, companionship, and a social life, brightened by all gracious influences. They need to be inspired by self-respect and a knowledge of their own powers and possibilities, by a consciousness that instead of pariahs they constitute the best elements of society, according to the excellence of their work, and the courage and constancy, the truth and honesty they put into it. It is an infinite pity that all the appeals that are made to them, nearly all the evidence of sympathy that reaches them, come in the shape of demands for wages and cutting down hours of labor, irrespective of the work or worker. This has a distinctly detrimental influence; antagonisms are fomented, there is no enthusiasm for work, or the attainment of excellence, and the only object set before any employé is to do as little and that as poorly as possible for the money he or she receives. This is immoral, and the results are as bad for employer as employed. It is not a question for the conscience of the worker whether the work is sufficiently paid; it is whether it is well done, and the conscience must be cultivated as a basis of permanent advancement, whether by those who labor or those who employ labor.

A Great Mission for Wealthy Women

Wealthy women, educated women, socially protected women have a great work lying right at their doors, in the isolation and needs of these armies of poor working girls and women. Is there any reason in what either do, or in what they are, why one should walk always in darkness, the other always in light; why one should enjoy all the sweet, the other all the bitter; why one should know the earth only as a driving wheel, the other as a garden of exquisite delights? The conditions are perhaps not so wide apart as they would seem, for in work there is compensation, in all idleness dreariness; but conscientious endeavor on the part of those who hold the power in their hands would build a bridge, over which angels might walk from one to another, bringing to one light and warmth, beauty and enlargement, and to the other elevation of character, which comes from a knowledge of duty and its performance.

ABRAHAM HUMMEL AND WILLIAM F. HOWE

"Store Girls: Their Fascinations, Foibles and Temptations"

William F. Howe (1828–1902), born in Boston and educated in London, returned to the United States in 1858 where he settled in New York City and was admitted to the bar the following year. His ambition, flamboyance, and skill for obtaining acquittals quickly garnered him an extensive clientele. In 1863, he hired a teenage Abraham Henry Hummel (1850–1926) as office boy. The two developed a good working relationship and an intimate friendship outside the workplace. Hummel had an aptitude for law, and Howe managed to have him admitted to the bar in 1869 when he was only nineteen. Howe then took Hummel into partnership, and together they became the two most widely known attorneys in the country. They specialized in sensational cases — murder, divorce, or those associated with celebrities of the stage — and usually succeeded due to Hummel's unscrupulous pretrial preparation and Howe's courtroom histrionics. The partnership flourished until Howe retired in 1900.

Together they published *In Danger! or, Life in New York* (New York: J. S. Ogilvie, 1888), a work that incorporated references to many of their famous cases. The following selection (31–39) incorporates many of the stereotypical attitudes toward the young workingwoman, including the idea

that reading the day's sentimental dime novels contributed to the downfall of many young woman by giving them exaggerated, romantic notions of the possibilities of life and love offered.

Since the time when Mary Rogers, the beautiful cigar girl of Broadway, met her sad fate over in Hoboken,[1] the pretty shop girls of New York have contributed more than their full quota to the city's contemporaneous history. They have figured in connection with many of its social romances and domestic infelicities, as well as with its scandals and its crimes — secret and revealed. In Gotham's grave and gay aspects — in its comedy, its tragedy, and its melo-drama, we are perpetually running across the charming face, graceful form, and easy, gay demeanor of the pretty shop girl.

As a rule, the temptress of the store is pretty — frequently quite beautiful, and almost invariably handsomer than those fortunate daughters of Mammon whom she is called upon to serve, and who often treat her with such top-lofty hauteur. And how stylish she frequently is, and how difficult it is to describe this incommunicable quality of *style,* which those artful setters of baits — the dealers in ready-made fabrics — understand so well! Who has not noticed how the tall, slender-framed girls, with their graceful movements and flexible spines, their long, smooth throats and curved waists, are drafted off to stand as veritable decoy-ducks? Who has not observed the grace and ease with which they wear risky patterns and unusual *façons,*[2] and so delude the arrogant but ungraceful customer into buying, in the belief that she will look just as well as the pretty model? The average well-to-do woman, with some pretensions to good looks, sees a beautiful young creature with Junoesque[3] air parading before her in bold color-combinations and doubtful harmonies, and she imagines she can venture the same thing with like effect. But alas! what a travesty the experiment frequently is!

[1] *Mary Rogers . . . Hoboken:* The 1841 murder of Mary Cecilia Rogers, a New York cigar girl, formed the basis for Edgar Allan Poe's tale, "The Mystery of Marie Roget." See John Walsh, *Poe the Detective: The Curious Circumstances Behind the Mystery of Marie Roget* (New Brunswick: Rutgers UP, 1968).

[2] *façons:* Fashions.

[3] *Junoesque:* The adjective refers to Juno, a mythical deity celebrated for her beauty.

Many of the New Yorkers who read this page will recall the Original Dollar Store on Broadway and its fascinating young salesladies. Some of these were perfect sirens with their loveliness of feature and delicacy of color; their luxuriant hair, made amenable to the discipline of the prevailing fashion; the gown stylish and perfect, and frequently not at all reticent in its revelations of form; the countenance calm, watchful and intelligent — frequently mischievous; the walk something akin to the serene consciousness of power which we are told that Phryne[4] exemplified before her judges, and accompanied with that grace which is the birthright of beauty in every age and under any circumstances.

For many reasons the tone of morality, in some instances, among store girls in this city is not high. A variety of obvious causes contribute to this result, among which may be mentioned their generally poor salaries; their natural levity, and the example of their companions; their love of dress and display, coupled with a natural desire for masculine attentions; long hours in close, impure air; sensational literature; frequent absence of healthy or adequate home influence; and the many temptations which beset an attractive girl in such a position.

Many of them enter stores as mere children in the capacity of cash girls. They are the children of poor parents, and as they grow up to young maidenhood, they acquire a sort of superficial polish in the store, and are brightened without being educated. Some grow up and take their places as full-blown salesladies, and begin to sigh for the gayety of the streets, for freedom from restraint, and for amusements that are not within their reach. Naturally *au fait* in style, with taste and clever fingers, they dress in an attractive manner, with the hope of beguiling the ideal hero they have constructed from the pages of the trashy story paper. It is a sort of voluntary species of sacrifice on their part — a kind of suicidal decking with flowers, and making preparation for immolation. Full of pernicious sentimentality, they are open to the first promising flirtation. They see elegantly-dressed and diamonded ladies, and their imagination is fed from the fountains of vulgar literature until they dream that they, too, are destined to be won by some splendid cavalier of fabulous wealth. Learning from the wishy-

[4] *Phryne:* The ancient Roman rhetorician Quintilian tells the story of a woman named Phyrne who was accused of impiety. When her lawyer saw that the issue was doubtful, he unveiled her bosom, which so influenced her judges that they acquitted her.

washy literature that their face is their fortune, and so, reading what happened to others, and how perfectly lovely and romantic it all was, they are ready for the wiles of the first gay deceiver. Waiting in vain for their god-like ideal, they are finally content to look a little lower, and favorably receive the immodest addresses of some clerk in their own store, or succeed in making a street "mash."

Sometimes the pretty girl rushes impetuously into marriage, repents and separates from her husband. She is still good looking, and her marital experience has given her an air of easy assurance, and she readily finds employment as a saleslady. Her influence afterwards, among girls comparatively innocent and without her experience, cannot but be pernicious, and at the same time must exert a certain formative and shaping process in determining the peculiar character of the whole class of girls in the store.

Very frequently she does not attain even to the questionable dignity of a marriage ceremony. Flattered by the attentions of some swell, the pretty shop girl will be induced to accompany him to the theatre and to supper in a concert saloon. Her vanity is kindled by his appearance. She rejoices in the style of his clothes, in the magnificence of his jewelry, and she thinks her mission in life is to walk beside the splendid swell, amid rose gardens, theatres and supper rooms, for the remainder of her life. Finally she yields to his soft solicitations, and her prospects are forever blighted. She becomes an incorrigible flirt, meets her "fellows" on the corner of the street near the store, spends a certain number of evenings and nights with them at hotels where no course of catechism takes place at the clerk's desk. She goes to Coney Island or local beer gardens on Sundays, manifesting a vivid animal pleasure in her enjoyment, with little manifestation of gratitude towards her escort who is supplying the money.

Sometimes, again, an exceptionally pretty girl will fall a victim to the proprietor, the manager or some of the superintendents of the store; and there have been cases of this kind heard in the courts, in one of which the proprietor not only seduced the girl, but married her, afterwards obtaining a divorce because of her incontinence. Sometimes the lapse of these girls from the paths of virtue is accompanied with exceptional hardships. The young lady is beautiful as well as good perhaps, and the pride of her idolizing parents, who have taught her that she is fit to be the wife of a duke. She attracts the eye of a man about town, and the process of courting and flattery — of sapping and mining — begins, with the result that he has had in view since the inception of the acquaintance. He is not a bad fellow as the world goes; but providence and society have made it very hard for single men to show kindness to single women in any way but one. He is sorry at her situation; but she is hardly the person for him to marry, even with her blooming, flower-like face. In such a situation — and such situations are far too common with the class — Byron's lines, slightly altered, seem peculiarly applicable to the pretty shop girl:

> "'Twas thine own beauty gave the fatal blow,
> And help'd to plant the wound that laid thee low."[5]

Sometimes it happens that the pretty girl, wearied of waiting for her knightly deliverer, comes across the advertisement of a gifted seeress — the seventh daughter of a seventh daughter, perchance, or "the only English prophetess who has the genuine Roman and Arabian talismans for love, good luck, and all business affairs;" or the wonderful clairvoyant who can be "consulted on absent friends, love, courtship and marriage." Not infrequently she falls into the toils of those advertising frauds, who frequently combine the vile trade of procuress with the ostensible trade of fortune-telling. When the girl is drawn to this den, the trump card offered her is, of course, the young gentleman, rich as Crœsus and handsome as Adonis, with whom she is to fall in love. He is generally described with considerable minuteness, and the time and place of meeting foretold. This may be fictitious, and it is fortunate for her if it is so. But the seeress too frequently needs no powers of clairvoyance or ratiocination to make these disclosures, for some *roue,*[6] who has exhausted the ordinary rounds of dissipation, or some fast young fellow seeking a change, has made a bargain with the prophetess for a new and innocent victim — the amount of the fee to depend on the means and liberality of the libertine and the attractiveness of the victim. The vain, silly girl is dazzled with the wily woman's story, and readily promises to call again. At her next visit the man inspects her from some place of concealment, and if she meets his views, either an introduction takes place or a rendezvous is perfected. Thus the acquaintance begins, with the result which every intelligent reader can see for himself. Sometimes the picture of the scamp is

[5] *"'Twas thine own beauty . . . that laid thee low":* George Gordon, Lord Byron, "English Bards and Scotch Reviewers," lines 839–40: "'Twas thine own genius gave the final blow, / And help'd to plant the wound that laid thee low."

[6] *roue:* A man who leads a life of pleasure and sensuality.

shown, but in every case there is but one end in view on the part of the seeress, and that end is almost invariably achieved. The girl thus becomes clandestinely "gay," and spreads the influence of her evil example and impure associations among her shopmates. Pope has told us in four immortal lines the effects of a constant contact with vice. In the second epistle of his Essay on Man, he writes:

> "Vice is a monster of so frightful mien,
> As, to be hated, needs but to be seen;
> Yet seen too oft, familiar with her face,
> We first endure, then pity, then embrace."[7]

In the case of the class of young girls under consideration this truth is peculiarly applicable. In consequence of their associations they hear and see things whose influence is almost wholly bad and pernicious. Those disguised advertisements in the newspapers called "Personals" are of this evil character. To young girls, with minds imperfectly disciplined, there is a fatal fascination in the mystery of surreptitious appointments and meetings. Mystery is so suggestive and romantic, and the young girl who, from piqued curiosity, is tempted to dally with a "Matrimonial" or a "Personal," is an object of commiseration. From dallying and reading and wondering, the step is easy to answer such notices. She believes that she has a chance of getting a rich and handsome husband, who will take her to Europe, and, in other respects, make her life a sort of earthly paradise. The men who write such advertisements know this besetting female weakness and bait their trap accordingly. And so a young girl, too frequently, walks alone and unadvised into the meshes of an acquaintanceship which leads to her ruin. It is perhaps as useless to ask the men who are base enough to conceive these things to refrain from publishing them, as it is to urge the mercenary proprietors of certain newspapers to refrain from printing them in their columns. Yet is must be perfectly clear to all right-thinking minds, that it is in vain for parents to warn, parsons to preach, friends to advise, for the good to deplore, and the ignorant to wonder, at the increasing deterioration of our metropolitan morals, while these tempting lures to feminine destruction are so alluringly displayed.

[7] *"Vice is a monster . . . embrace"*: Alexander Pope, "Essay on Man," epistle II, lines 217–20.

It would be doing very imperfect justice to this theme did we fail to record our conviction that some of the salesladies and shop girls of the city are thoroughly good, virtuous, honest and respectable. Many of them, amid unhealthy influences and corroding associations, preserve the white flower of a blameless life, and become the honored wives of respectable citizens. But these are a small minority. At the same time it is useless to disguise the fact that there are others whose character needs stronger colors for proper delineation than have hitherto been employed. There are those among pretty shop girls who simply give up their leisure time to surreptitious appointments. This is the worst and most dangerous form in which this prevalent vice stalks abroad, and it more clearly stamps the character of a community than does its more open and brazen manifestations. Many causes may lead to a woman's becoming a professional harlot, but if a girl "goes wrong" without any very cogent reason for so doing, there must be something radically unsound in her composition and inherently bad in her nature to lead her to abandon her person to the other sex, who are at all times ready to take advantage of a woman's weakness and a woman's love. Seduction and clandestine prostitution have made enormous strides in New York, and especially among the young women and girls connected with stores, within the last decade.

Not long ago a woman, who then occupied a prominent position in a Sixth avenue store, was met up-town in the evening. She is very good looking — strong and lithe and tall, with a cloud of handsome hair that glistens like bronze; large dreamy eyes that flash and scintillate witchingly; a handsome, pouting, ruddy mouth; while her neck, white and statuesque, crowns the full bosom of a goddess. She said that she came out evenings occasionally to make money, not for the purpose of subsistence, but to meet debts that her extravagance had caused her to contract. She said in substance: "You see my appetite is fastidious, and I like good eating and drinking. I have the most expensive suppers sometimes. I am engaged to be married to a young fellow who works on a daily newspaper and who is busy at night. We shall be married some day, I suppose. He does not suspect me to be 'fast,' and you don't suppose I am going to take the trouble to undeceive him. This is not a frequent practice of mine; I only come out when I want money, and I always have an appointment before I come out. I always dress well of course, and can pick up a gentleman anywhere when I like. Yes, I know I have good feet, and I know how to use them. I have hooked many a fifty dollars by showing a couple of inches of my ankle. Of course, I hate being in the store, but my fellow is rather jealous, and I keep going there as a blind. Will I reform when I am married? Perhaps so — if he gives me heaps of money. I am no worse than thousands of girls, single and married, who put on airs of purity and church-going. I know plenty of ladies who pay five hundred dollars at the store for silks and finery, which they persuade their husbands they bought for one-fourth of the price. And, for my part, I am going to eat well, dress well, and enjoy myself as long as ever I can get the money, by hook or by crook."

EDGAR FAWCETT

"The Woes of the New York Working-Girl"

Edgar Fawcett (1847–1904), was born in New York, educated in the city public school system, and attended Columbia College, where he obtained the A.B. in 1867 and the M.A. three years later. Independently wealthy, he devoted his time after college to writing fiction, verse, and drama. His first novel, *Purple and Fine Linen* (1873), attacks New York's high society, an attack that he continued in several later novels. The following article, which appeared in *Arena* 5 (December 1891): 26–35, furthers Fawcett's condemnation of the rich as it indicts their neglect of the city's poor.

It is not long ago since I stood within the reception-room of a well-known lodging-house for working-girls, only a few hundred yards from what one might call the ugly sanctity of Cooper Union.[1] I had brought a letter to the lady in charge of the establishment, and a sweet-mannered proprietress I found her, with a smile that some sculptor might profitably have stolen for a statue of Benevolence, and with eyes

[1] *Cooper Union:* Founded in 1857–59 by Peter Cooper (1791–1883), the Cooper Union for the Advancement of Science and Art was established to provide a free education to gifted students from the working class.

that must have beamed like stars of hope to many a wayfarer whose feet had paused at her threshold. She treated me almost as hospitably as I am sure she always treats the poor waifs that seek her welcome. She scarcely glanced at the letter I had brought her; it was enough that I had come to see and learn about the lodging-house, dear and tender home of mercy which it is. I was shown the clean though plain chambers where the girls ate and slept, the laundry where they washed and ironed, the dressmaking department where they did their hand-sewing, machine-sewing, and (if capable of labor in this line, more skilled than that of certain less apt sisters) their cutting out and fitting on of garments for the feminine customers who patronized them. I learned that the "home" was nearly always crowded, and could not accommodate more than fifty girls; that many had to be turned away from lack of room in the dormitories and dining-halls; that order and discipline prevailed here as the sane and wholesome consorts of compassion and help; that a chance of securing some sort of employment was held out to all who could be received, but that no recommendations were ever given except after a very long trial; and that while expulsion would only result from a defiance of rules, those rules meant in all cases a faithful adherence to work.

I went away from this peaceful and thrifty asylum with the sense that it was indeed a charming protest against that reckless and wholesale cynicism through which the professional pessimist would too often attack society; and I was destined later to become convinced that it stood only for one of many similar institutions in our vast town where humanity gives proof of not having ossified into a very adamant of selfishness. True, there are some places of refuge touched by the shame of religious preferment, yet now that all sects visibly weaken with each new decade, and creed threatens in a little while to become merely the trivial coral on which an unimportant fanaticism cuts its harmless gums, this minor evil merits but passing heed. He who carelessly denied the worth of women's lodging-houses would be a caviller of most trifling quality. And yet, to assert that they are in any manner potent forces of aid to the thousands whom poverty and degradation bind with mordant shackles, would be almost equally to err from the actual facts. In using that little word "thousands," one may hide behind a comfortable vagueness. The hideous skeleton of want is almost prettily draped by it, and not until we realize that about seventy thousand women live in New York by their needles alone, do we begin to perceive how enormous a figure this careless generalization may cover. That shelter can be given to no more than a meager minority of these drudges, all must grant; but it is easy to perceive, as well, that working-women's homes must of necessity protect those who are, for the most part, least worthy of protection. Innumerable are the cases in which fragile girls would rather die than live without those they love. It is the clinging to bedridden or invalid parents, to weak little brothers and sisters, which deepens the terrible pathos of their struggles. Often it is the clinging to kindred who deserve no fondness whatever; and many a sot who reels into some liquor saloon, bloated and thick of speech, has wrung from a toiling daughter's too lenient hand the coin he wastes there in drink. To girls bound by these ties of blood the lodging-house is like a prison. Their tiny rooms in filthy tenement-houses, with the kettle of tea on the stove at their elbow, and perhaps with a sick mother groaning from her mattress in the corner, are sweet liberty by contrast. But sweetest of all is the companionship of their own people, though this not seldom takes for them acrid and domineering forms. I heard but lately of two girls who work all day in one of our largest factories and support a lazy, intemperate father, who is forever pelting them with abuse. All they ask of the ichor-blooded[2] ruffian is that he shall "mind" their two poor little rooms during the day; but owing to his complete and drunken neglect of this simple task, the rooms were recently entered by thieves, and two plain but hard-earned suits of mourning (worn by the girls on Sundays in memory of the mother who died a few months ago) were ruthlessly stolen. It is only to be expected that the parents and other relations of girls thus forced to fight their way in the world should belong among the lower and lawless classes. One of the most horrible features of their fate is to be found in the piteous ignorance which is born with them, and tracks them like their own shadows ever afterward; for even when they should be learning at the public schools to read and write, the puny strength of their little limbs too often is demanded and taxed "at home." Chapters might be written on the raw diabolism of the working-girl's kith and kin, who may heartily wish to save her virtue from ruin, but who are willing, for all that, to freight her days with leaden discomforts. I know of two other sisters, dwelling at present in East Broadway, who are cloak-makers for a firm of good repute, and by

[2] *ichor-blooded:* Animal-like.

using every effort, and often sitting up after their return from the shop until a late hour with extra work, are enabled decently to support their elderly mother. They give her a certain sum each day for expenditures in the line of general house-keeping; but, to use their own phrase, she constantly "knocks down on them," and either buys them inferior food to allay her thirst for liquor, or else openly deprives them of food that she may drink her greedy fill. And yet these two stern-burdened children are very gentle and patient with her, possibly accepting the existence of her horrid appetite as naturally as they do the color of her eyes or hair. They were doubtless born to take one for granted just as they take the other. And there are so many family horrors like this that the working-girl must take for granted, besides staggering as best she can beneath the onus of her other torments!

It is marvellous how she does manage to stagger without actually sinking. The prices paid her are often a disgrace to her employers, who literally mount from mercantile obscurity to prominence on the bodies of herself and her dead or dying fellow-slaves. The "tailor-made" clothes for women are rendering her position, when she is a sewing-girl, even more sad than it once was. A New York dressmaker with a prosperous custom, will employ ten girls out of thirty at the niggardly pittance of three dollars a week, the other half getting more, but only a little over a dollar a day at the very utmost; and in certain cases which I could cite, those who receive five dollars a week must make long journeys every morning from places like Yonkers or Jersey City, reaching their destinations at eight in the morning and stitching away till six in the evening, with only the intermission of a half hour at mid-day for luncheon. More than once I have stood in Hyde Park of a Sunday and heard passionate and often eloquent harangues delivered there against the "sweating" processes of London capitalists. But on these occasions I have always felt that no harsher comment could be spoken concerning the complete social failure of our own high-vaunted republic than those to be noted in the unconscious analogy drawn between working-girls of the two greatest cities which England and America contain. For if there these ill-starved creatures reel broken-hearted by multitudes to nameless graves, do they not find here, by like if not equal multitudes, the same forlorn and undeserved goals? In New York it has been clearly shown that there are two hundred thousand working-women, of all ages, and he who runs may read of the despicable wages they receive. Shirts paid for by the very most liberal establishments, two dollars and a half per dozen! In firms of a less generous trend, one dollar and seventy-five cents for the same amount of work, prices of cotton being deducted from the sum earned. Between eight or nine dollars a week for women's chemises of elaborate sort, with tuckings and puffings in lace and cambric accompaniment. Seventy-five cents a dozen for babies' slips, with price of cotton also deducted. Forty cents a dozen for corset-covers. Eighty-five cents a dozen for ruffled skirts, and seven cents an hour for flannel underwear! . . . So the revolting chronicle goes on, with its odious monotony varied merely by the sort of work performed, which is not always even as endurable as that of the needle, packed with bane though the lot of the seamstress may strike us.

In cigar-factories the horror is still more accentuated, for here the vile scents of rankest tobacco are a stealthy and perpetual poison. Five dollars a week are doled out to the "strippers" and "bunchers" at this nauseous and asphyxiating toil, and rarely more. Their clothes and hair reek with the hateful odors of the weed they manipulate; their hands are forever stained with its noisome juices. What wonder that their morals, like their clothes and fingers, are often sadly stained as well? Haggard and jaded, they are not seldom robbed of even the physical chance to seek ease through sin. Their rouge and cosmetics are of so baleful a quality that you might fancy some bedraggled Quasimodo of the slums would alone feel a thrill of response at their pathetic lures. In the vanities of noxious rags, you might dream, they should deck themselves for ghastly coquetries, with but pools of gutter-slime as their mirrors, and as jewelry but the hideous ulcers here and there on their brutally envenomed flesh. . . . With the cigarette-makers it is even more dreadful; with the girls in book-binderies it is almost as bad, their fingers being blistered and often bleeding, so that every movement of the hand means pain. For the laundresses there is less suffering, though acutest fatigue; for the flower-makers there is death in the petals of the false blooms they fabricate; and thus from trade to trade it is always the same story, one in which the degree of sorrow and distress may vary, yet never their essential kind.

The more one observes the joyless lives of working-girls the more he wonders that so many of them should be jealous of their good names. In losing these they not only relinquish a possession about which no one, for the most part, cares much whether it be lost or kept, but they obtain material comforts which must fall on their jaded spirits like some magic mantle woven of starlight and sea-winds. The road to ruin, as we call it, is so fatally easy to them; who shall dare to blame them if they take it? And in so many cases to call it a road to ruin is so purely preposterous. It is indeed nothing of the kind; it is a road out of cold, famine, neglect, despair, into warmth, plenty, protection, hope. The protection may prove transient, and the hope a flower of frailest leaf; but even when seduction has ceded sway to its heir-presumptive, disillusion, the world continues far kinder than it was in the days of freezing attic and stony cot. No attempt is here made to uphold prostitution on the part of these wretched starvelings. This course, indeed, is nearly always with them an ultimate change for the worse; they live their little rosy hour of luxury, and then comes either the malady engendered by drunkenness or that produced from a worse cause. Drunkenness overtakes them, for the reason that excess and surfeit are apt to follow on abstinence, and that the fatal ignorance by which they are all cursed makes them see only with dulled vision the coarseness that inebriety begets. But cases could be quoted where years of extreme moral laxity, unaccompanied by intemperance, have ended in no hospital tragedy whatever, and sometimes have ended in decent, even happy marriages. For the girls who thus escape the enmity of that savage destiny which seems to enwrap all their kind like the folds of one monstrous serpent, too ardent gratulations cannot be framed. It would be hard to think of any class of human beings more dismally handicapped from their births than are these same poor helpless victims of our massive social mistakes. In countless instances all real education is denied them. They come into a world whose greeting might be symbolled by the threat of a bayonet-point thrust at their faces. They are the Cinderellas of society; their sisters go to the ball, and they must bide at home, but alas! with no fairy godmother, no miraculous mice, no necromantic pumpkin. The "prince," in a certain sense, comes to them, and here is the horrible part of it. As if their hapless feet were not already beset by enough deep and miry pitfalls, a certain human (or shall we say "inhuman?") antagonism forever assails them. Only the very ugly are free from this continual temptation, this persistent, Mephistophelean assault. As youth and maidenhood are seldom an ugly combination, these grief-doomed beings rarely reach the age of sixteen without having sensuality somehow address them, in tones either stealthy or bold. The charge would indeed be unfair if it were brought against all men of our metropolis under five-and-thirty; but how large is the male throng that regards working-girls as fair prey for passions at once conscienceless and deliberate? Men of actually the best repute regard what they would call an "adventure," in this connection, as lightly, as jocosely as they would regard having quaffed a bottle of champagne too much at their favorite club, or having ridden a thoroughbred in a "gentlemen's" steeplechase. I do not for an instant mean that they boast of their *bonnes fortunes* publicly among their intimates, after the fashion of characters in Bulwer's worst novels, and in some of Ouida's worst as well.[3] The growth of good taste has luckily delivered us from such vulgarities. But has it delivered us from the silent and secret raids upon feminine virtue which add a new anguish to the working-girl's already heart-breaking fate? I recall, years ago, hearing a man of position and excellent so-called character, say carelessly in a fashionable club that if A. T. Stewart[4] really carried out his idea of a great lodging-house for working-girls (he alluded to that yet incomplete edifice which is now the Park Avenue Hotel) it would swiftly become one of the most notorious brothels in town. This gentleman spoke quite at random; he doubtless had not the least vicious personal feeling; he merely expressed a current sentiment with respect to "men about town" in their estimate and general treatment of the working-girl. And now, ten years later, she is fair game for the lewd fervors of the opposite sex, just as she was ten, twenty, thirty years before. The whole deplorable drama goes on to-day precisely as it went when our grandfathers were urchins. The difference is merely a numerical one; more toil-dulled eyes are made to sparkle now than then by gifts of trinkets, little clandestine feasts, rolls of coveted bank-notes, and all those insidious details of enticement which the tired child of labor cannot but prize as a wilted weed prizes the dew. Of course the whole question is a complicated one. Innocence is by no means always on the feminine side of these intimacies, and coyness is often assumed by Daphne[5] toward a pursuer who could teach her nothing in the art of deceit. But the weight of blame should not, for this reason, be shifted from those hardier shoulders where it properly belongs. The fault of the alleged "gentleman" is in all circumstances far greater than that of the unlettered working-girl, with the stars in

[3] *Bulwer's worst novels, and . . . Ouida's worst as well:* Edward Bulwer-Lytton (1803–1873) was a prolific British novelist, widely popular in nineteenth-century America. Ouida was the pen name of another prolific British novelist, Marie Louise de la Ramée (1839–1908). Crane would publish an appreciation of Ouida's *Under Two Flags* in *Book Buyer* 13 (1897): 968–69.

[4] *A. T. Stewart:* Alexander Turney Stewart (ca. 1802–1876) developed his reputation and made his fortune in retail. At the time of his death, he was erecting a building to house and board working women at cost.

[5] *Daphne:* The mythical woman pursued by Apollo.

their courses, as it would seem, her bounden foes! She has the immense excuse of a guideless, a spurned individualism; at the great, strange banquet of life, she is like a kicked scullion who crouches by a crevice in the grand hall of revellers and scents with greedy nostrils the odors of meats that she is not deemed even worthy to serve. He, on the other hand, has the immense drawback of an existence passed amid enlightenment and good repute; in the commission of any such wanton act he is like a wayfarer who willfully quits a clean road and plunges his feet in mirk and bog, of which he already knows the black taint, even to its utmost filth of soilure.

That great and good woman, Mrs. Annie Besant,[6] not long ago declared that much potential aid for those of the starving and agonized among their own sex was possessed by the London ladies of wealth and consequent high standing, but that all their movements in this direction were drearily lukewarm. With the ladies of New York is it not precisely a similar state of affairs? The chief trouble with their charities always has lain in their making them a pastime rather than a pursuit. They visit Baxter Street and Bayard Street very much as if they were going to a menagerie. The few earnest women (or shall we call them angels?) who have almost devoted their lives, here in New York, to the succor and amelioration of this woe-begone sisterhood, will tell you, if you ask them, that in the large masses of cultivated matrons and damsels there is a mournful weakness of eleemosynary[7] purpose. But if men are selfish, why take to task the wives and daughters who as yet are content admiringly to reflect them and no more? It seems to me that we touch the very horrible centre of this unassuaged social sore when we state that most of our well-placed women, who could aid their kind, will not really aid them, and that they are bored unspeakably by even the small, dainty profferings of time and pin-money which the modish churches they attend demand that they shall exploit. The ordinary woman of New York society is a good wife and an excellent mother. She has her aims to get her growing-up maidens into the "right set," and her darling sophomore son out of collegiate perils at Columbia. She has an eye upon the real or menaced infidelities of her dear husband, and her curtain-lectures to prepare from week to week,

[6] *Mrs. Anne Besant:* Annie Besant (1847–1933) was a liberal British political activist.
[7] *eleemosynary:* Of or pertaining to alms or almsgiving.

whether they be hot philippics[8] or only tepid reprimands. There is her visiting-book to be watched and weeded; there is Mrs. Manhattan to be lunched, or Mrs. Moneypenny to be glared down. There is the opera, and gossip with one's intimates, and a scheme to make Mrs. Schenectady send herself and lord cards for the next Patriarch's ball without really having seemed to cringe for them, and that new "very select" dancing class to which Carrie and Fannie haven't yet been asked, but to which they must be asked, *ruat cœlum,*[9] or else when they come out they will be horrid little wall-flowers (don't you know?), not on speaking terms with a single soul among the real "swagger" gentry. She has all these "duties," this charming New York wife and mother, and myriad more of a like sort. How should it concern her that girls of the same age as her Carrie and Fannie are starving, slaving, coughing up blood, dragging themselves from dirty vermin-thronged beds at five in the morning, being blackguarded and beaten by drunken parents, being tempted by rakes whose very lust seems a heaven of refuge to them? How should all these grisly things concern *her?* "Of course, you know," she will tell you, "I'm interested in the Skin and Cancer Hospital, and I contribute to the Woman's Protective Union on Clinton Place. Dear Doctor Silverspeech thinks I ought to, and I do just as he tells me. I make the girls follow his advice, too. They belong to the Amsterdam Sewing Society. It's just too lovely. Mrs. Amsterdam is so sweet, and so genuinely religious. And the girls, although they're sometimes thrown with poor creatures from Avenues A, B, C, and all such frightful places, don't mind it a bit, because Mamie Van Corlear belongs, and Lottie Van Dam, and . . . oh, well, don't you know, just the kind of girls that I want mine to grow up with and go out with into society."

And what of Doctor Silverspeech? He is simply one of the many clergymen who smile upon this terrible species of hypocrisy. If the working-girl of New York has any arch foe it is that sad fraud which to-day is termed Christianity. If to-day there is any class of men who entirely desert the requirements of their avowed profession it is the class of the clergy. They draw salaries — and some of them very large salaries — for preaching the doctrines of the Galilean to people whom Christ himself, if he were alive this hour, would cover with invective.

[8] *philippics:* Bitter verbal attacks.
[9] *ruat cœlum:* Though the heavens fall.

For even Christ, we must remember, sometimes lost his temper, sometimes got fearfully out of patience; as he did, for instance, when he overthrew the tables of the money-changers in the temple, and again when he "looked round about on them with anger" in the synagogue at Capernaum.

The clergy of our time and town are just as Christian as expediency permits them to be, and not a jot more. Perhaps it is cruel to blame them, for if they took up fiercely and devoutly such a charitable cause as that of the oppressed and stamped-on working-girl, their congregations would begin to yawn, and in that yawn they would see an omen of empty pews, and empty pews would mean curtailment of their apostolic incomes. If I may humbly write so, it seems to me that these "divine" gentlemen (with certain happy exceptions) think quite too little of their "divinity." They appear to spend a great deal of their time in squabbling like testy old women among themselves, and to waste a great deal more in orotund insults to the lofty and perhaps the only true thinkers of our century, Darwin, Huxley, and (of all men!) Herbert Spencer![10] Yes, even Herbert Spencer, the shadow of whose mind, now cast as it is for a little while longer upon earth, spreads there so vastly that when he dies the world will miss it as Switzerland would miss an Alp!

Ah, gentlemen of the clergy — and of the New York clergy in particular — two hundred thousand wretched New York working-women need your help far more than these noble scientific regenerators of the age need your anathemas! Cleave a little closer, pray ye, reverend gentlemen, to your alleged "Christianity," and accord us a kindly dearth of your fifteenth century polemics. Mankind will be the better for it, and (I dare swear) the poor working-girl as well!

[10] *Darwin, Huxley, and . . . Spencer:* Like that of Charles Darwin (1809–1882), the scientific research of Thomas Huxley (1825–1895) and the sociological research of Herbert Spencer (1820–1903) scrutinized the fundamental basis of human behavior.

Working Women Tell Their Own Stories

"In Employment Offices: Trials of Women Who Look There for Work"

"Counter Trials: The Shop-Girl Speaks"

"The Song of the Shirt: How the Finisher Lives"

"The Embroiderer: Quick Work and Scanty Wages"

"The Scarf-Maker"

"The Lace-Maker"

"The Black-Borderer: A Solemn Kind of Work"

During the 1880s, the *New-York Tribune* began devoting greater space for female readers. By the end of the decade, the *Sunday Tribune* contained several columns devoted to fashion, homemaking, charity, and other aspects of women's life and work. Many helped inform the well-to-do woman about her less privileged sisters. Each of the following articles appeared on the woman's page of the *Sunday Tribune.* All are the result of interviews between a *Tribune* reporter and several young working women. Purportedly, all are transcriptions of the actual words of the working women, but their words often seem a little too polished. Reading them, one can see the reporter's hand shaping the details of each. The first, which appeared in the *Sunday Tribune* on December 29, 1889, provides a general overview of the difficulties involved with obtaining employment. The following six articles were part of a series of interviews with working women in 1890. They appeared in the *Sunday Tribune* on March 16 and 23, April 13 and 27, and May 11 and 25.

"In Employment Offices: Trials of Women Who Look There for Work"

Sitting for Weeks Waiting for Something to Do — Some of the Experiences

Is it difficult for a girl to find work in New-York? There are many of them sitting in so-called intelligence offices who will tell you that it is, and that when a woman gets an offer of employment in many of these places, she is unable to seize the chance, for the reason that she cannot live on the wages offered to her. But almost all of them will tell you that even opportunities entailing the hardest work and the smallest remuneration rarely come to many of these employment bureaus. What is the experience of a woman who looks for work at these offices? If you can stand a close room, packed full of discouraged humanity, a red-hot stove blazing in the middle of the floor, hard wooden benches, absolute lack of ventilation, and a stifling inactivity for half a day, you will have no trouble in finding a place where you may see for yourself. But you will not enjoy your investigation. An employment office is not a cheerful place at any time. Often it is worse — dirty, greasy, foul-aired and dispiriting.

Pay Your Dollar First

A young woman who wanted to find employment as a nurse went to one of these places, which is not a block away from Forty-third-st. nor yet a mile away from Sixth-ave. There was one large room, with long benches arranged around the walls, a glowing stove in the centre and a desk in one corner, behind which stood a short, fat man, with an oily face. He was not well dressed, but he had a sleek, comfortable look, as if the world treated him fairly well, however it dealt out its favors to the women who sat listlessly around on the benches. These women paid little attention to the newcomer. In fact, they seemed to notice nothing in particular, unless it was the dust on the walls, the ashes about the stove or the faint tracings of cobwebs on the ceiling. They were simply waiting. A good many of them had been waiting just that way for several weeks.

"Can I get work?" asked the girl.

"You must register and pay your dollar first," said the fat man, looking up from his desk.

"I don't want to register until I know whether I can get my place."

"And you can't know till you pay your dollar — there."

"Can I get my place if I pay my dollar?"

"You must register first."

"I want to know, honestly."

"I won't talk to you till you pay your dollar."

So the girl tried another place in Thirty-fourth-st. This office was more imposing. It had two entrances and two desks. There were even a few chairs in it. But the air was quite as foul as in the other one. There were more women waiting here. A tall, slim young man with a dark mustache, answered her questions. He was a good talker, and conveyed the impression that he took a personal interest in the applicant and that he could easily secure work for "his customers." He assured the girl that he could get her work.

"Just register, please — and now one dollar," he added, politely.

"Now," he went on, smoothly, "if you will step in this room I will soon get you something."

"This afternoon?"

"Oh, undoubtedly."

The waiting-room was not so luxurious as the reception-room. It was adorned with the customary benches and the everlasting stove blushed deeply. The windows were closed tightly. Three women were talking loudly in a corner. One was stout, with a red face, sharp black eye, and hair pushed straight back from her forehead.

"Well," she said, "I had a place with a young married woman who didn't know anything about keeping house. On Thanksgiving Day I had some brandy for a pudding and I drank it. I had to drink it, for I was sick and needed it."

"Of course," put in a sympathetic chorus.

"When my young mistress scolded I got up and left her with her company on her hands. That will teach her how to treat her servants. That's my way," she added, her eyes snapping.

"I wanted to be boss of my own kitchen," said another, "and I haven't had a place since I left the last one on that account.

"Mary Brady," bawled a loud voice, as a head came up over the partition.

Three girls got up to answer the call. Each one insisted that she was wanted. Finally a shock-haired man came in and picked out one. The other two got into a dispute with him, each declaring that he had promised that she was the one to have the first place. The one selected went out, but came back in a few minutes.

"They only get us places we can't take," she said, in disappointed tones. "Cooking, housework and washing, $8 a month and live at home."

"You won't get work here," said one woman to the girl. "I've been sitting here six weeks.

Her dress was shabby, and her collars and cuffs were badly soiled.

"Don't you think you might do better if you took better care of your dress," said the young girl, looking at her own cuffs, scrupulously white.

"Wait till you've been sitting in this dirty room six weeks without money, and see how you'll be dressed," answered the woman. "And you have to pay a dollar every four weeks, too."

Offering Sympathy

A quiet little woman with a white face came over and sat down beside the girl.

"I hope you've got enough money to get along with for awhile," she said. "I'm afraid you'll have to wait a long time. I've been here several weeks. They don't pay much attention to us after we pay our dollar — until it gets time to pay the next one. Then we get plenty of promises, but no work."

During the afternoon one woman got work in a hotel. There were not less than sixty in the room, and all seemed eager, with, perhaps, the exception of the woman who drank brandy for her health, and one or two others, to get anything to do which would afford them a decent living.

A day spent in another agency in Fourth-ave. showed the same condition of affairs. The front office, a room just big enough for a desk, was presided over by a bustling young man with a red beard, who made himself very familiar with every one who came in. Behind his office was a long waiting-room, filled with women of all nationalities. The room, as usual, was stuffy and the air depressing.

A neatly dressed girl came in and applied for work. After she had registered her name, the young man said: "Now, Annie, give me a dollar, and I will have work for you as soon as possible."

The girl took her seat on one of the benches, when a young woman beside her began to hum "Where Did You Get That Hat?"

"What kind of work do you want?" said the girl, biting off her song suddenly.

"I am a nurse: I take care of children."

"You won't get anything here," consolingly. "Besides, you aren't strong enough to tag around after children and carry them. No one will think you can do it."

"Oh, yes I can. I could take care of two, for I had three before; but it was too hard for me."

"There!" triumphantly, "didn't I say you weren't strong. You'll never get anything. I took care of one brat once, and he nearly killed me. I couldn't live a day with three. How long did you stand it?"

"Two years."

"Two years! My!" and then she took up her humming of "Where Did You Get That Hat?" chewing gum at times.

A man came in and said: "Ann, you've got to get out of here. You won't get any work here, and you can't come back."

"Ann" got up from one corner and protested.

"You let her alone," said another woman. "You told her you had a place for her, and when she went there she found that she was to get $10 a month, when you had promised her $14."

"You can get out, Ann," he repeated. "You'll get no more chance here."

"Give me back my dollar, then."

"Give you back what? Didn't we get you a place?"

"Yes, but it wasn't what you said it was. I won't go unless you give me my dollar."

The man went out whistling.

"She's foolish," said one girl. "She'll never get anything now. You see, when anybody hires a servant he has to pay $2 to the office, so it makes money both ways. And when the girls take places and find that they've been cheated and leave the house, that makes the office mad, because the man who has paid his $2 for a servant won't go there again."

One girl sat on a bench with her arm bent to hold her head. Her face was white, as if she was ill. She scarcely moved during the afternoon. When the man came in and said that a girl was wanted to do kitchen work at $4 a week she jumped up eagerly.

"I'll take that," she said quickly.

"You'll have to live at home."

"I have no home," she said wearily, sitting down again. "I'd have to rent a room. But I'll take it. I must do something. What kind of work is it?"

"Restaurant kitchen, 6 o'clock in the morning until 8 at night."

"Oh, I'll take it," weakly.

Why She Took the Work

The frail young girl who was willing to take care of two children went over and spoke to her kindly.

"I have had no work for two months," said the woman, buttoning her shabby coat with nerveless fingers. "I broke down working for a family of nine. I have no home, but a woman whom I know lets me sleep at her house. I haven't had money to get enough to eat for days, and, besides, I've been sick. But I would suffer anything rather than leave a place again as long as I could stand on my feet. I know now how hard it is to get a place."

Some one wanted a chambermaid in the country for $8 a month.

"I'd take it," said a very big woman, "but I'm too fat. It's all a cheat, a cheat, a cheat," she whimpered. "They don't get us anything we can do."

"Well, don't we all know it's a cheat?" flashed a black-eyed young woman. "What's the use of crying about it? As for the country, I know what that means. When you get there you don't get any money for two or three weeks because it's held out for your fare. And after you are there you have got to stay, because it costs so much to get back. And you never know what kind of a place they will send us to. No country in mine, unless I know something about the people."

The women did not go out for luncheon. Perhaps few of them could spare the money, but most of them were afraid of missing an opportunity. An appleman came in, and stray pennies were fished up from rusty pocketbooks, and then the women munched away, discussing their woes.

Bits of Conversation

"I don't like to work in a hotel," declared one stout woman, who looked as if she had no fickle appetite. "The stuff you get to eat there is terrible."

"That depends on the hotel," put in a voice.

"Oh, you're too smart."

"The best place I ever had," said a young woman who seemed not to have been out of work long, "was with Mrs. B———, in Lexington-ave. I was treated well there, but I couldn't stand one thing. Mrs. B———'s daughter had to have her breakfast in bed every morning. I wasn't going to encourage laziness."

"You were a fool," said the woman who did not like to work in hotels.

"I begin to think I was."

"I should think you were," said an honest-faced young German woman. "I did all the work for a family of seven and helped to take care of the baby. I had to do my washing at night, and I wouldn't have left that place, but they couldn't afford to keep a girl."

"You were a fool, too."

"Well, I'd like the chance again. I've been looking for one for eight days."

"You won't get it here."

One Woman Succeeds

There were ample grounds for doubts, for at evening only one woman had left the place with work assured — the pale-faced, weak creature who had gone to work in the restaurant because she was starving to death. The courageous little nurse-girl looked brave as ever when she went home for the night, but the stories of the trials she had heard and the close room and hot, foul air had flushed her face till the freshness had turned into a red like that of the poor woman who was "too fat" to be a chambermaid.

Still another visit was made on the next day to a bureau in Sixth-ave., below Fourteenth-st. This place was in charge of a thin, angular little woman, whose sharp black eyes, keen features and lively activity denoted a business woman. She was more crisp and abrupt in her manner toward the women than were the men in the other offices. Almost the first thing heard there was a complaint about broken promises.

The Same Old Story

A girl who went in to register was advised by some one who had a grievance not to throw away her money. "You won't get any work," said the woman. "I'm as willing as anybody, but I didn't succeed."

One of the most striking features about the "intelligence offices" visited was the utter indifference to the comfort and health of the women. In no case was the air fit for a human being to breathe. The furniture was of the rudest kind. There was enough dirt about the rooms to furnish each applicant work for a day, and everything was lacking which would be likely to make a favorable impression on the person who was looking for a servant. Even the girls appeared in a worse light than was necessary, owing to the gloomy and unfavorable surroundings. While there are many respectable labor bureaus in the city, there are a great many more which are precisely like those described, and many more which are even worse. The working girl who has to trust to securing employment through a system of "checking off" applicants has a hard path to tread in the city of New York.

"Counter Trials: The Shop-Girl Speaks"

"Ah-h! I feel as if my feet had gone to Heaven!" said the sales-girl, as she exhibited a number three foot in a number five slipper. "I get solid comfort out of these old slippers when I come home in the evening with my feet hurting so I can hardly stand. Shoes too tight? Not a bit of it! My shoes are always too large for me, but standing all day swells my feet and makes them painful.

"The first day I went into the store I thought I'd die before 6 o'clock, but one can get used to anything except being scolded — that I never can get used to, no matter how long I may stay in the store. Why am I so naughty as to need scolding? It aint me that's naughty, it's the customers. I do get so mad sometimes that I could jump over the counter and beat 'em! They aint got no notion of buying when they come in, but all the same they want to see this, that and the other thing, and so they make me take down box after box, and then they stand there, worrying the life out of me, asking no end of questions, and crowding out other folks who might buy if they could only get a chance; and when all's said and done they walk away without my making a sale, and then I get a scolding from one of the bosses. Which one? Well, most always it's our buyer. I spoke back to him the other day! Says I, "It takes a smarter girl than me to make folks buy when they aint got no money to buy with." Why, bless your heart! Sometimes women, wearing hats and wraps that I'd be ashamed to be seen in will come to my counter and make me show 'em my highest priced laces. They aint the only kind, though, that only come to look. There's them in sealskin sacques and $50 bonnets who will spend hours in a store and go out without buying a penny's worth. Them's the sort that make me nervous when they come around because I get my awfullest scoldings on their account. I s'pose it's fun for them to come and see what we've got, and then go on to see what they've got in other stores, but it aint no fun for the girls who are expected to make sales.

"Oh, me! If here aint a place on my sleeve that's most worn through! I'll have to get a new dress for the store. If I had my way about it I'd get one for Sunday and wear my blue cashmere in the store, but I'd have to get it dyed first, and I hate to have a dress dyed when there aint nothing the matter with it. Wear it as it is? Oh, dear, no! That wouldn't be allowed. We are expected to dress either in black or in some grave color that don't make no show. Black is what they like best, and that is why I have lived in this old thing so long. In common stores the young ladies may wear all the colors of the rainbow, but it's different in a first-class establishment. At one time there was talk of our having to give up our bangs, but I guess they thought better of it.

"Are we expected to lie professionally? What do you mean by that? Oh, to 'get off goods on customers.' Well, some of the young ladies will do it because it's in 'em and they can't help it, but lies aint paid extra for, as in some stores where a girl gets a commission on her sales of out-of-style wraps and things. Employers Christian men? I don't know as our bosses are extra Christian, but I heard one of 'em say once that lying wasn't business-like. You see our store has a reputation to keep up. Talk about sales-ladies saying what aint so, you ought to hear the customers! When I was in the cloak department last winter I was every day sending off goods C. O. D. that did nothing but come back to me again. The ladies, when they said they'd take them, had no more idea of paying for them than I had.

"As for sass, let people say what they will, I'm sure there's them that come to buy that's sassier than any of them that sell. My, how sick and tired I used to get of trying sacque after sacque on a customer, and when I'd tell her it was a perfect fit having her ask me if I called wrinkles a fit! Of course a sacque will be full of wrinkles when a woman is twisting around in front of the mirror, trying to see her own back. On my sacred word of honor, I've tried thirty sacques on one customer, and then, after her taking up my time for two hours or more, and maybe preventing my getting my lunch till half past 3 in the afternoon, she'd go away saying that probably she'd come in again the next day, and then she hoped she would have a more obliging young woman to wait on her. Just the other day I was stooping down behind the counter, putting away some lace, when, the first thing I knew, somebody poked me in the back with a great umbrella, and when I looked up the customer that done it began business by threatening to report me for inattention.

"I tell you, with all the work I have to do, and all I have to put up with, nobody ought to grudge me my $10 a week. A fair salary, you say? Yes, I s'pose it is. I know it is twice as large as what some girls get in common stores, but then in common stores they employ common girls."

"The Song of the Shirt: How the Finisher Lives"

"I don't know what I am going to do about it," said the shirt-finisher. "My room-mates vow they won't consent to have the alarm clock go off at half-past 5 in the morning. You see all three of 'em are salesladies, and so they can afford to lie abed till nearly 7, while I ought to be up as soon as I can see to work, though, for the life of me, I cannot wake without the alarm. I'd rather sew late at night, so as to sleep the next morning, but my room-mates won't agree to my having a candle, as they say the light keeps them awake; so I am sometimes left in the dark, in the middle of a button-hole, when the gas is turned off at half-past 10. I don't sew on buttons in my dreams, as that woman done in the 'Song of the Shirt' (I heard it read at a club meeting), but it would be a great saving of time if I could only sew 'em on in the dark. By working early and late I can't finish more than five shirts a day — when they are custom shirts I do four — and as the highest pay at the place where I work is a dollar and a half a dozen, I consider myself in luck when I can pay my board, three dollars, at the end of the week. For stock shirts I get only a dollar a dozen, and when business ain't brisk, of course I get more stocks than anything else.

"I could do an awful lot more if I was allowed to slight my work, like girls who finish cheap jerseys. The buttons drop off my jerseys the first time I fasten 'em, but no such work as that is put on shirts. It ain't poor work, poor pay with me, but good work, poor pay. The over-looker at our place, though she wears glasses, has got the eyes of a hawk, and in the button-holes the stitches has to be just so close together, and the hemming in the gussets has to be almost as nice as what would be put on a pocket handkerchief; the buttons must be sewed on hard and tight; and as for the eyelets, they are just the torment of my life, they have to be worked so awfully round and smooth. I always dread to get a set of shirts with eyelets in 'em because then I am sure to be found fault with when I hand my work in — sometimes

I get 'em too large, and sometimes too small. I don't have that trouble with button-holes, because they are cut for me.

"Supplied with thread and needles? That we ain't! We buy our own thread and needles, and it counts up in the end I can tell you. That's all I do buy nowadays, and it often comes out of my board money. As for having my washing done, I told the superintendent of the home where I live that I just couldn't afford it, and so there has been an exception made in my favor, and I am allowed to go down into the laundry and wash my own duds. It's against the rules for the other boarders to do it, though they'd all like to, no matter how much they may earn. The girls in my room are always washing their handkerchiefs and such things, and hanging them behind their washstands where they think they won't be seen.

"Why do I work for so little? Well, three dollars a week is better than no dollar a week, and as I ain't got nobody to look to for support, I have to catch on to any work that comes along. Relations? None nearer than cousins, and they don't amount to much — at least, mine don't. Why , a cousin of mine, whose husband earns his three dollars a day as a painter, invited me to do my washing at her house (that was before I got permission to do it at the home) and afterward fell out with me, and made me pay for the coal I had burned in heating my irons.

"The girls in my room tell me that I could never get a place in a store because I am not tall enough; but if I could once lay up money enough to pay my board for a week or two, I'd tramp up and down the city till I found a place where they would take me. I was in a store once around the holidays, when there was a lot of extra hands needed, and I tell you it was a satisfaction to bring home my five dollars every Saturday night! It's an awful misfortune to be so short when a girl has her own living to make. If I was six inches taller I'd be earning almost half as much again as I am now, for I ain't one of the stuck-up sort who think they are put behind a counter only to show off their bangs and their bangles. I made lots of sales when I worked as an extra, and I could do it again as a regular hand if the store-keepers could only get over my being so short. I get out of all patience when I hear a girl whining and saying she don't see why she has to work. Why, work is what I want, and the only thing that I do want. If I could earn five dollars a week at shirt-finishing, I'd be willing to keep at it till I was too old to thread a needle. It's only work that doesn't pay enough to keep me that I don't like. On five dollars a week I could live like a lady:

three for my board; one to put aside for dress; and one for my other expenses. Any girl that's got the knack of fixing things can dress decent on fifty dollars a year, but it's awful hard to do it on nothing."

"The Embroiderer: Quick Work and Scanty Wages"

"Time money?" said the pale-faced woman, as she shook the many-colored ends of floss silk from the white apron she had taken off preparatory to donning her hat and sacque. "Well, my time don't amount to much money — not more than 5 cents an hour, no matter how steady I work; and that don't take in the hour I spend in going and coming between here and the shop. I was six hours working this (unrolling a piece of cotton velvet that lay on the table), and all I'll get for it will be 30 cents, and maybe a little jawing because the stitches ain't so near together as they might be on that rose in the corner, though that ain't my fault, as the girl who deals out the materials gave me a scant supply of pink silk. A double carnation you say it is? Well, I call all the pink flowers roses; it don't matter to me what they are, just so they are made like them in the pattern.

"You needn't count the flower-clusters; I can tell you how many they are — just two dozen. One dozen pair of slippers will be cut out of this velvet, and though they will be sold cheap, being cotton, you may be sure there will be money made on them, and lots of it. You see the embroidery on one pair comes only to a little over one penny, and the material is bought wholesale. No, these ain't the only kind of slippers I embroider. There's them I get 50 cents a dozen for. They are of a better quality of velvet, and the flowers has to have more pains taken with them. A dozen pair of that kind will keep me busy from half-past 7 in the morning till 6 in the evening, allowing me half an hour for lunch. The best kind — them I get 75 cents a dozen for — are worked on silk velvet, and to get a dozen of them finished in a day I have to begin as soon as I can see in the morning, and keep at 'em till 10 o'clock at night. Often, though, I don't get 'em done because the girl don't give me enough of one kind of silk, and I have to lose an hour by going back for it. Hope I scold her well? Why, bless you! if I say a word to her she only gives me sass. She will say I waste the silk; and she has even insinuated that I hook it for my own use — just as if I had the time to make fancy pin-cushions or any of them sort of things! I can never get used to the impudence of these young girls who are all bangs and boot-heels. You just ought to hear this one talk if I give in my work unpressed, though they have got pressers in the shop, and if the cook here don't happen to be in a good humor when I go into the kitchen, she is sure not to have any irons hot.

"No, the madame herself ain't so cranky as you might suppose a woman would be, who had a grown-up no-'count son to support. I remember the first work I done for her didn't begin to come up to the mark; a lot of it had to be ripped out and done over again in the shop; they couldn't give it to me to do, as they were in a hurry to send it off. Well, after speaking her mind pretty sharply, the Madame said 'she would give me another trial.' 'What the dickens are you going to do that for?' says the son, who was hanging around the work-room like a tame cat. 'It don't pay to give out work to botchers. Keep it for them as can do it right.' 'She will do it right next time,' says Madame; and so, though I came home with 20 cents instead of the 30 I had expected, I brought home another piece of velvet, and none of my work had to be ripped out next time. Women are awful skins when it comes to money matters — not to save your life would they pay you a penny more than your due, and they are always ready to dock you if you get a stitch wrong — but they have a lot more patience than men — that is, when they are old enough to have sense.

"No, there ain't no money to speak of in embroidery of this kind. That the Sisters have to do on altar clothes and such things is different, and a good worker can earn $10 a week at it. But all that is given to them that have been to the Sisters' schools, and learned to do the best kind of work; what they do is what might be called skilled labor. I learned in one morning how to do this kind, and if I was to tell madame to-day that I wouldn't do no more, she would only be at the trouble of putting an advertisement in tomorrow's paper: that would bring her all the embroiderers she wanted and a lot more girls who live at home, and would like to make that sort of work an excuse to get off from washing dishes and ironing clothes.

"Why do I make this my profession? My! do you suppose I'd work at this regular, year in and year out? I'm only doing it now to keep from getting into debt for my board while I am out of a place. I've got my name down at an agency, and made 'em understand there just what I want. Folks say I am a patient woman, and I am looking out for work that can't be done by nobody but a patient woman — nursing a fractious invalid that ain't sick enough to require a trained nurse; or looking after a weak-minded child that's too much for one of the sort of girls that's usually engaged to mind children. A place of that kind will turn up, sooner or later — sooner, I hope, as I am tired to death of these roses, or double carnations, or whatever you may call 'em."

"The Scarf-Maker"

"Yes, kerosene is awful cheap just now," observed the industrious virgin as she carefully filled her lamp: "and a good thing it is for me, as I burn such a lot of it. There's hardly a night that I put out my light before 12 o'clock, and in winter I am sewing by lamplight for nearly an hour every morning.

"I needn't go to the shop till 9 or 10 o'clock, but the work has to be done all the same, as I ain't got no notion of starving at my time of life; it looks sort o' like it though, working for $4 a week, and, as true as I live, there are weeks when I don't make that, no matter how steady I stick to my work.

"Now that my lamp is lighted I'll have to sew and talk together, as these scarfs must be finished to-night. Why don't I cut my thread, instead of breaking it? Because it takes too much time to pick up and put down my scissors. I didn't know enough to do this till they told me down at the shop that I was wasting an awful lot of time by stopping to cut my thread.

"Look tired, do I? Then my looks ain't telling no lie. I've been tired for two years past, and I hate the sight of these scarfs as if they was snakes. The folks who sent me to the seaside cottage last summer thought I had a good rest from scarfs; they didn't know I took a lot of 'em with me. We had a big order to fill about that time, and I couldn't afford to hold my hands.

"So you think I am a first-rate worker? I wish our forelady thought as you do, but she don't, and I'll tell you why. There's a lot of girls working down at our place who live in their own homes, and they fetch and carry great piles of scarfs, letting her believe that they make 'em all themselves, instead of telling her of the help they get from their folks. Now I ain't got nobody to help me, and the forelady thinks I am poky. I ain't much of a favorite with her, and if there are any of us bounced when the slack season comes I shouldn't wonder if I was one, though she knows I ain't got no folks and have my room-rent to pay, besides feeding myself and buying my duds. That's just the way it is in them scarf places; in the busy season they work you to death, and in the dull season you may go and jump off Brooklyn Bridge for anything they care. If I thought I'd come out alive, I'd go and do it to-morrow, for then I'd get my $30 a week in a Bowery museum as the Great American Female Jumper. Sometimes I feel like throwing down my work and going and offering myself to Barnum for an Early Christian or something of that kind. You know, he is continually advertising just now for supes[1] to fill up the scenes in the Destruction of Rome. If his season wasn't so short, I'd do it as true as I live. Afraid of the wild beasts? No, I ain't afraid of nothing but of falling behind with my room-rent and being turned out on the street.

"What do you suppose I am paid a dozen for the kind of scarfs I am making now? A dollar? You missed it a long way that time! Just forty-two cents, and if I get this dozen finished by bedtime I'll be doing well. These are the cheapest kind that are made at our place, and it seems to me I get more than my share of them, though I s'pose the forelady tries to do the right thing by us all. Sinfully cheap? Yes, that is just it. They are sinfully cheap. And yet the men who buy 'em will talk about the hardness that women show to women when they run after the cheap things that working-girls make. A friend of mine runs a machine in a slop-shop that puts cheap underwear in the market, and she earns more than I do, for she is paid by the week — $5 a week all the year round. All such shops ain't like that though. In some of them it's piece-work, and the operators are paid only 15 cents a dozen for some garments. Fifteen cents 'apiece,' don't I mean? No, I mean just exactly what I say. A girl told me so who works in such a place, and though girls may lie about other things, you can depend on what they say when they tell you how little they get for their work. It's awful hard, though, to hammer the truth into some folks' heads! The other evening I attended a meeting of our Girls' Friendly Society, and my old Sunday-school teacher came up to me, bringing another uptown lady along with her; 'Here's a poor dear girl, Mrs. A.———', says she, 'who supports herself by making scarfs, and sometimes she is paid only 10 cents apiece for them. Isn't it too bad?' I didn't like to contradict her, but I says to myself, 'No, it's too good.' If I could get 10 cents

[1] *supes:* Slang for supernumerary, a *supe* basically meant to the theater what the term *extra* means to the motion pictures today.

apiece for scarfs, I'd put money in the bank. I told my teacher as plain as I told you just now that for some scarfs, I was paid only three and a half cents apiece — and that's how she remembered it.

"No, mine isn't all hand-work what I am doing now is slip-stitching, and that is always done by hand, and in the shop we machine some of our work, and there's where the trouble comes in. There's only one machine for half a dozen girls, and sometimes there's two or three after wanting it an once, and then some complain that they fix it when it is out of order, while others only wait for it to be fixed for them. The forelady has to settle it the best way she can, and poor thing! I get sorry for her sometimes, for she hates a row, and scolds awful seldom, considering it's only her that stands between us and the boss, who sometimes swears at her like forty. I have a hard enough time, you may think, but, all the same, I'd rather be me than her."

"The Lace-Maker"

"Yes, I am a stay-at-home boarder," said the Swiss girl. "My room-mate she go out to her work, and so I am alone all day while I am here waiting for de position I expect to get as travelling companion. Feel lonely? No, I feel good. Dere never was such a woman before as my room-mate for keeping de windows shut up tight, cold night or hot night, but when she is gone out in de day I open wide de windows and haf all de air I want. You know I am accustomed to air dat comes down de Alps, over de snow — air dat invalids come to Switzerland for in summer time and winter time. In my own country, when my housework was done, I used to take my lace-pillow and my alpenstock and climb up de hillside, and dere I would stay till I make half a yard of lace. Haf I not show you my lace-pillow? I keep it before me all day now, and dere is companionship in de rattling of de bobbins. How many bobbins? Dat is as de lace is narrow or wide. For dis pattern I am making now I use twenty-four. Yes, I seem to trow dem dis way and dat way, widout looking, but all de time I am working out a difficult pattern; and if I was to stick one of dis multitude of pins in de wrong place, among de crossing treads, dere would be much work to undo. I haf patterns of lace wide enough only for edging on baby-clothes, and patterns wide enough for trimming skirts.

"Ready sale for it? Umph! Not so ready as I could wish. De last order I had kept me working steady for a week, and den I was told dat de lady who ordered it was gone away, not to return for tree months, and had left no directions about it. Here is de lace — six yards. It is so wide dat it take me a whole day to weave a yard. A dollar a yard? No, half-a-dollar I was to get. I did ask 75 cents, but dat she refuse to give, saying she could get machine-made lace just as good to look at for 25 cents, but would give me 50, as mine seemed more strong. Strong it should be when I haf to buy pure linen tread with which to make it. It surprise me, in dis country, to hear ladies talk mit respect of machine-made lace. One accustomed to hand-woven lace would never look at dat what comes from a machine. I read in de newspaper one day, as a joke on some poor shop girls, dat when dey see de lace in de museum, dey say, 'Umph! We see better dan dat down on Grand-st.' Many fine ladies would say de same ting. Dey can no more tell imitation lace from real dan dey could tell dis pin from dat pin. In a store one day I hear one lady in a sealskin sacque scold anoder lady in a sealskin sacque for paying dirty dollars for a lace veil, when she could get one dat look just as well for fife dollars. What for you smile? You say 'dirty dollars' sound like 'fildy lucre'? Well, I cannot pronounce your th till I try two, tree times, and you know what I mean. I am German Swiss, and it said dey haf more trouble mit de th dan do the French Swiss. Yes, I speak French quite as well as I speach English, but German is my native language, and dat is de language in which I tink, and in which I pray.

"How did I learn to make lace? I was taught at home by my aunt. All de time she can spare from housework she is making lace, and dat is de way mit many of de housekeepers. Dey can hardly make it fast enough for de woman who goes from house to house mit a basket buying de lace which she takes to de hotels, and sells to de tourists. Yes, I do trow de bobbins quick, but all de same, such a difficult pattern of lace as dis makes itself but slowly. Dis is to go into de showcase of a friend of mine who keeps a fancy store, and dere I am afraid it will stay. She has many rich ladies among her customers, but dey are not too rich to buy machine-made lace. Try de Women's Exchange? Dat is what so many people ask me! How would my poor little rolls of lace look up dere among all dat finery? I was in dere one day, mit a little package of lace in my hand, but such piles upon piles of bright-colored fancy-work did I see dat I lose all heart, and so, when de lady come up and ask me what was my business, I slip my package into my pocket, and say I haf no business; I merely come to look.

"Too easily discouraged? I should not be if I was going to make lace all de time, but dat is not my profession except as — as — what you call it? A stop-gap. Soon dere will be a number of families going to Europe, people who speak no language but dere own, and den my tongue will be wanted more dan my lace is now; for I can make myself understood in tree languages besides English, and dat, to be sure, will make me a very useful companion indeed."

"The Black-Borderer: A Solemn Kind of Work"

"You can't tell me there is no good in novels," said the girl in glasses. "Some years ago my father read in a novel which was then coming out in one of the magazines — I don't remember the name of it now, nor in what magazine it was coming out — that every girl should know how to do one thing well enough to earn her living by it; and he was so much taken with the idea that he determined then and there that I should learn a trade. I was his only daughter, only child in fact, and there were but two of us in family, as my mother had died some years before. Well, I agreed to it just to satisfy him, though I didn't see the use of it, as he was making money as a real estate agent, and had always given me everything I wanted.

"First I tried telegraphy, and found I was no good at that, and then I tried typewriting, and found that wasn't my forte either. I gave up then, and told papa that if he couldn't support me I'd either have to get married or go to the Free Home for Destitute Young Girls; but he had lately visited a paper-mill, and so he suggested another trade, the one you see me working at now. Do I call myself a paper-decorator? No, I am a black-borderer; but, all the same, I am just as much of a skilled laborer as any girl who paints flowers and newly hatched chickens on Easter cards, and such things, though I do nothing but blacken the borders of mourning-paper. Thought that was machine-work? No, that it isn't! It is woman's work. They have tried, time and again, to invent machines that would do it, but they have all turned out flat failures, and I am glad of it, as this is the only kind of work by which I have been able to earn my bread and butter for the last three years — since papa was taken off suddenly by heart failure, and it was found that he had laid up no money, and his insurance policy had lapsed.

"Profession much crowded? No, thank Heaven! This is one profession in which there is a little elbow-room. One reason, I suppose, is that it is not so well-paid as it might be, and another is (though I say it that shouldn't say it) that it isn't every girl who could learn the art. You thought yourself that it was done by a machine, so you must have noticed how very smoothly the black is laid on. Of course I do not take such pains with the inferior qualities as with the best linen, such as I am doing now; but the worst must be done well or it won't pass master. My employer, when he advertised for a girl to do this work, didn't get more than a dozen answers, and most of them were from girls who didn't know anything whatever about it; they only thought they could do it, as they had taken lessons in water colors. They didn't know that this kind of painting is a trade by itself. It is a great satisfaction to me to feel assured that as long as there is work at this place I can get it, and that if I were to be 'bounced' to-day another worker as good as myself could not be found to-morrow.

"Trying to the eyes? Yes, when one's eyes are not very strong to begin with; but if I find my eyes giving out I can always stop and rest awhile. It is my own loss, you know, as mine is piece-work, paid for by the ream. How much a ream? For fine qualities $1.50 a ream; for the very commonest, 45 cents; but there is very little of that kind to be done, for which mercy I am thankful. Unless we are awfully busy, I can come when I like and go when I like. Usually I get here at half-past 8 in the morning and leave at 5 in the afternoon. In that time, if I don't give my eyes too many resting spells, I can border a ream and a half.

"Lonely? No, I am too busy to feel lonely, though I sit here all day by myself, seeing nobody but the man who brings in my paper. That is, he brings it to me in the busy season, but if the work is anyways slack I have to go for it myself. The slack season begins in June, and work does not pick up again till some time in September. In July I live in daily fear of being laid off for awhile, though that hasn't happened to me yet. I am not like girls who live at home and work for 'fun.' When a crowd of them have been addressing circulars in a novelty establishment they don't mind it a bit when the superintendent announces that the job is finished and the mob is to be dispersed. If I were to be laid off I'd have to draw on my savings, instead of adding to them, as I do every week now, except when there is a new dress to be bought, or some other expense of that kind to be incurred.

"Treated as a lady should be? Yes, my employer is as polite as pie whenever I see him, as indeed he ought to be to a girl who he knows by this time is working for all she is worth. At first it seemed to me to be an awfully solemn kind of work, and I used to think about deaths and funerals, and sometimes would get to crying, though I took good care not to damage the paper by letting my tears fall on it; but one can get used to anything, and now I don't any more mind painting black borders than I'd mind painting red ones."

3

"The Painted Cohorts"

In "A Harlot's Progress," William Hogarth illustrates the fallen woman's characteristic journey from innocence to death. She begins as an attractive young woman who is brought into a house of ill repute where she loses her innocence and descends into the outwardly lavish yet inwardly degrading world of the eighteenth-century bawdy house. Before long, she is sentenced to prison, but after being released she returns to prostitution, contracts venereal disease, and dies. Journalists and novelists of Stephen Crane's time, a century and a half after Hogarth, depicted the prostitute in much the same way. Crane's contemporaries described her as a young woman seduced by a man who had promised her the world only to cast her by the wayside after corrupting her. Believing that no self-respecting man would want an impure woman, she turns to a life of prostitution. She first joins one of the finer parlor houses, but as her degradation continues, she descends to concert halls and, eventually, the streets. Finally, she kills herself.

In the essentials, Maggie Johnson follows this pattern. The way Crane tells her story, however, lifts it above earlier stories of prostitution. Crane's careful choice of detail makes Maggie's downward spiral original. He restricts his account of Maggie's prostitution to a single chapter, Chapter XVII, a minor masterpiece in itself. As the previous chapter ended, Maggie's mother had thrown her out and Maggie, having seen Pete for the last time, is left alone to wander the streets with nowhere to go. As Chapter XVII begins, several months have elapsed.

***Will God Forgive Her?* The results of being seduced and abandoned. From J. W. Buel, *Metropolitan Life Unveiled; or the Mysteries and Miseries of America's Great Cities* (St. Louis: Historical Publishing, 1882).**

It is a rainy night and a "girl of the painted cohorts" appears amidst a throng of people exiting the theaters. The girl's name is not mentioned, and although it is obviously Maggie, the generic label allows her to represent all young women in similar situations. Her looks and her behavior indicate how she has spent the intervening months. She has learned to wear makeup, something that nice girls of the period did not do. She has also learned to analyze prospective clients instantly and to shape her approach accordingly. Crane's description of her looks and behavior obviate the need for him to describe Maggie's life during the intervening months.

Though the seduced-and-abandoned pattern remained well established in the popular imagination, its basis in fact was questionable. In the *History of Prostitution*, William W. Sanger classifies the reasons why women turned to prostitution based on statistics gathered from interviews with prostitutes. The percentage of women Sanger interviewed who said they turned to prostitution after being seduced and abandoned is relatively small, yet Sanger himself contradicts his own statistics and reasserts the importance of the traditional pattern. Better to do that than to have to admit women could have strong sexual desires or that they might place their lives above their chastity. Charles Loring Brace also asserts the idea of feminine purity and maintains that a woman held the responsibility to maintain her purity against all odds. Edward Crapsey blames men and women equally for prostitution's existence, citing an anecdote of a woman seduced and abandoned who turns to prostitution because she lacked the nerve to kill herself. Also contradicting Sanger's statistics, Crapsey blatantly asserts that seduction caused "all the prostitution with which we are cursed" (282). J. W. Buel further perpetuates this idea. His essay begins with realistic description but quickly becomes a fanciful, sentimental, even cloying episode describing how a well-to-do young woman, seduced and abandoned, ultimately meets death at her own hands.

"Prostitution and the Working Woman," an excerpt from a lengthy U.S. Department of Labor report, presents and analyzes additional statistics with the purpose of dispelling the stereotypes about the loose morals of "shop girls" that authors such as Abraham Hummel and William F. Howe perpetuated (see pp. 226–233). Correlating the previous occupations of prostitutes, the report authoritatively disproves any connection between morality and occupation.

The final essay in this section, an extended anecdote by Helen Stuart Campbell, shows how poverty could turn a woman to prostitution. Campbell's sympathetic account portrays Rose Haggerty as a working woman going about her business in order to support her needy family. Unlike so many of the other works, Campbell accepts the reality that poverty alone, without seduction, could lead a woman to prostitution. The story of the sober and industrious Rose Haggerty dispels another stereotype, that of the prostitute as a woman so ashamed of herself that she must immerse herself in a sea of liquor to avoid facing her shame. When the account first appeared in the *New-York Tribune* (October 31, 1896), the editor introduced it with a remark that suggests that public attitudes toward prostitutes were beginning to change: "In a graphic sketch of the sad career of a young working-girl she shows the nature and extent of the trials and temptation to which thousands of unfortunate women are subjected. The desperate battle for honest living, the gradual defeat through the combined pressure of competition and greed, the hopeless yielding at the last, and the sullen acceptance of shame as the one means by which starvation for the dependent helpless ones can be fended off — all illustrate not, unfortunately, rare and exceptional misfortunes, but peril and suffering which are the bitter lot of whole classes. . . . Society, arraigned more and more seriously and menacingly for its apathy, heartlessness and self-absorption, has hitherto adjudged the blame, with perverse obstinacy, to the victims of its own mismanagement. To remedy this injustice the facts must first be readjusted and presented plainly as they actually are."

Taken as a whole the documents in this section provide a good indication of the attitudes toward prostitution in Stephen Crane's day and thus provide a touchstone for better understanding Maggie's plight and seeing how Crane affirmed yet challenged the prevailing stereotypes of the prostitute. Crane's contemporary readers did not need another story of the prostitute's stereotypical descent from seduction to the parlor house to the street; they already knew it. The details Crane does provide show that the stereotypical story was oversimplistic. While other stories depicted streetwalking as the prostitute's lowest point, Crane shows that, on the street, the descent continued through many more subtle gradations. Taking Maggie from childhood to her last night when, on a dark street, she encounters the fat, greasy man who precipitates her suicide, Stephen Crane made the life of the street girl poignant, sympathetic, and memorable.

WILLIAM W. SANGER

From The History of Prostitution

William W. Sanger, resident physician at the correctional facility on Blackwell's Island, prepared *The History of Prostitution: Its Extent, Causes, and Effects throughout the World* (New York: Harper, 1859) as an official report to the Governors of the Alms-House of the City of New York. Upon its publication, Sanger's work became the standard authority on the subject, and it was reprinted many times well into the twentieth century. The first two-thirds of the nearly seven-hundred-page volume, largely compiled from other sources, trace the history of prostitution around the world. The last third of the book, from which the following selection has been taken (488–94), is based on primary research and thus is much more important than the rest of the book. At the time of its publication, it was recognized as the most important sociological study of prostitution ever conducted. Sanger interviewed two thousand prostitutes on Blackwell's Island and gathered information about age, nationality, marital status, parents, religion, and literacy. (He discovered that three-quarters of the prostitutes interviewed could read, and 60 percent could read and write.) The results of the one question Sanger called the most important — the cause of the women's turn to prostitution — are reprinted here.

Question. WHAT WAS THE CAUSE OF YOUR BECOMING A PROSTITUTE?

Causes.	Numbers.
Inclination	513
Destitution	525
Seduced and abandoned	258
Drink, and the desire to drink	181
Ill-treatment of parents, relatives, or husbands	164
As an easy life	124
Bad company	84
Persuaded by prostitutes	71
Too idle to work	29
Violated	27
Seduced on board emigrant ships	16
" in emigrant boarding houses	8
Total	2000

This question is probably the most important of the series, as the replies lay open to a considerable extent those hidden springs of evil which have hitherto been known only from their results. First in order stands the reply "Inclination," which can only be understood as meaning a voluntary resort to prostitution in order to gratify the sexual passions. Five hundred and thirteen women, more than one fourth of the gross number, give this as their reason. If their representations were borne out by facts, it would make the task of grappling with the vice a most arduous one, and afford very slight grounds to hope for any amelioration; but it is imagined that the circumstances which induced the ruin of most of those who gave the answer will prove that, if a positive inclination to vice was the proximate cause of the fall, it was but the result of other and controlling influences. In itself such an answer would imply an innate depravity, a want of true womanly feeling, which is actually incredible. The force of desire can neither be denied nor disputed, but still in the bosoms of most females that force exists in a slumbering state until aroused by some outside influences. No woman can understand its power until some positive cause of excitement exists. What is sufficient to awaken the dormant passion is a question that admits innumerable answers. Acquaintance with the opposite sex, particularly if extended so far as to become a reciprocal affection, will tend to this; so will the companionship of females who have yielded to its power; and so will the excitement of intoxication. But it must be repeated, and most decidedly, that without these or some other equally stimulating cause, the full force of sexual desire is seldom known to a virtuous woman. In the male sex nature has provided a more susceptible organization than in females, apparently with the beneficent design of repressing those evils which must result from mutual appetite equally felt by both. In other words, man is the *aggressive* animal, so far as sexual desire is involved. Were it otherwise, and the passions in both sexes equal, illegitimacy and prostitution would be far more rife in our midst than at present.

Some few of the cases in which the reply "Inclination" was given are herewith submitted, with the explanation which accompanied each return. C. M.: while virtuous, this girl had visited dance-houses, where she became acquainted with prostitutes, who persuaded her that they led an easy, merry life; her inclination was the result of female persuasion. E. C. left her husband, and became a prostitute willingly, in order to obtain intoxicating liquors which had been refused her at home. E. R. was deserted by her husband because she drank to excess, and became a prostitute in order to obtain liquor. In this and the preceding case, inclination was the result solely of intemperance. A. J. willingly sacrificed her virtue to a man she loved. C. L.: her inclination was swayed by the advice of women already on the town. J. J. continued this course from inclination after having been seduced by her lover. S. C.: this girl's inclination arose from a love of liquor. Enough has been quoted to prove that, in many of the cases, what is called willing prostitution is the sequel of some communication or circumstances which undermine the principles of virtue and arouse the latent passions.

Destitution is assigned as a reason in five hundred and twenty-five cases. In many of these it is unquestionably true that positive, actual want, the apparent and dreaded approach of starvation, was the real cause of degradation. The following instances of this imperative necessity will appeal to the understanding and the heart more forcibly than any arguments that could be used. As in all the selections already made, or that may be made hereafter, these cases are taken indiscriminately from the replies received, and might be indefinitely extended.

During the progress of this investigation in one of the lower wards of the city, attention was drawn to a pale but interesting-looking girl, about seventeen years of age, from whose replies the following narrative is condensed, retaining her own words as nearly as possible.

"I have been leading this life from about the middle of last January (1856). It was absolute want that drove me to it. My sister, who was about three years older than I am, lived with me. She was deformed and a cripple from a fall she had while a child, and could not do any hard work. She could do a little sewing, and when we both were able to get work we could just make a living. When the heavy snow-storm came our work stopped, and we were in want of food and coals. One very cold morning, just after I had been to the store, the landlord's agent called for some rent we owed, and told us that, if we could not pay it, we should have to move. The agent was a kind man, and gave us a little money to buy some coals. We did not know what we were to do, and were both crying about it, when the woman who keeps this house (where she was then living) came in and brought some sewing for us to do that day. She said that she had been recommended to us by a woman who lived in the same house, but I found out since that she had watched me, and only said this for an excuse. When the work was done I brought it home here. I had heard of such places before, but had never been inside one. I was very cold, and she made me sit down by the fire, and began to talk to me, saying how much better off I should be if I would come and live with her. I told her I could not leave my sister, who was the only relation I had, and could not help herself; but she said I should be able to help my sister, and that she would find some light sewing for her to do, so that she should not want. She talked a good deal more, and I felt inclined to do as she wanted me, but then I thought how wicked it would be, and at last I told her I would think about it. When I got home and saw my sister so sick as she was, and wanting many little things that we had no money to buy, and no friends to help us to, my heart almost broke. However, I said nothing to her then. I laid awake all night thinking, and in the morning I made up my mind to come here. I told her what I was going to do, and she begged me not, but my mind was made up. She said it would be sin, and I told her that I should have to answer for that, and that I was forced to do it because there was no other way to keep myself and help her, and I knew she could not work much for herself, and I was sure she would not live a day if we were turned into the streets. She tried all she could to persuade me not, but I was determined, and so I came here. I hated the thoughts of such a life, and my only reason for coming was that I might help her. I thought that, if I had been alone, I would sooner have starved, but I could not bear to see her suffering. She only lived a few weeks after I came here. I broke her heart. I do not like the life. I would do almost any thing to get out of it; but, now that I have *once done wrong,* I can not get any one to give me work, and I must stop here unless I wish to be starved to death."

This plain and affecting narrative needs no comment. It reveals the history of many an unfortunate woman in this city, and while it must appeal to every sensitive heart, it argues most forcibly for some intervention in such cases. The following statements of other women who have suffered and fallen in a similar manner will show that the preceding is not an isolated case. M. M., a widow with one child, earned $1.50 per week as a tailoress. J. Y., a servant, was taken sick while in a situation, spent all her money, and could get no employment when she recovered. M. T. (quoting her own words) "had no work, no money, and no home." S. F., a widow with three children, could earn two dollars weekly at cap-making, but could not obtain steady employment even at those prices. M. F. had been out of place for some time, and had no money. E. H. earned from two to three dollars per week as tailoress, but had been out of employment for some time. L. C. G.: the examining officer reports in this case, "This girl (a tailoress) is a stranger, without any relations. She received a dollar and a half a week, which would not maintain her." M. C., a servant, was receiving five dollars a month. She sent all her earnings to her mother, and soon after lost her situation, when she had no means to support herself. M. S., also a servant, received *one dollar a month wages.* A. B. landed in Baltimore from Germany, and was robbed of all her money the very day she reached the shore. M. F., a shirt-maker, earned one dollar a week. E. M. G.: the captain of police in the district where this woman resides says, "This girl struggled hard with the world before she became a prostitute, sleeping in station-houses at night, and living on bread and water during the day." He adds: "In my experience of three years, I have known *over fifty cases* whose history would be similar to hers, and who are now prostitutes."

These details give some insight into the under-current of city life. The most prominent fact is that a large number of females, both operatives and domestics, earn so small wages that a temporary cessation of their business, or being a short time out of a situation, is sufficient to reduce them to absolute distress. Provident habits are useless in their cases; for, much as they may feel the necessity, *they have nothing to save,* and the very day that they encounter a reverse sees them penniless. The struggle a virtuous girl will wage against fate in such circumstances may be conceived: it is a literal battle for life, and in the result life is too often preserved only by the sacrifice of virtue.

"Seduced and abandoned." Two hundred and fifty-eight women make this reply. These numbers give but a faint idea of the actual total that should be recorded under the designation, as many who are included in other classes should doubtless have been returned in this. It has already been shown that under the answer "Inclination" are comprised the responses of many who were the victims of seduction before such inclination existed, and there can be no question that among those who assign "Drink, and the desire to drink" as the cause of their becoming prostitutes, may be found many whose first departure from the rules of sobriety was actuated by a desire to drive from their memories all recollections of their seducers' falsehoods. Of the number who were persuaded by women, themselves already fallen, to become public courtesans, it is but reasonable to conclude that many had previously yielded their honor to some lover under false protestations of attachment and fidelity.

It is needless to resort to argument to prove that seduction is a vast social wrong, involving in its consequences not only the entire loss of female character, but also totally destroying the consciousness of integrity on the part of the male sex. It matters not under what circumstances the crime may be perpetrated, none can be found that will exonerate the active offender from the imputation of fraud and treachery. A woman's heart longs for a reciprocal affection, and, to insure this, she will occasionally yield her honor to her lover's importunities, but only when her attachment has become so concentrated upon its object as to invest him with every attribute of perfection, to find in every word he utters and every action he performs but some token of his devotion to her. Love is then literally a passion, an idolatry, and its power is universally acknowledged.

But this passion can not be the growth of an hour. Its developments are gradual. From the first stage of mere acquaintance, it ripens progressively under the influence of tender words and solemn vows, frequently sincere, but often simulated, until the woman owns to herself and admits to her lover that she regards him with affection. Such an acknowledgment, virtually placing her future life in his custody, should inspire him with the high resolve to protect her name and fame, to justify the confidence she has reposed. But not unfrequently is it made the medium for dishonorable exactions, and for a momentary gratification, valueless to him except as a proof of her fervent adoration, and fatal in its consequences to her, he tramples on the priceless jewel of her honor, confidingly surrendered to this love and truth.

It should be remembered that, in order to accomplish this base end, he must have resorted to base means; must either have professed a love he did not feel, or have allowed his affection to cool as he approached its consummation. Pure and sincere attachment would effectually prevent the lover from performing any act which could possibly compromise the woman he adores. None but an unmitigated ruffian can calmly and deliberately wrong an unsuspecting female who has acknowledged a tender sentiment toward him, thus placing herself so entirely in his power. The crime of seduction can be viewed only as a mean and atrocious perjury, and strangely callous must he be whose conscience in after life does not pursue him with scorpion stings and fiery tortures.

But how account for the participation of the female in the crime? Simply by viewing it as an idolatry of devotion which is willing to surrender all to the demands of him she worships; to the intensity of her affections, which absorbs all other considerations; to a perfect insanity of love, excited and sustained by a supposed equal devotion to herself. As soon as this conviction of a mutual love possesses her mind, as soon as her heart responds to its magic touch, she lives in a new atmosphere; her individuality is lost; her thoughts revert only to her lover. Devoted to the promotion of his happiness, she thinks not of her own; and only when it is too late does she awake from the spell that lures her to destruction. In such a case as this, a woman does not merit the contempt with which her conduct is visited. She has sinned from weakness, not from vice; she has been made the victim of her own unbounded love, her heart's richest and purest affections.

Moralists say that all human passions should be held in check by reason and virtue, and none can deny the truthfulness of the assertion. But while they apply the sentiment to the weaker party, who is the sufferer, would it not be advisable to recommend the same restraining influences to him who is the inflictor? No woman possessed of the smallest share of decency or the slightest appreciation of virtue would voluntarily surrender herself without some powerful motive, not preexistent in herself, but imparted by her destroyer. Well aware of the world's opinion, she would not recklessly defy it, and precipitate herself into an abyss of degradation and shame unless some overruling influence had urged her forward. This motive and this influence, it is believed, may be uniformly traced to her weak but truly feminine dependence upon another's vows. Naturally unsuspicious herself, she can not believe that the being whom she has almost deified can be aught but good, and noble, and trustworthy. Sincere in her own professions, she believes there is equal sincerity in his protestations. Willing to sacrifice all to him, she feels implicitly assured that he will protect her from harm. Thus there can be little doubt that, in most cases of seduction, female virtue is trustingly surrendered to the specious arguments and false promises of dishonorable men.

CHARLES LORING BRACE

"Street Girls"

Charles Loring Brace (1826–1890), born in Litchfield, Connecticut, graduated from Yale in 1846. He worked briefly as a schoolteacher, considered a career in the ministry, travelled widely, but found his true calling when he began doing philanthropic work among impoverished New Yorkers. In 1853, he helped found the Children's Aid Society, an organization that assisted poor children, mostly immigrants, by establishing industrial schools, night schools, summer camps, and sanitariums. The notion of self-help formed the basis of his efforts. Brace wrote prolifically as well, but *The Dangerous Classes of New York* (1872) is generally considered his most important work. Indeed, the phrase "dangerous classes" received some currency after the book's publication. In *How the Other Half Lives,* Jacob A. Riis reused the phrase and kept it current. The following selection is from the third edition of *The Dangerous Classes* (New York: Wynkoop & Hallenback, 1880), 114–19.

A girl street-rover is to my mind the most painful figure in all the unfortunate crowd of a large city. With a boy, "Arab of the streets," one always has the consolation that, despite his ragged clothes and bed in a box or hay-barge, he often has a rather good time of it, and enjoys many of the delicious pleasures of a child's roving life, and that a fortunate turn of events may at any time make an honest, industrious fellow of him. At heart we cannot say that he is much corrupted; his sins belong to his ignorance and his condition, and are often easily corrected by a radical change of circumstances. The oaths, tobacco-spitting, and slang, and even the fighting and stealing of a street-boy, are not so bad as they look. Refined influences, the checks of religion, and a fairer chance for existence without incessant struggle, will often utterly eradicate these evil habits, and the rough, thieving New York vagrant make an honest, hard-working Western pioneer. It is true that sometimes the habit of vagrancy and idling may be too deeply worked in him for his character to speedily reform; but, if of tender years, a change of circumstances will nearly always bring a change of character.

With a girl-vagrant it is different. She feels homelessness and friendlessness more; she has more of the feminine dependence on affection;

the street-trades, too, are harder for her, and the return at night to some lonely cellar or tenement-room, crowded with dirty people of all ages and sexes, is more dreary. She develops body and mind earlier than the boy, and the habits of vagabondism stamped on her in childhood are more difficult to wear off.

Then the strange and mysterious subject of sexual vice comes in. It has often seemed to me one of the most dark arrangements of this singular world that a female child of the poor should be permitted to start on its immortal career with almost every influence about it degrading, its inherited tendencies overwhelming toward the indulgence of passion, its examples all of crime or lust, its lower nature awake long before its higher, and then that it should be allowed to soil and degrade its soul before the maturity of reason, and beyond all human possibility of cleansing!

For there is no reality in the sentimental assertion that the sexual sins of the lad are as degrading as those of the girl. The instinct of the female is more toward the preservation of purity, and therefore her fall is deeper — an instinct grounded in the desire of preserving a stock, or even the necessity of perpetuating our race.

Still, were the indulgences of the two sexes of a similar character — as in savage races — were they both following passion alone, the moral effect would not perhaps be so different in the two cases. But the sin of the girl soon becomes what the Bible calls "a sin against one's own body," the most debasing of all sins. She soon learns to offer for sale that which is in its nature beyond all price, and to feign the most sacred affections, and barter with the most delicate instincts. She no longer merely follows blindly and excessively an instinct; she perverts a passion and sells herself. The only parallel case with the male sex would be that in some Eastern communities which are rotting and falling to pieces from their debasing and unnatural crimes. When we hear of such disgusting offenses under any form of civilization, whether it be under the Rome of the Empire, or the Turkey of to-day, we know that disaster, ruin, and death, are near the State and the people.

This crime, with the girl, seems to sap and rot the whole nature. She loses self-respect, without which every human being soon sinks to the lowest depths; she loses the habit of industry, and cannot be taught to work. Having won her food at the table of Nature by unnatural means, Nature seems to cast her out, and henceforth she cannot labor. Living in a state of unnatural excitement, often worked up to a high pitch of nervous tension by stimulants, becoming weak in body and mind, her character loses fixedness of purpose and tenacity and true energy. The diabolical women who support and plunder her, the vile society she keeps, the literature she reads, the business she has chosen or fallen into, serve continually more and more to degrade and defile her. If, in a moment of remorse, she flee away and take honest work, her weakness and bad habits follow her; she is inefficient, careless, unsteady, and lazy; she craves the stimulus and hollow gayety of the wild life she has led; her ill name dogs her; all the wicked have an instinct of her former evil courses; the world and herself are against reform, and, unless she chance to have a higher moral nature or stronger will than most of her class, or unless Religion should touch even her polluted soul, she soon falls back, and gives one more sad illustration of the immense difficulty of a fallen woman rising again.

The great majority of prostitutes, it must be remembered, have had no romantic or sensational history, though they always affect this. They usually relate, and perhaps even imagine, that they have been seduced from the paths of virtue suddenly and by the wiles of some heartless seducer. Often they describe themselves as belonging to some virtuous, respectable, and even wealthy family. Their real history, however, is much more commonplace and matter-of-fact. They have been poor women's daughters, and did not want to work as their mothers did; or they have grown up in a tenement-room, crowded with boys and men, and lost purity before they knew what it was; or they have liked gay company, and have had no good influences around them, and sought pleasure in criminal indulgences; or they have been street-children, poor, neglected, and ignorant, and thus naturally and inevitably have become depraved women. Their sad life and debased character are the natural outgrowth of poverty, ignorance, and laziness. The number among them who have "seen better days," or have fallen from heights of virtue, is incredibly small. They show what fruits neglect in childhood, and want of education and of the habit of labor, and the absence of pure examples, will inevitably bear. Yet in their low estate they always show some of the divine qualities of their sex. The physicians in the Blackwell's Island Hospital say that there are no nurses so tender and devoted to the sick and dying as these girls. And the honesty of their dealings with the washerwomen and shopkeepers, who trust them while in their vile houses, has often been noted.

The words of sympathy and religion always touch their hearts, though the effect passes like the April cloud. On a broad scale, probably no remedy that man could apply would ever cure this fatal disease of society. It may, however, be diminished in its ravages, and prevented in a large measure. The check to its devastations in a laboring or poor class will be the facility of marriage, the opening of new channels of female work, but, above all, the influences of education and Religion. . . .

EDWARD CRAPSEY

"Prostitution"

Edward Crapsey wrote *The Man of Two Lives!: Being an Authentic History of Edward Howard Rulloff*[*son*], *Philologist & Murderer* (1871) and *The Nether Side of New York, or, The Vice, Crime, and Poverty of the Great Metropolis* (New York: Sheldon and Company, 1872). The second work, from which the following is excerpted (138–46), devotes separate chapters to confidence games, thievery, gambling, and abortion.

Take the lowest type first, and find it in the middle of any night by merely sauntering through Broadway from Grand to Fourteenth street, or again from Twenty-third to Thirtieth street, or in some of the side streets. The type is the night-walker, and gradations of the class are almost as numerous as its representatives. To meet the worst, Greene, Wooster, Houston, Bleecker, or Amity streets must be traversed. There was a time, and it is not long past, when only the Fourth Ward[1] could show the prowling prostitute in her most abject degradation, but it is not necessary now to get lost in the tortuous mazes of the old town, to find the most repulsive phases of female frailty. The Eighth Ward has taken the place of the Fourth, and the stranger need only turn three hundred feet out of Broadway anywhere between Grand and Amity streets, to encounter the most startling evidence of the possibility of total depravity.

[1] *Fourth Ward:* In the late seventeenth century, Manhattan was subdivided into political divisions known as wards. By the mid–nineteenth century, the ward system had become rife with corruption and was abolished in 1857. Though political divisions were gone, New Yorkers continued to refer to different parts of the city as wards through the nineteenth century.

To see the worst, stand for the hour before midnight on the corner of Houston and Greene streets. In that time a hundred women apparently will pass, but the close observer will notice that each woman passes the spot on an average of about twice, so that in fact there are not more than fifty of them. This frequency of appearance leads to the supposition that they do not go far, which is the fact. Each set of prostitutes has its metes and bounds laid down by an unwritten code of its own enactment, which is rarely violated. The set now under consideration travels Houston, Bleecker, Wooster, and Greene streets, with occasional forays upon Broadway, which is the common property of all. But these poor fallen creatures rarely go there to put themselves in fruitless competition with more attractive sin. They are poorly dressed, have nothing of beauty in form or face, and are always uncouth or brazenly vulgar in manner. They are miserably poor, herding in garrets or cellars, and are driven by their necessities to accost every stranger they meet with what the silly law of New York calls "Soliciting for the purpose of prostitution." When a woman offers to sell her body to a man she never saw before, for fifty cents, she has fallen low indeed, and this offer will be made at least a dozen times within the hour to any observer at the spot mentioned, whose appearance does not absolutely forbid advances.

Next stand for the same period at Amity and Greene streets. As many women will pass, and in about the same ratio as to reappearances. They are a shade better in appearance as to dress, and some of them have the faint remnants of former personal beauty. They are vulgar yet, but are a vast improvement on the set first seen. All of them will so look at you as to invite advances, but only about one in five will speak first. When they do, it is merely to say "Good evening" or "How are you, my dear," instead of a direct invitation to go home with them, which is the first greeting of the other set. These Amity street women are, as a rule, better housed and fed than the first set, as they live in the houses bordering their tramping-ground, which are all well built and finished. Some of the women have attained to, or more correctly speaking, have not fallen below the prosperity of occupying a room in one of these houses alone, and none of them have more than one female room-mate. Instead of the rough pine furniture of Houston street, the rooms here are given an almost decent appearance by imitation oak, or else are filled up with those strainings for respectable adornments known as "cottage furniture." Another decided proof of better condition is the absence of the cooking stove, for these girls either board with the "Madam" or obtain their food at restaurants.

This class, which is thus better housed, better dressed, better behaved, has the middle rank, and contains the majority of all women plying their vocation in the public streets. Although I have mentioned only Amity and Green streets as a post of observation, it can be seen at many other points, and notably so at Twelfth street and University Place, which latter stately thoroughfare has lately become a chief tramping-ground for abandoned females.

There is yet another grade of these night-walkers, and it can be best seen at Broadway and Washington Place, or Broadway and Twenty-fourth street. But whoever wishes to observe this class must go earlier, as these women have nearly all retired from the tramp by ten o'clock, and can be seen in greatest numbers only between eight and nine in winter or nine and ten in summer. Almost without exception, they seem in the faint light of the streets to be dressed with elegance and taste, to be handsome in feature and form, and to have left in them something of womanly reserve and modesty. True, they are out in the streets at unseemly hours without male escort, but walking quickly as they do, without looking to the right or the left, the unpractised observer doubts that they belong to the *demi-monde*,[2] and charitably supposes that they have been compelled to leave the shelter of their homes by sudden sickness in the family or by some equally urgent necessity. If the stranger is bold enough to accost one of them, he is even less sure than before of her character. She does not exactly repel his advances, but she does not invite them, and is sufficiently adroit to assume a maidenly reserve that perplexes while it allures him. She will not stop to talk with him, but if he walks beside her she will converse on ordinary topics and use language to which no exceptions can be taken. There is nothing essentially vulgar, much less indelicate, about her words, demeanor, or appearance, and by the time a novice has walked a block in her company he is in a tremor of apprehension that he has committed the grave indiscretion of speaking to a lady who happened to be unprotected, and who is luring him on to be cowhided by her brother or husband. If she succeeds in getting him to her home he finds it a house of respectable exterior and furnished within with some pretensions to elegance. As there was nothing indecent about the woman herself, so there is nothing bawdyish about her home. The pictures which adorn the walls are not, as in houses of a lower grade, sug-

[2] *demi-monde:* "The class of women of doubtful reputation and social standing, upon the outskirts of 'society'" (*OED*).

gestive of the vile lives of the inmates; the furniture is handsome and of a kind to give an impression of a quiet, reputable life. . . .

In all the great cities of the United States, so far as my personal observation extends — and I have been in all of them — the walking of the streets after nightfall by prostitutes has become an alarming evil; but New York is entitled, I am afraid, to preëminence in this respect. Not only is the city first in the number of its street-walkers, but nowhere else has the class become so degraded. I have hinted something of the profanity and obscenity of the women who can be found after midnight in any of the side streets, but it is not possible to describe in detail the scenes which will be forced upon the observer any night in Houston, Bleecker, Amity, and Fourth streets, as well as in the lower Bowery, Chatham street, and some other east-side thoroughfares. Singly, in couples, or groups, these girls, many of whom are mere children and very few of whom have scarcely passed maturity, plunge along the sidewalks, accosting every man they meet, or, stopping at the street corners, annoy all passers until they are driven away by the police. Many of them are under the influence of liquor, and not a night passes but some of these degraded creatures are carried into the station-houses helplessly or furiously drunk. Until within the past two years I never saw any of these women drinking in public bar-rooms, but now it has become so common that it has ceased to be remarked; it is true there are few of the saloons which will serve them, but there is always one on each route of the tramps which will sell to any one, and here these poor painted wrecks of womanhood can be seen standing at the bar, drinking vile liquors until they have won the beatitude of stupefaction, or until they reel out into the streets indecently drunk. If the unconsciousness of inebriety is ever a blessing, it is such in the case of these lost women, as it permits them for the time to forget what they are and must be always. Often suffering for the necessaries of life, burdened almost without exception with "lovers" who despoil them of the pittance they receive for moral and physical death, harassed by the police, shunned by their more prosperous sisters in sin, corroded morally and physically with the leprosy of their vice, no class needs so much of pity, none has less of it, and none is so little aware that it needs commiseration. Calloused by crime which is unnatural and bestializing, the street-walkers have forgotten that they were ever undefiled and lost all desire to be other than they are. Numbering about two thousand, constantly infesting the public thoroughfares, inoculated and inoculating with loathsome diseases, they are the great danger and shame of civilization found in all cities, but here more numerous more dangerous and more shameful than anywhere else on the continent.

It has not been from any wish to pander to a morbid desire for the repulsive that I have set this type of prostitution in the foreground, Palpable facts cannot be ignored, and a vice that is thus obtruded upon every passer through the public streets cannot be too soon or too fully described; but having presented the facts in such plain terms that they cannot be misunderstood, I gladly take leave of this lowest type of metropolitan prostitution. It is hardly more agreeable to speak of the next grade, which is found in the lowest of what are known as "parlor houses." The chief difference between the inmates and the street-walkers is that the former do not cruise the streets to entice strangers to their dens. If this is a comparative virtue it is the only one these women can boast, as they are fully as bestial in every other respect as their sisters of the pave. The houses in which they live and ply their infamous vocation are always unmistakable even to the novice. In Greene and Wooster streets several blocks are almost wholly taken up by such houses, but others but little less open and degraded can be found in many other quarters of the city. In many of these houses there is a public bar; in all of them the orgies are indecent to such a degree that they cannot be described. Next above these dens are houses a shade more sufferable, which attempt to hide their infamy behind cigar stores or some other kind of shop, and are filled with women who do not shock at the first glance. Above these again are houses which really have parlors, and in which the women make a pretence to decency in their demeanor while in public. After these come the grand saloons where the evil is painted in the most alluring colors. The houses are of the largest and stateliest, the furniture the most elegant, the inmates beautiful, accomplished, captivating in dress and manner, who, with woman's only priceless jewel, would adorn any circle. It is difficult to persuade one who has no personal knowledge of the matter, that, taken into the parlors of one of these houses and meeting the inmates without a previous intimation of the character of the place, he would believe himself in a pure, refined home. Yet such is the fact. Such houses as these can be found in every desirable neighborhood, and no man can be sure that he has not one of the sepulchres next door.

But the vice has taken in New York a more insidious if less alluring form than this. For some years past a most deplorable change has been going on which has had the effect of greatly decreasing the number of parlor houses, while houses of assignation have multiplied in the same ratio. The effect has been to intrude prostitution into circles and places where its presence is never suspected. Hundreds of houses are thus defiled, and the corroding vice creeps into families of every social grade. Women of high position and culture, no less than the unlettered shop girls, resort to the houses of assignation, which are of every grade, from the palaces in the most aristocratic quarters of the city to the frowsy rooms in the slums. Many of the frequenters of these houses are married women, who are driven by an insane desire for display to thus add to a scanty income; others are young girls led astray by faulty education, and yet others are driven by starvation to sell their virtue to any casual buyer. Many of these cases have come to the knowledge of the police, and there is nothing which pleads so strongly against the flagrant injustice which has closed the doors of productive industry against women, as the fact that when forced to fall back upon their own resources, so many of them have been compelled to choose between prostitution and destitution. But for this fact the chief evil of the age could not have become so prevalent as it is. Woman is naturally chaste, and if those who have fallen can be induced to tell the cause, it will be found that at least six in every ten are forced by sheer necessity to become confirmed prostitutes. I do not mean to say that they will plead this as the cause of the first lapse from virtue, as nine cases out of ten of them will charge that to their betrayal by men whom they loved. But after that first lapse, and after their desertion by these men, they claim they had no choice between their way of life and death from starvation. The story is told by types of every class of prostitutes, from the adroit adventuress who lays her snares in the great hotels, to the poor drunken creature who tramps the streets, and there is so little variation that one will answer for all; that one was told a few nights ago at a station-house desk by a young girl of rare beauty, who had been taken from a house which had been seized by the police. Having given her name, age, and birth-place, she was asked the usual question as to her occupation, and answered, "I am what men have made me." Then she went on speaking rapidly, as if to have her revenge upon society before her resolution failed her:

"Sir, only a year ago I was a happy innocent girl in my father's house in a town near this city. I had a lover everybody thought an honorable man, and we were engaged to be married. I adored him, father trusted him, all my friends envied me because of him. Well, I was weak, he was mean, and he betrayed me. After that he deserted me; the consequences of my sin after a time could not be hidden. Then my father cast me off, and all who had ever known me shunned me. In that town, sir, there was no human being who would shelter me or give me a crust; but *he,* mind you, was received among them all just as before, except father. Of course I had to starve or come here. I hadn't nerve enough to kill myself, so I went to the house where you found me to-night. That's a 'disorderly house' you say — perhaps it is — I know it's vile enough, but ain't the men you found in it as bad as the women? You don't seem to think so, for you've let the men go and we women are to be locked up. I'm young yet, sir, but I'm old enough to have found out that all the sin and shame of this thing falls on us; the men get none of it, yet they cause the whole of it."

With some variations in immaterial details, I have heard this story so often, and from women of every type, that I am convinced that they believe it to be the cause of all the prostitution with which we are cursed. There is the corroborative fact that the majority of all the public prostitutes in the city are from the country, and drift hither from the towns and villages within two hundred miles, when they enter upon their career of open shame. Many are from the New England States, as many from New York State, a few from the other Middle and the Western States, and the others are produced by the city itself, and, as they claim, by like causes. From all I have seen of the vice I am constrained to believe that the perfidy of men is chargeable with the greater part of it, and of nearly all of that which is most open and bestial. There is enough in the mental and moral characteristics of these weak, thoughtless, hapless women to sustain the indictments which they present against their betrayers. All prostitution cannot of course be thus explained, for there are women who have deliberately chosen it, as there are others who have been led astray by love of dress and other equally unworthy motives. For the first there should only be boundless pity, for the others only measureless scorn.

From the commencement of this chapter I have written under manifold difficulties. I have striven to present only general facts and to present them so that the full measure of the shame of the metropolis in this respect can be seen, without rudely shocking the false modesty which is so prevalent. Of the dangers which threaten us from this cause it is hardly necessary to speak in a work of this character, as they are the same here as in other cities and not much greater here than elsewhere in proportion to numbers, unless it is in the corrupting influence of vice upon the domestic life. It may be that here the evil in its covert form is more general in its ramifications through all circles of society, and is thus corroding where its presence is not suspected. I believe such to be the case, and if it is, the moral stamina of the community is being undermined, and it is impossible to imagine what will be the debasement it will entail upon the next generation. It is a terrible state of affairs when the chastity of men is hooted at as an absurdity, and the virtue of women seeming to be virtuous is suspected; yet such is the condition of New York. That such things can be asserted and believed is in itself a proof of a profligacy that has become ominous; that houses of assignation into which women can steal from reputable homes have gradually replaced the houses of prostitution, is a startling evidence that there is too much foundation for these assertions and this belief. As I look upon it in the light of many facts which are of such character that they cannot be hinted at, much less mentioned, the chief danger that threatens the city from the social evil does not come from the street-walkers, nor the inmates of public houses of prostitution. These are women known to be unchaste, who are without home ties and without influence except to a very limited extent. On the contrary, those women who are unsuspected prostitutes occupy and defile the holiest positions of domestic life, and there is no limit to the evil which their crime produces. And this form of the plague is more deplorable because it is one which no law can cure, although it might be mitigated to some extent by statutory remedies.

But as yet there has been no attempt to apply any remedy to any form of this vice. Chiefly because of a senseless delicacy the subject has not been sufficiently agitated to compel the Legislature to attend to it, and the consequence has been that prostitution has gone on unchecked until it threatens to ulcerate the whole body politic. What little demand has been made for putting it under legal restraints, has been for such expedients as have been adopted in some European cities with a view to ameliorate the physical consequences of debauchery, and has been met by the Pecksniffian[3] objection that such laws are viler than the vice they regulate. These cavillers declare that every man should be compelled to take the possible consequences of his sin, and that to lessen the chances of these consequences occurring is to increase rather than diminish the evil. It is useless to tell them that prostitution has always existed and always will in every segregation of mankind, and there is little sense in attempting to ignore a thing that every one knows to exist. Yet it has been ignored by the law-makers of New York, with the exception that they have declared that it shall be "unlawful to solicit men on the public streets for the purposes of prostitution." No Legislature has, however, ever been induced to enact that

[3] *Pecksniffian:* A reference to Mr. Pecksniff, a character in Charles Dickens's *Martin Chuzzlewit* (1844), used to suggest hypocrisy.

all houses of prostitution or assignation shall be licensed and a register kept of the inmates and frequenters, although it is evident such a law would materially decrease both the number of houses and of prostitutes. There is nothing so certain as that New York must do something to check this evil, if it does not desire to be known the world over as a marvel of lechery within the next twenty years, and it is as sure that while statutory restrictions may assist to that end they cannot alone accomplish it. The moral character of this people needs to be rebuilt from the foundation. Public writers and speakers must make unceasing warfare upon this form of iniquity until unchastity in males as well as females is once more an outrage upon the social code. The truth as to the facts must be plainly told, and no prudish delicacy must be allowed to prevent the adoption, much less the suggestion of the needed remedies. I may be all wrong, but I have always had a fancy that when profligate men are socially outlawed, a great deal will have been done to exterminate profligate women. It would not, as nothing else will, exterminate prostitution, but it will do more to mitigate it than anything else.

The full measure of our danger from prostitution cannot be seen from the meagre statistics on the subject which are attainable. A census of all public prostitutes is taken each year by the police, and the last enumeration shows 351 houses of prostitution with 1,223 inmates, and 113 houses of assignation. Two years before there were over 600 of these houses, and when some one complained to Mr. John A. Kennedy during the last of the ten years that he worthily filled the office of Superintendent of Police, that this was too small an exhibit to be a true one, he answered that if all these houses were together they would line both sides of Broadway from the Battery to Houston street, and he thought New York ought not to have more than three miles of houses of ill-fame. That is true, but there is vastly more of prostitution than these figures show, for the reason that they do not give any of that covert class to which I have alluded. This class to a large extent are domiciled in tenement houses where two women live together occupying a suite of rooms; but there are others who belong to a higher social grade and living unsuspected as members of respectable families are frequent visitors to houses of assignation. It is evident that none of these get into the police statistics. The full measure of our shame exceeds this showing as to the number of women who sell their virtue in the open market, at least three times, and to a far larger extent as to those who go astray from less sordid but equally unworthy motives.

What we must acknowledge we have of the vice is enough to move us as a community from our apathy regarding it; what we have good reason to dread that we have is enough to startle a people more debauched and debased than we have yet become.

J. W. BUEL

From Metropolitan Life Unveiled

The following selection from *Metropolitan Life Unveiled; or the Mysteries and Miseries of America's Great Cities* (St. Louis: Historical Publishing, 1882), 61–69, begins as a description of prostitution in New York, yet quickly becomes a cloying, sentimental, imaginary description of a presumably typical fallen woman. The excerpt is worth including here not for its literary value, but because it represents an attitude toward the prostitute that continued to predominate through the nineteenth century.

Prostitution, in New York, is one of the sciences, as well as the evils, with which legislation has had much to do in efforts for its suppression, while society has been even more active in its extension. It is a sin which, though the most blighting and debasing, is so general and infectious that we see it flaunting its scarlet colors in every by-way and aristocratic walk of our Metropolis. An evil that subsists solely by reason of the gain it secures or the lust it satiates, must thrive where society recognizes its prevalence and throws sweet morsels, like dainty pastries, to its votaries. Since the custom prevails of blaming only the weaker vessel for being broken by the stronger, the ministration of angels could not abate the crime of prostitution, and until the time shall come when the morals of men shall be held as sacred as those of women, we can entertain no hope for its eradication, or even curtailment.

As New York is the great centre of aggrandizement, so is she a swirling charybdis, forever drawing the hopes and ambitions of mankind to her vortex, and then scattering the debris throughout the realm of her influence.

By recent estimate, based on reports submitted from various charitable and municipal officers, Gotham has fifteen thousand registered prostitutes, and perhaps four times that number of women who are secret followers of the destructive vice. By comparison, this aggregate is not so astounding, for it gives but one lewd woman to every eight, estimating the population of New York at 1,200,000, and one-half that number females. But the total of secret bawds cannot be approximated, and any estimate is therefore liable to be greatly exaggerated or underrated; hence the prevalence of prostitution must be considered as it appears on the surface, always having in view the surroundings which nourish it. . . .

Let us take an example and follow its subject to the borders of that mysterious realm which spirits only can explore:

A beautiful girl with the dawning flush of approaching womanhood mantling her innocent soul, a heart swelling with the rich and roseate romance of life, an eye that wistfully follows the lengthening, but laughing shadows of gathering years. On this picture I could gaze forever without tiring, for it brings before me a vision of my own young days; those halcyon hours when budding ambition just peeped above the purple horizon of life and left its first mark like a dew-drop in a bed of blushing roses. I can even now see a young girl turning the meridian of her sweetest years, clad in a celestial robe of purity, that glintens and gathers like the changing iris on the feathered throat of a dove. Even her hair reflects a brilliant sheen, rippling in an early sunlight, as her foot is lifted to touch the shore, the floating isle of her morning dream. Could the god of ruthless and unpitying conquest, with devilish power, despoil that bank of violets whereon so guileless a child would rest her feet, and make the bewitching isle of love's first dream a shore of misery?

Looking back over the flowery pathway she has so lithesomely trodden, no dark clouds are visible; nothing but a bright vision of her blessed mother teaching an infant prayer:

"Oh, Lord, I am but a 'ittle child,
But teach me all dy ways to know;
'Teep me f'om sin, b'ess mamma too,
Dat in all doodness we may drow."

In a moment, as it were, this beautiful girl traverses the highway of youth to adolescence; she leaves a mother's sheltering wing to try her pinions, and make a short flight into the empyrean. Society now recognizes her as a *debutante;* young men seek her company and find it charming; but one of her admirers was born on a Friday, and nature, while giving him Hyperion's form,[1] destroyed her work by endowing him with a villain's heart. But the heart is hid, and love never thinks of a magic looking-glass. Following society's practices, the young and trusting birdling becomes a frequenter of theaters, operas, social gatherings, receptions, always with the Hyperion-Satyr, who whispers in her willing ears soft stories, so thrilling when first heard, but frequently changing from honeyed phrases to the poisonous gall of wretched experience.

Familiarity succeeds the first protestations of love, and carriage drives, trips to Coney Island, or Long Branch, soon follow, far removed from parental care, which gives a feeling of freedom from restraint. Late suppers, a little wine, charades, jests with double meanings, and other hidden snares gather around our lovable, beautiful and confiding girl, and bind her unsuspectingly.

Months of time may be spent devising schemes for the certain ruin of this innocent child — two hearts interchanging, one giving trust and holy devotion, the other nothing but lees of blackest villainy. At last the prey has been caught in silken meshes; at a ball, turning from the lascivious waltz to a wine table, while under the deadly effects of this social combination, devised by hell for the spoliation of virtue, unconscious of her own acts and flushed by warm embraces, she yields that priceless gem of womanhood to him who, gaining nothing, has left her poor indeed.

Where are the fires of heaven, or the thunderbolts of Nemesis?[2] Can it be that there is no purgatory to punish this cannibal, this foulest vulture that ever tore flesh from a dead body? There must be a hell, else eternal life, the law of cause and effect, the adage "Time makes all things even," are all travesties and delusions, or the work of creation is yet incomplete.

Having gained his infamous purpose, this beast in human form gluts his sensuous maw and eulogises his own cunning, while the poor, trembling victim, with soul rolling in the fires of remorse, hugs her bosom lest that terrible secret should escape. In her own bed-room, she locks the door against intrusion and yet images of a threatening future arise in hideous phantom shape, and with fingers of scorn point

[1] *Hyperion's form:* Hyperion was a mythical god often taken for the sun itself.

[2] *Nemesis:* One of the infernal deities, Nemesis, the goddess of vengeance, severely punished wrongdoers.

to her withering ambitions and proclaim her obloquy. Still she trusts the leper that has infected her and looks anxiously forward to the promised marriage day. Poor thing! the roses you saw yester'en are dead; the sun of your dreamy morning is veiled behind a black cloud; the beds of violets on that beautiful little isle have all faded, while the zephyrs of your golden summer day have changed to a simoom, blowing dead leaves in your face.

A letter comes to her; it is from the one she still loves most; how greedily she breaks the seal and reads! I see her face blanching, her fingers relax; look! help! she is gasping; why, heaven help us, the expected bride of a week has fallen in a swoon which is an excellent counterfeit of death; water, quick! and brandy; send for a physician, let no time be lost!

"My God! speak to me, child, oh, what shall I do? Darling, look up into your mother's face; my heart is breaking; heaven minister to me and my child." The agony of that fond mother who, unconscious of her daughter's shame, wrings her hands over the prostrate form, would make dumb pity speak, and patience weep tears like scalding rain.

The letter is discovered and the contents made known to her doting parents; they then look no further than a broken heart for unrequited love, reeking nothing of the priceless sacrifice their trusting daughter had rendered up with all that soul of obedience shown by Abraham; yea, even greater, for she herself lay on the pyre.

Life came back again to the simple child, but it was a living without heart, an existence groveling in the shadows of despair; a life walking in eternal night; reading her epitaph in every flower, hearing no sound save the curses of leering imps, and looking nowhere save into phantom faces mocking her dishonor.

Alone, in the room where mother's love taught her that little prayer beside a sister who, too, was a pupil, she thinks of all those roseate hours, those beautiful jewels in life's garniture; of a mother's bounteous love and care; of all the bright visions thrown out so vividly on the canvas of her youth, and then — of the bitter dregs; the searing conscience; the blasted hopes, and withered life. There, before her, lies that cruel letter, the seal of obloquy, and gazing on it through the windows of her tears she grasps the thought which steals into her brain like a wraith of murder — Lethe's waters roll at her feet and under its waves the soul reposes in forgetfulness. Once conceived, no hand can stay that dreadful purpose; softly she withdraws from a shelter prepared by mother's hands; out from the shades of honeysuckles and the perfumes of May flowers, she flies swiftly to the rolling beach or some towering cliff, and ends the beautiful life begun with cymbal and song and dying in tears and shadows.

U.S. DEPARTMENT OF LABOR

"Report on Prostitution and the Working Woman"

The following text is excerpted from the introduction to the U.S. Department of Labor's fourth annual report, *Working Women in Large Cities* (Washington, D.C.: GPO, 1889), 72–76. While the bulk of the report consists of tables outlining the demographics of the working woman, the following selection seeks to refute the traditional stereotypes, which, in general, linked work to morality and specifically associated working women and prostitutes.

The statistical method is not the best under which to determine moral conditions; yet, as an auxiliary, it has great efficiency. Statistical science can only be employed to show the results of the lives of the people; it can not show the inner motives which lead to results. So, in this investigation relating to working women, it is comparatively easy to determine, through statistics, their home surroundings, their wages and general conditions, their income and expenses, and all that belongs to their material environment. But their characters, as to personal integrity or virtuous conduct, can not well be ascertained through a statistical investigation. Yet it is necessary to consider this side of the question.

Observation is not sufficient, and personal interviews might lead to difficulties greater than those belonging to observation alone. The force of statistics in such conditions is rather negative than positive, and this negative quality is brought into use here.

It is often flippantly asserted that the shop girls, those comprising the class under investigation, recruit the ranks of prostitution. It would be a relief, of course, to all honest citizens to have this charge removed entirely, and further, to have the facts warrant its removal. Of course, such a charge can not be entirely removed when applied to any class. The only question here is, Does it apply to the class against which it is brought?

A few statistics of a negative character have been collected, relating to prostitution. This partial investigation has been made as to how far the ranks of prostitution are recruited from girls belonging to the industrial classes. It should be distinctly borne in mind that this partial investigation was applied only to what may be called professional prostitutes; for no statistical investigation can disclose the amount of immoral conduct of any class of people. So that quiet, unobtrusive, and unobserved prostitution, which exists in all communities, has no place in the present consideration.

Original investigation was made in the following cities: Brooklyn, Buffalo, Chicago, Cincinnati, Cleveland, Indianapolis, Louisville, Newark, New Orleans, New York, Philadelphia, Richmond, Saint Louis, and San Francisco.

In 1884 the Massachusetts Bureau of Statistics of Labor made a report as to the previous occupations of one hundred and seventy professional prostitutes in the city of Boston, and the facts then given have been incorporated in this report.

The number of prostitutes as stated in the following table, for any one of the cities named, falls far below the total number of prostitutes in that city, but the number and variety of those from whom information has been received are sufficient to insure representative results. Thus in Chicago, for example, there are, or were at the time of the investigation of the Department, 302 houses of ill-fame, assignation houses, and "rooming" houses, known to the police, containing 1,097 inmates. This investigation involved 557 of this number. In some of the other cities, Philadelphia and Brooklyn notably, the proportion of prostitutes interviewed was not so large as in Chicago, but a sufficient number of reports were obtained to afford a basis for a fair conclusion as to the part played, if any, by the working women in swelling the ranks of these unfortunates.

In certain of the cities in which this subject was investigated return was made of the number of women who had been married before entering on a life of shame. Some of these women were married before engaging in industrial work, some between periods of industrial employment; some after working at various employments were married, and then entered upon a life of prostitution, and some of the married prostitutes had never been industrially employed.

The facts as to marriage are shown by the following brief table, which gives the number of prostitutes furnishing information in the cities referred to, the number reporting themselves as having been married, and the per cent. of the total number who were married:

City.	Number of prostitutes furnishing information.	Number of prostitutes reported as having been married.	Per cent. of married of total number.
Boston	170	13	7.65
Chicago	557	143	25.67
Cincinnati	382	77	20.16
Louisville	263	70	26.62
New Orleans	167	4	2.40
Philadelphia	100	3	3.00
San Francisco	323	81	25.08

It is hardly worth while to take space at this time to give the occupations of all those who have entered prostitution from the different cities involved.

The number of prostitutes giving information was 3,866, and the following summary exhibits the occupations of this number preceding their entry upon their present life. For this purpose, occupations similar in character have been grouped, and no occupation or group containing less than ten persons has been included in the classification — those numbering under ten being put into the general classification of "various occupations":

Occupation	Number
Actresses, ballet girls, circus performers, singers, etc	52
Bead-trimming makers, embroiderers, lace workers	21
Bookbinderies	18
Bookkeepers, clerks, copyists, stenographers, typewriters, etc.	31
Candy factories	10
Cigarette, cigar, and tobacco factories	78
Corset factories	16
Dressmakers, seamstresses, employés of cloak and shirt factories, button-hole makers, etc.	505
Hairdressers and hair workers	15
House work, hotel work, table work, and cooking	1,155
Laundry work	70
Milliners and hat trimmers	71
No previous occupation (home)	1,236
Nurses (hospital and house), and nurse girls	22
Paper box factories	32
Rope and cordage factories	12
Saleswomen and cashiers	126
Shoe factories	43
Students (at schools or convents)	14
Teachers, governesses, etc	23
Telegraph and telephone operators	11
Textile factories	94
Various occupations	211

The following list shows the character of the more important occupations omitted from the foregoing summary, with the number of women who had been employed in each, and which are included under "various occupations":

> Artificial flower makers, 9; button factories, 9; farm work, 9; canning establishments, 8; necktie makers, 8; housekeepers, 7; straw sewers, 7; hat and cap factories, 6; bag factories, 5; canvassers, 5; clock and watch factories, 5; box factories (wooden), 4; chewing-gum factories, 4; florists, 4; feather curlers and sewers, 4; restaurant, 4.

The preceding figures are exceedingly instructive. By them it will be seen that the largest number coming from any occupation has been taken from those doing house work, hotel work, and cooking; this number, 1,155, being 29.88 per cent. of the whole number comprehended in the statement.

The next largest number, so far as occupation is concerned, ranks with the seamstresses, including the dressmakers, employés of cloak and shirt factories, etc., this number being 505.

A fact which strikes one sadly is the large number who enter prostitution directly from their homes. This number is 1,236, being 31.97 per cent. of the whole number comprehended.

It can not be said, therefore, so far as this investigation shows, that the employés in workshops are to be burdened with the charge of furnishing the chief source whence the ranks of prostitution are recruited.

The experience of the writer in making an examination in many cities, both in the United States and in Europe, sustains the statement, but more strongly than the figures here given, that working women do not recruit the houses of prostitution.

Nor does the investigation show that employers of labor are guilty of reducing their employés to a condition of prostitution, as is often alleged. Only in the rarest cases can one meet with a whisper that this is the case. And these whispers, followed to their source, have rarely disclosed any facts which would lead to the conclusion that employers make bargains based on the loss of character of their employés.

From all that can be learned one need not hesitate in asserting that the working women of the country are as honest and as virtuous as any class of our citizens.

All the facts are against the idea that they are not virtuous women. The statistics given show that a very large percentage of them are living at home. They are living in whatever moral atmosphere there is in their homes.

And it is true that they are not corrupted by their employers, nor do their employers seek to corrupt them. All such impressions originate in the idea that girls can not dress well upon the small wages they receive unless they lead immoral lives, in which they receive pecuniary assistance. But all the testimony that the writer has ever been able to collect upon this point is against such a sentiment, which prevails in too great a degree. The testimony of capable and honest women — of the heads of departments in great stores and millinery establishments and shops, forewomen of shops, the matrons of homes, and of all those best informed and in the best positions to give testimony on this point — is that the working women are as respectable, as moral, and as virtuous as any class of women in the country.

Of course there are exceptions in this class, as in all, but the grand fact must stand out plainly that the working women are not to be burdened with a charge that belongs to others as well as to them, so far as it lies at all.

Working women are not street-walkers. They could not carry on their daily toil and walk the streets too. A captain of police expressed the matter well when he said that people who charge the working women with walking the streets at night for evil purposes do not know what they are talking about. Night-walkers are all of them hardened convicts. The prostitutes, some of them, may have been hard-working women, but no working woman ever walks the streets as a prostitute. This captain said that when a girl falls from virtue she has first to graduate as a "parlor" girl, and then serve some time in a still lower house, before she is hardened enough to take to the streets.

The fact that here and there a case of depravity comes to public attention can not be considered as conclusive evidence that the class to which the depraved case belongs is the cause of the depravity, or that the class itself is depraved.

Virtue and integrity belong to the individual. Either may be stimulated by surroundings, or destroyed by them. But when it is known that women are willing to work from morning till night for the paltry sum of $5 per week, out of which they must, and do, assist their friends, contribute, as a rule, to the general expenses of the household, to a large extent pay for their sewing, and in various ways help on the family, it can not be assumed, with any reasonableness, that they enter lives of prostitution, even in that private way which it is alleged often accompanies their lives. Certainly the houses of prostitution do not contain them.

Many professional prostitutes, when finding a new acquaintance, are very apt to state that they are saleswomen in this or that well-known house. This attracts the victim, and gives him to understand that he is in company with some half-respectable woman, and not in company with a professional prostitute. His conscience, what little he has left, might rebel at associating with a professional prostitute, when it would allow him to continue in the company of a woman in a respectable calling.

The virtuous character of our working women is all the more attractive when the cost of their virtue is recognized. With their poor pay, if they continue virtuous they are the more entitled to our applause, and certainly one must recognize the heroic struggle they make to sustain life, to appear fairly well, and to remove what every honorable-minded man and woman seeks to remove, the appearance of poverty.

All the helps that are thrown around them in our great cities, all the kindness and the care of benevolent employers, all the influences of church and school, must be increased, and these, joined with the heroic efforts of the women themselves, must gradually deepen their characters, strengthen their purposes, and help them to gain a more generous livelihood.

HELEN STUART CAMPBELL

"The Case of Rose Haggerty"

Helen Stuart Campbell (1839–1918), born in upstate New York, moved to New York City with her family at an early age. Well educated in public and private schools, she began writing during the 1860s, publishing children's stories and adult fiction, both magazine pieces and novels. During the late 1870s and early 1880s she became involved in the home economics movement, taught cooking, and wrote a cookbook. Also during this time she became active with a city mission, delineating her experience in *The Problem of the Poor* (1882). Several articles that appeared in the *New-York Tribune* in late 1886 and early 1887 were collected as *Prisoners of Poverty: Women Wage-Workers, Their Trades and Their Lives* (Boston: Roberts Brothers, 1887). "The Case of Rose Haggerty" appeared in the

New-York Tribune, October 31, 1886: 13, and later as part of *Prisoners of Poverty* (18–29), the source of the following text.

"The case of Rose Haggerty." So it stands on the little record-book in which long ago certain facts began to have place, each one a count in the indictment of the civilization of to-day, and each one the story not only of Rose but of many another in like case. For the student of conditions among working-women soon discovers that workers divide themselves naturally into four classes: (1) those who have made deliberate choice of a trade, fitted themselves carefully for it, and in time become experts, certain of employment and often of becoming themselves employers; (2) those who by death of relatives or other accident of fortune have been thrown upon their own resources and accept blindly the first means of support that offers, sometimes developing unexpected power and meeting with the same success as the first class; (3) those who have known no other life but that of work, and who accept that to which they most incline with neither energy nor ability enough to rise beyond a certain level; and (4) those who would not work at all save for the pressure of poverty, and who make no effort to gain more knowledge or to improve conditions. But the ebb and flow in this great sea of toiling humanity wipes out all dividing lines, and each class so shades into the next that formal division becomes impossible, but is rather a series of interchanges with no confinement to fixed limits. Often in passing from one trade to another, chance brings about much the same result for each class, and no energy or patience of effort is sufficient to check the inevitable descent into the valley of the shadow, where despair walks forever hand in hand with endeavor.

This time had by no means come for Rose, with just enough of her happy-go-lucky father's nature to make her essentially optimistic. Born in a Cherry Street tenement-house, she had refused to be killed by semi-starvation or foul smells, or dirt of any nature whatsoever. Dennis Haggerty, longshoreman professionally, and doer of all odd jobs in the intervals of his discharges and re-engagements, explained the situation to his own satisfaction, if not to that of Rose and the five other small Haggertys remaining from the brood of twelve.

"If a man wants his dhrink that bad that no matter what he's said overnight he'd sell his soul by the time mornin' comes for even a thimbleful, he's got jist to go to destruction, an' there's no sthoppin' him. An' I've small call to be blamin' Norah whin she comforts herself a bit in the same manner of way, nor will I so long's me name's Dennis Haggerty. But you, Rose, you look out an' get any money you'll find in me pockets, an' keep the children straight, an' all the saints'll see you through the job."

Rose listened, the laugh in her blue eyes shadowed by the sense of responsibility that by seven was fully developed. She did not wonder that her mother drank. Why not, when there was no fire in the stove, and nothing to cook if there had been, and the children counted it a gala day when they had a scraping of butter on the bread? But, as often happens in these cases, the disgust at smell and taste of liquor grew with every month of her life, and two at least of the children shared it. They were never beaten; for Haggerty at his worst remained good-natured, and when sober wept maudlin tears over his flock and swore that no drop should ever pass his lips again; and Norah echoed every word, and for days perhaps washed and scrubbed and scoured, earning fair wages, and gradually redeeming the clothes or furniture pledged round the corner. Rose went to school when she had anything to wear, and learned in time, when she saw the first symptoms of another debauch, to bundle every wearable thing together and take them and all small properties to the old shoemaker on the first floor, where they remained in hiding till it was safe to produce them again. She had learned this and many another method before the fever which suddenly appeared in early spring took not only her father and mother, but the small Dennis whose career as newsboy had been her pride and delight, and who had been relied upon as half at least of their future dependence. There remained, then, Norah, hopelessly incurable of spinal disease and helpless to move save as Rose lifted her, and the three little ones, as to whose special gifts there was as yet no definite knowledge. In the mean time they were simply three very clamorous mouths to be stopped with such food as might be; and Rose entered a bag-factory a block away, leaving bread and knife and molasses-pitcher by Norah's bed, and trusting the saints to avert disaster from the three experimenting babies. She earned the first month ten dollars, or two and a half a week, but being exceptionally quick, was promoted in the second to four dollars weekly. The rent was six dollars a month; and during the first one the old shoemaker came to the rescue, had an occasional eye to the children, and himself paid the rent, telling Rose to return it when she could. When the ten hours' labor ended, the child, barely fourteen, rushed home to cook something warm for supper, and when the children were comforted and tucked away in the wretched old bed, that still was clean and decent, washed and mended their rags of clothes, and brought such order as she could into the forlorn room.

It was the old shoemaker, a patient, sad-eyed old Scotchman, who also had his story, who settled for her at last that a machine must be had in order that she might work at home. The woman in the room back of his took in shirts from a manufacturer on Division Street, and made often seven and eight dollars a week. She was ready to teach, and in two or three evenings Rose had practically mastered details, and settled that, as she was so young, she would not apply for work in person, but take it through Mrs. Moloney, who would be supposed to have gone into business on her own account as a "sweater." Whatever temptations Mrs. Moloney may have had to make a little profit as "middle-man," she resisted, and herself saw that the machine selected was a good one; that no advantage was taken of Rose's inexperience; and that the agent had no opportunity to follow out what had now and then been his method, and hint to the girl that her pretty face entitled her to concessions that would be best made in a private interview. Shame in every possible form and phase had been part of the girl's knowledge since babyhood, but it had slipped away from her, as a foul garment might fall from the fair statue over which it had chanced to be thrown. It was not the innocence of ignorance, — a poor possession at best. It was an ingrained repulsion, born Heaven knows how, and growing as mysteriously with her growth, an invisible yet most potent armor, recognized by every dweller in the swarming tenement. She had her father's quick tongue and laughing eyes, but they could flash as well, and the few who tried a coarse jest shrunk back from both look and scorching word.

Thus far all went well with the poor little fortunes. She worked always ten and twelve, sometimes fourteen, hours a day, yet her strength did not fail, and there was no dearth of work. It was in 1880, and prices were nearly double the present rates. To-day work from the same establishment means not over $4.50 per week, and has even fallen as low as $3.50. In 1880 the shirts were given out by the dozen as at present, going back to the factory to pass through the hands of the finisher and buttonhole maker. The machine operator could make nine of the best class of shirts in a day of ten hours, being paid for them at the rate of $1.75 per dozen. Four spools of cotton, two hundred yards each, were required for a dozen, the price of which must be deducted from the receipts; but the firm preferred to supply twenty-four-hundred-yard spools, at fifty cents for six-cord cotton used for the upper thread, and thirty cents for the three-cord cotton used as under thread, the present prices for same quality and size being respectively forty-five and twenty-five cents. Making nine a day, the week's wages would be for the four dozen and a half $7.87, or $7.50 deducting thread; but Rose averaged five dozen weekly, and for nearly two years counted herself as certain of not less than thirty dollars per month and often thirty-five. The machine had been paid for. The room took on as comfortable a look as its dingy walls and narrow windows would allow; and Bridget, age five, had developed distinct genius for housekeeping, and washed dishes and faces with equal energy and enthusiasm. She did all errands also, and could not be cheated in the matter of change. She knew where the largest loaves were to be had, and sniffed suspiciously at the packets of tea.

"By the time she's seven, she'll do all but the washing," Rose said with pride, and Bridget reverted to childhood for an instant, and spun round on one foot as she made answer: —

"Shure, I could now, if you'd only be lettin' me."

"There's women on the west side that'll earn $2.50 a dozen, for work no better than you're doing now," some one who had come from that quarter said to her one day, but Rose shook her head. There is a curious conservatism among these workers, who cling to familiar haunts and regard unknown regions with suspicion and even terror.

"I've no time for change," Rose said. "It might not be as certain when I'd got it. I'll run no risks;" and she tugged her great bundle of work up the stairs, rejoicing that living so near saved just so much on expressage, a charge paid by the workers themselves.

There were signs well known to the old hands of a probable reduction of prices, weeks before the first cut came. More fault was found. A slipped stitch or a break in the thread was pounced upon with even more promptness than had been their usual portion. Some hands were discharged, and at last came the general cut, resented by some, wailed over by all, but accepted as inevitable. Another, and another, and another followed. Too much production; too many Jew firms competing and underbidding; more and more foreigners coming in ready to take the work at half price. These reasons and a dozen others of the same order were given glibly, and at first with a certain show of kindliness and attempt to soften harsh facts as much as possible. But the patience of diplomacy soon failed, and questioners of all orders were told that if they did not like it they had nothing to do but to leave and allow a crowd of waiting substitutes to take their places at half rates. The shirt that had sold for seventy-five cents and one dollar had gone down to forty-five and sixty cents respectively, and as cottons and linens had fallen in the same proportion, there was still profit for all but the worker. Here and there were places on Grand or Division Streets where they might even be bought for thirty and forty cents, the price per dozen to the worker being at last from fifty to sixty cents. In the factories it was still possible to earn some approximation to the old rate, but employers had found that it was far cheaper to give out the work; some choosing to give the entire shirt at so much per dozen; others preferring to send out what is known as "team work," flaps being done by one, bosoms by another, and so on.

For a time Rose hemmed shirt-flaps at four cents a dozen, then took first one form and then another of underclothing, the rates on which had fallen in the same proportion, to find each as sure a means of starvation as the last. She had no knowledge of ordinary family sewing, and no means of obtaining such work, had any training fitted her for it; domestic service was equally impossible for the same reason, and the added one that the children must not be left, and she struggled on, growing a little more haggard and worn with every week, but the pretty eyes still holding a gleam of the old merriment. Even that went at last. It was a hard winter. The steadiest work could not give them food enough or warmth enough. The children cried with hunger and shivered with cold. There was no refuge save in Norah's bed, under the ragged quilts; and they cowered there till late in the day, watching Rose as she sat silent at the sewing-machine. There was small help for them in the house. The workers were all in like case, and for the most part drowned their troubles in stale beer from the bucket-shop below.

"Put the children in an asylum, and then you can marry Mike Rooney and be comfortable enough," they said to her, but Rose shook her head.

"I've mothered 'em so far, and I'll see 'em through," she said, "but the saints only knows how. If I can't do it by honest work, there's one way left that's sure, an' I'll try that."

There came a Saturday night when she took her bundle of work, shirts again, and now eighty-five cents a dozen. There were five dozen, and when the $1.50 was laid aside for rent it was easy to see what remained for food, coal, and light. Clothing had ceased to be part of the question. The children were barefoot. They had a bit of meat on Sundays, but for the rest, bread, potatoes, and tea were the diet, with a cabbage and bit of pork now and then for luxuries. Norah had been failing, and to-night Rose planned to buy her "something with a taste to it," and looked at the sausages hanging in long links with a sudden reckless determination to get enough for all. She was faint with hunger, and staggered as she passed a basement restaurant, from which came savory smells, snuffed longingly by some half-starved children. Her turn was long in coming, and as she laid her bundle on the counter she saw suddenly that her needle had "jumped," and that half an inch or so of a band required resewing. As she looked the foreman's knife slipped under the place, and in a moment half the band had been ripped.

"That's no good," he said. "You're getting botchier all the time."

"Give it to me," Rose pleaded. "I'll do it over."

"Take it if you like," he said indifferently, "but there's no pay for that kind o' work."

He had counted her money as he spoke, and Rose cried out as she saw the sum.

"Do you mean you'll cheat me of the whole dozen because half an inch on one is gone wrong?"

"Call it what you like," he said. "R. & Co. ain't going to send out anything but first-class work. Stand out of the way and let the next have a chance. There's your three dollars and forty cents."

Rose went out silently, choking down rash words that would have lost her work altogether, but as she left the dark stairs and felt again the cutting wind from the river, she stood still, something more than despair on her face. The children could hardly fare worse without her than with her. The river could not be colder than this cold world that gave her no chance, and that had no place for anything but rascals. She turned toward it as the thought came, but some one had her arm, and she cried out suddenly and tried to wrench away.

"Easy now," a voice said. "You're breakin' your heart for trouble, an' here I am in the nick o' time. Come with me an' you'll have no more of it, for my pocket's full to-night, an' that's more'n it'll be in the mornin' if you don't take me in tow."

It was a sailor from a merchantman just in, and Rose looked at him for a moment. Then she took his arm and walked with him toward Roosevelt Street.

It might be dishonor, but it was certainly food and warmth for the children, and what did it matter? She had fought her fight for twenty years, and it had been a vain struggle. She took his money when morning came, and went home with the look that is on her face to-day.

"I'll marry you out of hand," the sailor said to her; but Rose answered, "No man alive'll ever marry me after this night," and she has kept her word. She has her trade, and it is a prosperous one, in which wages never fail. The children are warm and have no need to cry for hunger any more.

"It's not a long life we live," Rose says quietly. "My kind die early, but the children will be well along, an' all the better when the time comes that they've full sense for not having to know what way the living comes. But let God Almighty judge who's to blame most — I that was driven, or them that drove me to the pass I'm in."

4

Realism and Beyond

When they go out together, Maggie and Pete frequently attend the theater and see "plays in which the brain-clutching heroine" is "rescued from the palatial home of her guardian, who is cruelly after her bonds, by the hero with the beautiful sentiments." Crane's ironic description of the New York stage during his day reflects his general disgust toward the sentimentality that pervaded popular culture. *Maggie,* like much realistic fiction, challenges sentimental novels in which the heroines are chaste and the heroes honorable. In those kinds of books, the protagonists may experience many trials and tribulations along the way, but in the end, the good are rewarded and the immoral punished. The sentimentalists presented an ideal world whereas the realist writers, recognizing that life was seldom so perfect, sought to create books that depicted life not as it should be, but as it was. While some American realists, most notably William Dean Howells, retained the moral basis of the sentimentalists and continued to reward the good and punish the wicked, others, like Stephen Crane, did not. What sentimentality there is in *Maggie* is present as a satiric target. Showing Maggie's susceptibility to sentimentalism, Crane mocks it and reveals its dangerous consequences for young women who take the world of sentimental fiction for reality. Imagining Pete to be akin to the hero of a popular melodrama, Maggie succumbs to his seduction.

Price, 50 Cents

MAGGIE

A Girl of the Streets

(A STORY OF NEW YORK)

By

JOHNSTON SMITH

Copyrighted

Hamlin Garland's inscribed copy of *Maggie* (1893). Courtesy of the Lilly Library, Indiana University. Crane was significantly influenced by Garland's approach to literature.

The documents included in this section help to situate *Maggie* within the literary milieu of the late nineteenth century. The first set of documents provides a variety of contemporary theoretical discussions treating several literary approaches. The comments of Eugène Véron, though mainly applied to painting and music, also suit literature. Véron emphasized the aesthetic importance of impressionism and stressed that the individual artist should represent reality as he sees it rather than seeking to create a more objective work. Though writing in the late 1860s and 1870s, Véron would not influence American literature until the 1890s. Literary realism dominated the best American novels of the 1870s and 1880s. Literary realism is an umbrella term that covers many other literary trends. Psychological realism, of which Henry James is the undisputed master, sought to capture the world as reflected in the mind. The local color movement, exemplified by such writers as Sarah Orne Jewett and Mary Wilkins Freeman, sought to capture the dialect, settings, and personality of a particular geographical region. Literary impressionism, like artistic impressionism, captured reality in a highly subjective manner. Naturalism, sometimes considered as a subgenre of realism, embodied the idea of determinism, and naturalist novels depict a realistic world that continues on its way with little regard for the people who fall in its path. To a greater or lesser extent, *Maggie* incorporates elements of each of these trends.

Two of the most significant progenitors of American literary realism, William Dean Howells and Henry James, are represented here. In his fiction and criticism, Howells exemplified and advocated a brand of realism that provided objective, true-to-life description containing both the good and the bad, the squalor and the beauty. Howells also held the long-standing notion that literature must instruct as well as delight, and his works have a firm moral basis. The Howells essay "New York Low Life in Fiction" expresses his enthusiasm for Crane's work, yet embodies his conservative moral stance. Though Henry James, for the most part, took the upper classes as his subject, he appreciated Crane's slum novels and, in the essay included here, admits that novelists could choose any subject they wished. In terms of critical approaches to fiction, no one came closer to Crane's than Hamlin Garland, for Garland advocated local color and literary impressionism, a dual approach to literature that he identifies in "The Future of Fiction." With *Maggie,* Crane fulfills Garland's critical principles, for he captures a real-life situation of New York street life, yet he does so using often highly subjective description and carefully selected detail. Rupert Hughes places *Maggie* within the context of other contemporary novels set among the poor.

The second set of documents presents fictional treatments of either the fallen woman or New York tenement life or both. All reflect the desire to treat low life in fiction, yet none is distanced from the sentimental tradition. The first selection comes from *The Two Detectives; or, The Fortunes of a Bowery Girl,* one of hundreds of works by the prolific dime novelist Albert W. Aiken. The selection brings the vernacular of the street life alive with a feisty Bowery girl unafraid to fight men to get her way, but the novel, as a whole, embodies a conservative morality — the girl remains pure, and she is rewarded with a marriage proposal. The excerpts from Harriet Beecher Stowe's and Edgar Fawcett's novels sympathetically address the causes that might lead a woman to prostitution, yet both reflect the sentimentality that prevailed among popular authors. Finally, J. W. Sullivan tells a story about a wedding that accepts conventional morality, yet is stripped of all tradition and sentiment. While helping to situate *Maggie* in its contemporary literary milieu, these documents, in comparison, also reveal Stephen Crane's originality.

Critics and Theorists

EUGÈNE VÉRON

From Aesthetics

L'esthetique by Eugène Véron (1825–1889) first appeared in 1878. The following year, W. H. Armstrong translated the work into English as _Aesthetics_ (London: Chapman and Hall, 1879), the source of the following text (v–xiv). Though largely forgotten now, Véron's work enjoyed a significant contemporary reputation. Before the end of the century, the work had gone through several additional French editions and been translated into Polish and Japanese. Véron's work significantly influenced Hamlin Garland and, through him, Crane. Garland wrote in his copy of Véron, "This work influenced me more than any other work on art. It entered into all I thought and spoke and read for many years after it fell into my hands about 1886" (quoted in Holloway 101). The selection closes with Véron's definition of art, which emphasizes the importance of environment and heredity to artistic production.

Art is nothing but a natural result of man's organization, which is of such a nature that he derives particular pleasure from certain combinations of forms, lines, colours, movements, sounds, rhythms, and images. But these combinations only give him pleasure when they express the sentiments and emotions of the human soul struggling with the accidents of life, or in presence of scenes of nature. The plastic arts, being addressed to the eye, manifest these impressions by the direct and more or less complete representation of objects, forms, attitudes, and of the real or imaginary scenes that they call up. The other arts, which are addressed to the ear, have for their domain, and also for their instruments, the infinite multiplicity of sounds.

The principles, upon which each of the two groups reposes, find their explanation, therefore, in the two sciences pertaining to the study of the organs of sight and hearing — namely, optics and acoustics. The explanation is far from being complete, for a large number of problems still remain unsolved; but from what we already know, we may be allowed to guess at future discoveries. And at least we can indicate the general directions with a great degree of certainty.

The explanation of the cerebral phenomena of what is commonly called the moral influence of art, is not so far advanced, and in most cases we are compelled to content ourselves with pure empiricism. Upon this point Æsthetics is perforce limited to the statement and registration of facts, and to their classification in the order most probable. So far, then, it ceases to be a science in the complete sense of the word.

However, we are able to deduce from the observation of these facts, a principle of the utmost importance; which is, that outside the material conditions that relate to optics and acoustics, that which dominates in a work of art and gives it its special character, is the personality of its author. Ontology[1] disappears to give place to man. The realisation of the eternal and unchanging Beauty of Plato[2] is cast aside. The value of the work of art rests entirely upon the degree of energy with which it manifests the intellectual character and aesthetic impressions of its author. The only rule imposed upon it, is the necessity for a certain conformity with the mode of thinking and feeling of the public to which it appeals. Not that such conformity can itself add to or take from the intrinsic value of the work. It is easy to understand that, in theory, a poem may express sentiments or ideas which, although they

[1] *Ontology:* The study of being.

[2] *Beauty of Plato:* Véron reflects Plato's conception of "forms," the notion that earthly things had a correlative ideal existence.

are incomprehensible to the contemporaries of the author, are not on that account the less worthy of the admiration of some more enlightened period or country. But as a matter of fact, it is certain that such a want of harmony often causes a work to fall rapidly into oblivion.

Happily cases of this nature are very uncommon, and the danger is much less to be feared by the artist than by the thinker. It is very rare, we may even say impossible, for an artist to be much before his time. Without going so far as to admit, as some have done, that he must necessarily be a simple echo, an Æolian harp played upon by every breath of contemporary emotion — it is certain that for a multitude of reasons which we have not space to enumerate, the artist and the poet, above all men, live the life of those among whom they are placed; and consequently it is only in exceptional cases that they are exposed to the danger which we have indicated.

An artist of true feeling has but to abandon himself to his emotion and it will become contagious, and the praise that he deserves will be awarded to him. So long as he shall observe the positive rules that spring from the physiological necessities of our organs, and which alone are certain and definitive, he need never trouble himself about academic traditions and receipts. He is free, absolutely free in his own province, on the one condition of absolute sincerity. He must seek only to express the ideas, sentiments, and emotions proper to himself, and must copy no one.

As there is no such thing as abstract art, *l'art en soi,*[3] because absolute beauty is a chimera,[4] so neither is there any definitive and final system of Æsthetics. All the various formulas by which at various times it has been imprisoned — idealism, naturalism, realism, and such like, are nothing but different ways of looking at art, which is not entirely contained in any one of them. Each of them may recommend itself to certain individual or national temperaments; but it is absurd to force them upon natures to which they are repugnant. It is quite as ridiculous to condemn Flemish or Dutch art in the name of Greek sculpture, as to go through the reverse process, and to refuse all praise to Phidias because he is not Rembrandt.[5] Courbet,[6] too, is legitimate.

[3] *l'art en soi:* Art in itself.

[4] We shall have to demonstrate that the principle of Beauty, absolute or relative, is quite insufficient to account for the complexity of artistic manifestations. [Véron's note.]

[5] *Phidias . . . Rembrandt:* Phidias (born ca. 490 B.C.), Athenian sculptor; Rembrandt Harmen van Rijn (1606–1669), Dutch painter.

[6] *Courbet:* Gustave Courbet (1819–1877), French painter.

We may be allowed to prefer one to the other, according to our natural attraction and affinities; but Æsthetics has no more right to exclude either the one or the other, than we have to import passion and partiality into a question of science.

Is this equal to saying, with certain philosophers, that the freedom of art is the freedom of indifference; that for it one system of direction is as good as another; and that it knows no law but the infinite variety of individual caprice? To answer this question in the affirmative, would be both an exaggeration and a mistake. The artist, as we have said before, lives the life of his own time and country, and so he is naturally led by the inspirations therein existing. Now, in spite of all the changes in human civilization — it is obvious that science, so long retarded by the pursuit of insoluble problems and the ontology of the theologians, has at last transferred its investigation from things of heaven to things of earth. It has substituted the direct study of things, facts, and living beings, for the fantastic explanations of metaphysicians, and of ancient and modern mythology. After wasting century after century in seeking for answers to the enigmas that puzzled it in the actions of gods and imaginary entities, it was obliged, in order to explain the physical and moral world, to take direct account of nature and of man. Man became a perpetual subject of observation for his own sake; and to have given this new direction to the investigation of science, is surely one of the chief glories of the nineteenth century.

Art, also, becomes ever more and more inclined to extend itself in the same direction. It is gradually withdrawing itself further and further from mythology and metaphysics, to which it was faithful so long as civilization set it an example of fidelity. This fact accounts for the ever-growing predominance of expression and of the portrayal of the passions and sentiments, so marked a characteristic of contemporary art. It also explains why landscape painting — that is, the painting of human emotions in the presence of the works of nature — has, for the last forty years, occupied a position of daily increasing importance. The same thing, again, is the cause of the transformation that life and movement have wrought in contemporary sculpture, making Carpeaux and Dalou[7] the chief favourites of the public.

All this is as much as to say that art, always human in its point of departure, which is the manifestation of the ideas and emotions of mankind, became equally so in its subjects and in its final aim. Instead

[7] *Carpeaux and Dalou:* Jean Baptiste Carpeaux (1827–1875) and Jules Dalou (1838–1902), French sculptors.

of representing the forms of the gods, or celebrating them in verse; instead of devoting itself to a symbolism that could never end in anything better than a dry subtlety: it applied itself, with visible effort, to re-enter the pure field of humanity, in which alone from that time it was able to awaken those sympathies without which neither talent nor genius are preserved from oblivion; and which, also, is the only one wherein the artist draws immediately upon the sincere and profound emotions that excite his own desire and power to create.

This movement has, of course, found an energetic and violent opponent in tradition, which, with us, possesses peculiar power on account of the organization of our academies and of our official teaching. This retrograde force exercises a most fatal influence over our art progress, especially as those of us who are subject to it, are, for the most part, unconscious of its existence. Young and without any philosophic education, these unconscious students find the schools and their surroundings impregnated with a multitude of academic prejudices, that taking hold of them, stereotype their ideas before they have ever thought in earnest about such things, or have formed any personal convictions. They unwittingly become enlisted from the very first in the official phalanx; and it is only the exceptionally independent and powerful intellects that are able either to resist this pressure at the beginning, or to escape from it at a later period. Our aim, then, is to denounce as strongly as our opportunities permit, this crushing of the future under the past, of liberty under dogmatism. We refuse to be bound by the narrow and antiquated rules which frustrate every attempt at emancipation; we repudiate the haughty and contemptuous criticism that, under pretext of protecting "good taste and calm doctrines," succeeds in discouraging every attempt at independence — a defensive criticism, which is, as M. Cuvillier-Fleury termed it, nothing but the open tyranny of academic doctrine and jealous impotence.

Our intention is to defend and uphold in every possible way the thesis which M. Eugène Viollet-le-Duc[8] has taken for the text of all his writings; namely, that without independence we can have neither art nor artists.

All the great art epochs have been epochs of liberty. In the time of Pericles as in that of Leo. X.,[9] in the France of the thirteenth century as

[8] *Eugène Viollet-le-Duc:* Eugene Emmanuel Viollet-Le-Duc (1814–1879), French architect and writer on archeology.

[9] *Pericles . . . Leo. X:* Pericles (ca. 495–429 B.C.) was an Athenian statesman; Giovanni d'Medici (1475–1521) served as Pope Leo X from 1513 until his death.

in the Holland of the seventeenth, artists were able to work after their own fancies. No æsthetic dogmas confused their imaginations, no official corporations claimed any art dictatorship, or thought themselves responsible for the direction taken by the national taste.

In these great epochs, Art was truly national. Men's intellects, when left to follow their own devices, naturally worked out the particular kinds of art with which they had most sympathy; or rather they found them without search, by their own spontaneous movement, without other guide or rule than the instinctive preferences of the race as a whole.

This peculiar similarity between instincts when left to themselves, explains the close sympathy that subsists between the works produced by different men during the great periods of art; whilst, at the same time, freedom is made manifest by the characteristic whose place nothing else can supply — namely, individual originality.

Men whose lives belong to the same period are generally influenced by the same set of facts. The sources of inspiration afford but little variety. Sometimes a single idea or sentiment is impressed upon a whole generation. But each man interprets it after his own fashion, after the fulness of his own personal inspiration, and according to the measure of his own genius.

This is the source of the infinite variety in unity — variety of expression in unity of sentiment — which is the mark of certain periods. In fact, the artist is never more powerful or more inspired, than when he finds himself in perfect accord with the age in which he lives; and art is never greater, than when it marches with the ideas and sentiments that influence a whole condition of society.

Now this universality of art and of artistic sentiment, at a certain moment in their intellectual evolution among the great majority of nations, is of the most capital importance in the history of the manifestations of human intelligence. The Egyptians, Assyrians, Greeks, Chinese, and Japanese, all possessed spontaneous forms of art, which, springing from the inmost feelings of the nation, have the appearance of being equally understood and appreciated by every individual of the race. Something of the same kind is to be found in France during the Middle Ages, and in Italy during the period of the Renaissance.

That such a statement should be true of so many different races of mankind, cannot be attributed to mere chance. Chance is far too convenient an explanation; besides, it has the disadvantage of really explaining nothing. Chance is not even an hypothesis; it is a mere negation. There is no chance in history. Every event, small or great, is but part of a continuous chain. Some of the links may escape our notice, but, nevertheless, the chain exists.

Art, considered from a psychological point of view, is nothing but the spontaneous expression of certain conceptions of things, which follow logically from the combination of the moral and physical influences to which different races are subject, with the original or acquired tendencies and aptitudes of each separate race.

It is an interpretation of the sentiments to which this *mélange* gives birth; a more or less literal, or more or less ideal interpretation, according as the nations in question give the first place to the material reality of things or to the habits and predilections of the race. But, whatever the result of such mixture may be, it is certain that the two primitive elements, reality and personality, are never wanting — in spite of the contrary theories that would reduce art either to the condition of photographic plagiarism, or to mere conjectural restoration of so-called ideal types.

HENRY JAMES

From The Art of Fiction

Nearly three decades older than Stephen Crane, Henry James (1843–1916) established his reputation as one of America's most important men of letters with several early works of fiction, including two extraordinary stories that took the American woman as their subject matter, *Daisy Miller* (1878) and *The Portrait of a Lady* (1881). While James preferred to set his own works among the European upper class, he was not necessarily averse to novels set in the tenement districts. As James states in the following essay, the novelist had no preset guidelines. He need not write from his own experience, but could choose whatever subject he wished. This essay is a response to an article by the prominent British man of letters Walter Besant, who maintained the long-standing notion that a literary work needed a moral basis. James eloquently answered Besant, arguing that the novel did not need a conscious moral purpose. Besant's and James's essays were published together in book form as *The Art of Fiction* (Boston: Cupples, Upham and Company, 1885), the source of the following selection (60–70). While living in England during the last years

of his life, Crane and James became good friends. James frequently bicycled to Crane's home at Brede Place, where the two engaged in lively literary conversations. They were on familiar enough terms that Crane once played a practical joke on James, soaking his firewood in chemicals to create unusual pyrotechnic effects.

A novel is in its broadest definition a personal impression of life; that, to begin with, constitutes its value, which is greater or less according to the intensity of the impression. But there will be no intensity at all, and therefore no value, unless there is freedom to feel and say. The tracing of a line to be followed, of a tone to be taken, of a form to be filled out, is a limitation of that freedom and a suppression of the very thing that we are most curious about. The form, it seems to me, is to be appreciated after the fact; then the author's choice has been made, his standard has been indicated; then we can follow lines and directions and compare tones. Then, in a word, we can enjoy one of the most charming of pleasures, we can estimate quality, we can apply the test of execution. The execution belongs to the author alone; it is what is most personal to him, and we measure him by that. The advantage, the luxury, as well as the torment and responsibility, of the novelist, is that there is no limit to what he may attempt as an executant — no limit to his possible experiments, efforts, discoveries, successes. Here it is especially that he works, step by step, like his brother of the brush, of whom we may always say that he has painted his picture in a manner best known to himself. His manner is his secret, not necessarily a deliberate one. He cannot disclose it, as a general thing, if he would; he would be at a loss to teach it to others. I say this with a due recollection of having insisted on the community of method of the artist who paints a picture and the artist who writes a novel. The painter *is* able to teach the rudiments of his practice, and it is possible, from the study of good work (granted the aptitude), both to learn how to paint and to learn how to write. Yet it remains true, without injury to the *rapprochement,* that the literary artist would be obliged to say to his pupil much more than the other, "Ah, well, you must do it as you can!" It is a question of degree, a matter of delicacy. If there are exact sciences there are also exact arts, and the grammar of painting is so much more definite that it makes the difference.

I ought to add, however, that if Mr. Besant says at the beginning of his essay that the "laws of fiction may be laid down and taught with as much precision and exactness as the laws of harmony, perspective, and proportion," he mitigates what might appear to be an over-statement by applying his remark to "general" laws, and by expressing most of these rules in a manner with which it would certainly be unaccommodating to disagree. That the novelist must write from his experience, that his "characters must be real and such as might be met with in actual life;" that "a young lady brought up in a quiet country village should avoid descriptions of garrison life," and "a writer whose friends and personal experiences belong to the lower middle-class should carefully avoid introducing his characters into Society;" that one should enter one's notes in a commonplace book; that one's figures should be clear in outline; that making them clear by some trick of speech or of carriage is a bad method, and "describing them at length" is a worse one; that English Fiction should have a "conscious moral purpose;" that "it is almost impossible to estimate too highly the value of careful workmanship — that is, of style;" that "the most important point of all is the story," that "the story is everything"— these are principles with most of which it is surely impossible not to sympathize. That remark about the lower middle-class writer and his knowing his place is perhaps rather chilling; but for the rest, I should find it difficult to dissent from any one of these recommendations. At the same time I should find it difficult positively to assent to them, with the exception, perhaps, of the injunction as to entering one's notes in a common place book. They scarcely seem to me to have the quality that Mr. Besant attributes to the rules of the novelist — the "precision and exactness" of "the laws of harmony, perspective, and proportion." They are suggestive, they are even inspiring, but they are not exact, though they are doubtless as much so as the case admits of; which is a proof of that liberty of interpretation for which I just contended. For the value of these different injunctions — so beautiful and so vague — is wholly in the meaning one attaches to them. The characters, the situation, which strike one as real, will be those that touch and interest one most, but the measure of reality is very difficult to fix. The reality of Don Quixote or Mr. Micawber[1] is a very delicate shade; it is a real-

[1] *Don Quixote or Mr. Micawber:* Miguel de Cervantes Saavedra (1547–1616) published *Don Quixote* in 1605. Mr. Micawber is a character from Charles Dickens's *David Copperfield* (1849–50).

ity so colored by the author's vision that, vivid as it may be, one would hesitate to propose it as a model; one would expose one's self to some very embarrassing questions on the part of a pupil. It goes without saying that you will not write a good novel unless you possess the sense of reality; but it will be difficult to give you a recipe for calling that sense into being. Humanity is immense, and reality has a myriad forms; the most one can affirm is that some of the flowers of fiction have the odor of it, and others have not; as for telling you in advance how your nosegay should be composed, that is another affair. It is equally excellent and inconclusive to say that one must write from experience; to our suppositious aspirant such a declaration might savor of mockery. What kind of experience is intended, and where does it begin and end? Experience is never limited, and it is never complete, it is an immense sensibility, a kind of huge spider-web, of the finest silken threads, suspended in the chamber of consciousness and catching every air-borne particle in its tissue. It is the very atmosphere of the mind; and when the mind is imaginative — much more when it happens to be that of a man of genius — it takes to itself the faintest hints of life, it converts the very pulses of the air into revelations. The young lady living in a village has only to be a damsel upon whom nothing is lost to make it quite unfair (as it seems to me) to declare to her that she shall have nothing to say about the military. Greater miracles have been seen than that, imagination assisting, she should speak the truth about some of these gentlemen. I remember an English novelist, a woman of genius, telling me that she was much commended for the impression she had managed to give in one of her tales of the nature and way of life of the French Protestant youth. She had been asked where she learned so much about this recondite being, she had been congratulated on her peculiar opportunities. These opportunities consisted in her having once, in Paris, as she ascended a staircase, passed an open door where, in the household of a *pasteur*, some of the young Protestants were seated at table round a finished meal. The glimpse made a picture; it lasted only a moment, but that moment was experience. She had got her impression, and she evolved her type. She knew what youth was, and what Protestantism; she also had the advantage of having seen what it was to be French; so that she converted these ideas into a concrete image, and produced a reality. Above all, however, she was blessed with the faculty which when you give it an inch takes an ell, and which for the artist is a much greater source of strength than any accident of residence or of place in the social scale. The power to guess the unseen from the seen, to trace the implication of things, to judge the whole piece by the pattern, the condition of feeling life, in general, so completely that you are well on your way to knowing any particular corner of it — this cluster of gifts may almost be said to constitute experience, and they occur in country and in town, and in the most differing stages of education. If experience consists of impressions, it may be said that impressions *are* experience, just as (have we not seen it?) they are the very air we breathe. Therefore, if I should certainly say to a novice, "Write from experience, and experience only," I should feel that this was a rather tantalizing monition if I were not careful immediately to add, "Try to be one of the people on whom nothing is lost!"

I am far from intending by this to minimize the importance of exactness — of truth of detail. One can speak best from one's own taste, and I may therefore venture to say that the air of reality (solidity of specification) seems to me to be the supreme virtue of a novel — the merit in which all its other merits (including that conscious moral purpose of which Mr. Besant speaks) helplessly and submissively depend. If it be not there, they are all as nothing, and if these be there they owe their effect to the success with which the author has produced the illusion of life. The cultivation of this success, the study of this exquisite process, form, to my taste, the beginning and the end of the art of the novelist. They are his inspiration, his despair, his reward, his torment, his delight. It is here, in very truth, that he competes with life; it is here that he competes with his brother, the painter, in *his* attempt to render the look of things, the look that conveys their meaning, to catch the color, the relief, the expression, the surface, the substance of the human spectacle. . . . I cannot imagine composition existing in a series of blocks, nor conceive, in any novel worth discussing at all, of a passage of description that is not in its intention narrative, a passage of dialogue that is not in its intention descriptive, a touch of truth of any sort that does not partake of the nature of incident, and an incident that derives its interest from any other source than the general and only source of the success of a work of art, — that of being illustrative. A novel is a living thing, all one and continuous, like every other organism, and in proportion as it lives will it be found, I think, that in each of the parts there is something of each of the other parts. The critic who over the close texture of a finished work will pretend to trace a geography of items will mark some frontiers as artificial, I fear, as any that have been known to history. There is an old-fashioned distinction between the novel of character and the novel of incident, which must have cost many a smile to the intending romancer who was keen about his work. It appears to me as little to the point as the equally celebrated distinction between the novel and the romance — to answer as little to any reality. There are bad novels and good novels, as there are bad pictures and good pictures; but that is the only distinction in which I see any meaning, and I can as little imagine speaking of a novel of character as I can imagine speaking of a picture of character. When one says picture, one says of character; when one says novel, one says of incident, and the terms may be transposed. What is character but the determination of incident? What is incident but the illustration of character? What is a picture or a novel that is *not* of character? What else do we seek in it and find in it? It is an incident for a woman to stand up with her hand resting on a table and look out at you in a certain way; or, if it be not an incident, I think it will be hard to say what it is. At the same time it is an expression of character. If you say you don't see it (character in *that* — *allons donc!*) this is exactly what the artist, who has reasons of his own for thinking he *does* see it, undertakes to show you. When a young man makes up his mind that he has not faith enough, after all, to enter the church, as he intended, that is an incident, though you may not hurry to the end of the chapter to see whether perhaps he does n't change once more. I do not say that these are extraordinary or startling incidents. I do not pretend to estimate the degree of interest proceeding from them, for this will depend upon the skill of the painter. It sounds almost puerile to say that some incidents are intrinsically much more important than others, and I need not take this precaution, after having professed my sympathy for the major ones, in remarking that the only classification of the novel that I can understand is into the interesting and the uninteresting.

HAMLIN GARLAND

The Future of Fiction

Few contemporary authors more significantly influenced Crane's early writing than Hamlin Garland (1860–1940). Born in Wisconsin, Garland moved with his family to Iowa and later to the Dakota Territory. Dissatisfied with farm life, yet unable to find other satisfying work, Garland

embarked on a program of self-education, reading, among others, Hippolyte Taine, Herbert Spencer, and Henry George. In the late 1880s, he began writing short stories set in the Middle Border, which he collected as *Main-Travelled Roads* (1891). Its publication marked a new advance in literary realism. In his critical writings, Garland began to define his approach to literature, which he called veritism. Stephen Crane met Garland during the summer of 1891, and by the next summer the two had become good friends, for both shared similar notions about literature. Indeed, Crane better fulfilled Garland's literary theories than Garland himself. The following essay, which appeared in *Arena* (April 1893): 513–24, contains the kernel of Garland's literary theory. He would expand his ideas the following year in *Crumbling Idols.* Eager to make a career out of writing, Garland retreated from avant-garde theories to write more conservative, popular, traditional works. His subsequent books did not achieve the quality of *Main-Travelled Roads* until he abandoned fiction for autobiography and produced a series of reminiscences starting with *A Son of the Middle Border* (1917).

It is interesting to observe that all literary movements in the past had little or no prevision. The question of their future, their permanence, did not disturb them. My reading does not disclose to me that the affectation of Euphues, or the glittering allegorical and bloodless pageantry of Spenser, or the thunderous mouthing of Marlowe[1] ever grew aware of its dark future. Each school lived for its day and time without disturbing prophecy.

Pope,[2] the monarch of the circumscribed, the emperor of lace and ruffles, so far as I have read, had no gloomy forebodings. His reign was the most absolutely despotic and long-continued reign the literary history of England has ever seen. He could be pardoned for never imagining that real flowers could come to be enjoyed better than gilt and scarlet paper roses — all alike. It is not to be wondered at that he

[1] *Euphues . . . Spenser . . . Marlowe:* John Lyly (1554?–1606), *Euphues* (1578–1580), a prose romance famous for its ornate style and excessive use of rhetorical tropes, especially antithesis and alliteration. Edmund Spenser (ca. 1552–1599), English poet. Christopher Marlowe (1564–1593), English dramatist and poet.

[2] *Pope:* Alexander Pope (1688–1744), British poet who exemplified the neoclassical style of the Augustan Age through the use of heroic couplets.

had no prevision of Whitman or Henley or Lanier,[3] in the joyous jog-trot of his couplets.

Take larger movements — the Reformation[4] for example. This movement in its day filled the whole religious history of Europe. It filled the horizon from sea to sea. It transformed empires, and planted new colonies in the wilderness west. It dominated art, literature, architecture, laws, and yet it was but a phase of intellectual development. Its order was transitory; and had an evolutionist been born into that austere time, he would have predicted the reaction to enjoyment of worldly things and the sure passing of the whole world as it was then colored and dominated by puritanic thought.

In art this narrowness and sincerity of faith in itself has been the principal source of power of every movement in the past. To question was to weaken. Had Spenser suspected the prosiness and hollow absurdity of his combats (wherein the hero always wins), had he suspected that there was something else in life better worth while than allegory and the endless recounting of tales of chivalry, he would have failed to embody as he does the glittering and caparisoned barbarism of his day. And the crown which Pope wore would have rested like a plat of thorns on his brow had he been visited by disturbing visions of a time when men would prefer their poetry in some other form than couplets or even quatrains, and would even question whether Pope wrote poetry at all!

Because Shakespeare and the group around him were feudalistic and did not believe in the common personality, because Dryden[5] believed Shakespeare was a savage, because Wordsworth[6] believed that God was in the round rim of ocean and in the wildling breeze, because each age believed in its art and in the world of thought around it, therefore has each real age of literature embodied more or less faith-

[3] *Whitman or Henley or Lanier:* Walt Whitman (1819–1892), American poet; W. E. Henley (1849–1903), British poet; Sidney Lanier (1842–1881), American poet. All three rejected neoclassical prosody.

[4] *Reformation:* The great Protestant religious movement of the sixteenth century whose leaders included John Calvin (1509–1564) and Martin Luther (1483–1546).

[5] *Dryden:* John Dryden (1631–1700) and Alexander Pope were the two foremost neoclassical poets in British literary history. Dryden significantly revised Shakespeare's plays to suit the tastes of his own era.

[6] *Wordsworth:* William Wordsworth (1770–1850) was the pioneer of the British Romantic poets.

fully its own outlook upon life, and gone peacefully, if not arrogantly, to its grave at last, in blessed ignorance of the devastation the future held in store.

But while each age can be said in general terms to have had no prevision, it is probable that some few of its artists in each generic movement caught some glimpse of coming change, and that this power of prophecy grew slowly until there came upon the world the splendid light of the development theory uttered by Spencer and Darwin. I think it is not too much to say that, previous to the writing of these men, definite prevision, even on the broadest lines, was impossible.

Until men came to see system and progression, and endless but definite succession in art and literature as in geologic change, until the law of progress was enunciated, no conception of the future, and no reasonable history of the past, could be formulated. Once prove literature and art subject to social conditions, to environment and social conformation, and dominance of the epic in one age and the drama in another became as easy to understand and to infer as any other fact of a people's history. It has made the present the most critical and self-analytical of all ages known to us.

Evolutionists explain the past by means of laws operative in the present, by survivals of change. In an analogous way, we may infer (broadly, of course) the future of society, and therefore its art, from embryonic changes just beginning to manifest themselves. The developed future is always prophesied in the struggling embryos of the present. In the mould of the present are the swelling acorns of future forests.

Fiction already commands the present in the form of the novel of life. It already out-ranks verse and the drama as a medium of expression. It is so flexible, admits of so many points of view, and comprehends so much, uniting painting, and rhythm to the drama and the pure narrative, that it has come to be the highest form of expression in Russia, Germany, Norway, and France. It occupies with easy tolerance the high seats in the synagogue, and felicitates the other arts on having got in, or rather stayed in at all. At its best it certainly is the most modern and unconventional of arts.

One of the questions most often asked the veritist is the question of his permanence. "Do you think you've reached the farther wall? Is your school final?"

Certainly not, the veritist replies. As students of comparative literature and of the development theory, we know perfectly well that the movement called realism, in so far as it expresses our characteristic outlook on life, will change, will become history like the Shakespearean literature, like the classic literature of Pope, like the romantic school of Scott and Hugo. If we are sincere and direct, we are making enduring literary history in precisely the same manner that Burns did in voicing the rising democracy of his age. All are links in an endless chain, forms in an endless procession. There is no farther wall; nothing but space before and after us.

All things are relative in literature and art as in science. We are relativists as a matter of fact, and not absolutists. We believe the phrase "writing for all time" is a figure of speech, for time is long, and art is fleeting; that is, in its special phases. We do not say Scott was greater than Shakespeare; we say he was *different,* and that his only justification for being is that he contributes something to English literature which Shakespeare failed to perceive and utter.

The phrase "Shakespeare and Eschylus[7] rise into the realms of the absolute in art," is excellent in an oration, but the relativists demand that phrases of this sort be boiled down to their solid residuum. Shakespeare was great among his fellows because he excelled them in work of the same time and place. He was valuable and interesting to all ages following because he embodied so well the life and thought of his time. Had he done his work with keener perception and with less regard for traditions, he would have been greater, because he would thus have embodied more of the loves and aspirations of his fellow-men and less of the intrigues and crimes of the crowned brigands, whose lives were crimes, whose deaths were public blessings, and he would have been greater, and his relationship to democratic America closer than it is.

This is said, not so much to cavil, as to illustrate the relativity of the schools of art. This question also involves the question of the immediate future of fiction. If the present novel is to change, what is it likely to be? If the development of literature involves the sloughing off of certain peculiarities, what will be carried forward from the present to the future? Is there anything permanent in a literature? These are the questions which pour in on the rash relativists who lay any claim to prophecy. Having entered in so far, the veritist might as well proceed boldly.

We are about to enter the dark. We need a light. This flaming thought from Whitman will do for the search-light of the profound

[7] *Eschylus:* Aeschylus (525–456 B.C.), Greek tragic dramatist.

deeps, "All the past was not, the future will be." If the past was bond, the future will be free.

If the past was feudalistic, the future will be democratic. If the past ignored and trampled upon women, the future will place them side by side with man. If the child of the past was ignored, the future will cherish him, and fiction will embody these facts.

If the past was dark and battleful and bloody and barbarous, the future will be peaceful and sunny. If the past celebrated lust and greed and love of power, the future will celebrate continence and humility and altruism. If the past was the history of a few titled personalities riding high on obscure waves of nameless suffering humanity, the future will be the day of high average personality, the abolition of all privilege, the peaceful walking together of brethren, equals before nature and before the law. And fiction will celebrate this life.

If the past was gross and materialistic in its religion, worshipping idols of wood and stone, demanding sacrifices to appease God, using creed as a club to make men conform to a single interpretation of man's relation to nature and his fellow, then the future will be high and pure and subtle in its religious interpretations; and there will be granted to individuals perfect freedom in the interpretation of nature's laws, a freedom in fact, as well as in name. And to fiction is the task of subtly embodying this splendid creed.

As we run swiftly over the development of literary history, we see certain elements being left behind while others are carried forward. Those that are carried forward are, however, extremely general and fundamental. They are the bones of art, not its curve of flesh, or flush of blood.

One of these central elements of unchanging power, always manifest in every really great literature, is sincerity in method. This produces contemporaneousness. Those great writers of the past did not write "for all time," not even for the future. They mainly were occupied in interesting some portion of their fellow-men. Shakespeare had no care and no thought of the eighteenth century in his writing; he was painfully anxious to please my Lord This or That, who could be of living and very material use to him, — or to the Queen, who could help him keep the wolf from his door. Witness his abject dedications.

He studied his time, and tried sincerely to state it in terms that would please those whom he considered his judicious friends. Thus he reflected (indirectly) the feudal age, for that was the dominant thought of his day. So Dryden and Pope, each at his best, portrayed his day, putting his sincere and original comment upon the life around him, flavoring every translation he made with the vice and lawlessness which he felt to be the prevailing elements of his immediate surroundings. In the main, they believed in themselves.

The romantic school of fiction, while it reigned, was self-justifiable, at least in great figures like Scott and Hugo. Because it was a sincere expression of their likings and dislikings, it reflected directly and indirectly their rebellion against the old, and put in evidence their conception of the office of literature. They did their work. It will never be done so well again, because all that follows their model will be imitative; theirs was the genuine romanticism.

The fiction of the future will not be romantic in any such sense as Scott or Hugo was romantic, because to do that would be to re-live the past, which is impossible; to imitate models, which is fatal. Reader and writer will both be wanting. The element of originality follows from the power of the element of sincerity. "All original art," says Taine,[8] "is self-regulative." It does not imitate. It does not follow models. It stands before life, and is accountant to life and self only. Therefore the fiction of the future must be original, and therefore self-regulative.

As fiction has come to deal more and more with men and less with abstractions, it will be safe to infer that this will continue. Eugène Véron covered the ground fully when he said, "We care no longer for gods or heroes; we care for men." This is true of the body of veritists, whose power and influence augments daily; even the romance writers feel its influence, and are abandoning their swiftly running love stories for studies of character. Like the romantic school of painting they are affected by the influence they fear.

It is safe to say that the fiction of the future will grow more democratic in outlook and more individualistic in method. Impressionism in its true sense means the statement of one's own individual perception of life and nature, guided by devotion to truth. Second to this great principle is the law that each impression must be worked out faithfully on separate canvasses, each work of art complete in itself. Literalism, the style that can be quoted in bits, is like a picture that can be cut into pieces. It lacks unity. The higher art would seem to be the art that perceives and states the relations of things, giving atmosphere and relative values as they appeal to the sight.

[8] *Taine:* Hippolyte Taine (1829–1893), French philosopher and literary historian.

The fiction of the future will not be so obvious in its method as it has been in the past. It will put its lessons into general effect rather than into epigrams. Discussion will be in the relations of its characters, not on quotable lines or paragraphs. It will teach, as all earnest literature has done, by effect; but it will not be by direct expression, but by placing before the reader the facts of life as they stand related to the artist.

Destructive criticism is the most characteristic literary expression of the present and of the immediate future, because of this slow rising of the literary mind to prevision of change in life and literature. Because the fictionist of to-day sees a more beautiful and peaceful future social life, and in consequence a more beautiful and peaceful literary life, therefore he is encouraged to deal truthfully and at close grapple with the facts of his immediate present. His comment virtually amounts to satire or prophecy or both. Because he is sustained by love and faith in the future, he can be mercilessly true. He strikes at thistles because he knows the unrotted seed of loveliness and peace needs but sun and the air of freedom to rise to flower and fragrance.

The realist or veritist is really an optimist, a dreamer. He sees life in terms of what it might be, as well as in terms of what it is; but he writes of what is, and suggests what is to be, by contrast. He aims to be perfectly truthful in his delineation of his relation to life, but there is a tone, a color, which comes unconsciously into his utterance like the sobbing stir of the muted violins beneath the frank, clear song of the clarionet; and this tone is one of sorrow that the future halts so lamely in its approach. He aims to hasten the age of beauty and peace by delineating the ugliness and warfare of the present. But ever the converse of his picture rises in the mind of the reader. He sighs for a lovelier life.

This element of sad severity will change as conditions change for the common man, but the larger element of sincerity, with resulting contemporaneousness, will remain. Fiction, to be important and successful, must be original and suited to its time. As the times change, fiction will change. This must always be remembered.

The surest way to write for all time is to embody the present in the finest form with the highest sincerity and with the frankest truthfulness. The surest way to write for other lands is to be true to our own land and true to the scenes and people we love, and love in a human and direct way without being educated up to it or down to it.

Thus it will be seen that the fiction of the immediate future will be the working out of plans already in hand. There is small prophecy in it after all. We have but to examine the ground closely, and we see the green shoots of the coming harvest beneath our very feet. We have but to examine closely the most naive and local of our novels, and the coming literature will be foreshadowed there. The local novelist seems to be the coming — woman! Local color is the royal robe.

The local novel seems to be the heir-apparent to the kingdom of poesy. It is already the most promising of all literary attempts to-day; certainly it is the most sincere. It seems but beginning its work. It is "hopelessly contemporaneous"; that is its strength. It is (at its best) unaffected, natural, emotional. It is sure to become all powerful. It will redeem American literature, as it has already redeemed the South from its conventional and highly wrought romanticism.

The fiction of the South has risen from the dead. It is now in the spring season of shooting wildling plants and timorous blades of sown grains. Its future is assured. Its soil is fertilized with the blood of true men. Its women are the repositories of great, vital, sincere, emotional experiences which will inevitably appear in their children, and at last in art, and especially in fiction. The Southern people are in the midst of a battle more momentous than the rebellion, because it is the result of the rebellion; that is, the battle of entrenched privilege against the swiftly spreading democratic idea of equality before the law and in the face of nature.

They have a terrible mightily dramatic race-problem on their hands. The South is the meeting-place of winds. It is the seat of swift and almost incalculable change; and this change, this battle, this strife of invisible powers, is about to enter their fiction; the negro has entered it. He has brought a musical speech to his masters, and to the new fiction. He has brought a strange and pleading song into music. The finest writers of the New South already find him a never-failing source of interest. He is not, of course, the only subject of Southern fiction, nor even the principal figure; but he is a necessary part, and a most absorbingly interesting part. The future of fiction in the South will also depict the unreconstructed rebel unreservedly, and the race-problem without hate or contempt or anger, for the highest art will be the most catholic in its sympathy. It will delineate vast contending forces, and it will be a great literature.

The negro will enter the fiction of the South first as subject, second as artist in his own right. His first attempts will be imitative, but he will yet utter himself, as surely as he lives. He will contribute a poetry and a novel as peculiarly his own as the songs he sings. He may appear also in a strange half-song, half-chant, and possibly in a drama peculiar to himself; but in some form of fiction he will surely utter the sombre and darkly florid genius for emotional utterance which characterizes them.

In the North the novel will continue local for some time to come. It will delineate the intimate life and speech of every section of our enormous and disparate republic. It will catch and fix unchangeably the changing, assimilating races, delineating the pathos and humor and the infinite drama of their swift adjustment to new conditions. California, New Mexico, Idaho, Utah, Oregon, each wonderful locality in our Nation of Nations will yet find its native utterance. The superficial work of the tourist and outsider will not do. The real novelist of these sections is walking behind the plow or trudging to school in these splendid potential environments.

This local movement will include the cities as well, and St. Louis, Chicago, San Francisco, will be delineated by artists born of each city, whose work will be so true that it could not have been written by any one from the outside. The real utterance of a city or a locality can only come when a writer is born out of its intimate heart. To such an one nothing will be "strange" or "picturesque"; all will be familiar and full of significance and beauty. The novel of the slums must be written by one who has played there as a child, and taken part in all its amusements; not out of curiosity, but out of pleasure seeking. It cannot be done from above nor from the outside. It must be done out of a full heart and without seeking for effect.

The contrast of city and country everywhere growing sharper, will find its reflection in this local novel of the immediate future, the same tragedies and comedies, with the essential difference called local color, and taking place all over the land, wherever cities arise like Fungii, unhealthy, yet absorbing as subjects of fictional art.

The drama will join the novel in this study of local conditions. The indications are already to be seen in the dramas of strong local flavor, now being put upon the stage. New England, Kentucky, California, Alabama, Virginia, have already received serious, if not altogether truthful representation. Others are to swiftly follow. They will be derived from fiction, and in many cases the dramatist and novelist will be the same person. In all cases the sincerity of the author's love for his scenes and characters will find expression in tender care for truth, and there will be made to pass before our eyes wonderfully suggestive pictures of other lives and landscapes. The drama will grow in dignity and importance along these lines.

Both drama and novel will be colloquial. This does not mean that they will be exclusively in the dialects, but the actual speech of the people of each locality will unquestionably be studied more closely than ever before. Dialect is the life of a language, precisely as the common people of the nation form the sustaining power of its social life and art.

And so in the novel, in the short story, and in the drama, — by the work of a multitude of loving artists, not by the work of an over-topping personality, will the intimate social individual life of the nation be depicted. Before this localism shall pass away, such a study will have been made of this land and people as has never been made by any other age or social group. A literature from the plain people, reflecting their unrestrained outlook on life. Subtle in speech and color, humane beyond precedent, humorous, varied, simple in means, lucid as water, searching as sunlight.

To one who believes each age to be its own best interpreter, the idea of "decay of fiction" never comes. That which its absolutist takes for decay is merely change. The conservative fears change, the radical welcomes it. The conservative tries to argue that fundamentals cannot change; that they are the same yesterday, to-day, and to-morrow. If that were true, then a sorrowful outlook on the future would be natural. Such permanency would be death. Life means change.

As a matter of fact, the minute differentiations of literature which the conservative calls its non-essentials, are really its essentials. Vitality and growth are in these "non-essentials." It is the difference in characters, not their similarity, which is forever interesting. It is the subtle coloring which individuality gives, that vitalizes landscape art, and so it is the subtle differences in the interpretation of life which each age gives, that vitalizes its literature and makes it its own.

The business of the present is not to express fundamentals, but to sincerely present its own minute and characteristic interpretation of life. This point cannot be too often insisted upon. Unless a writer add something to the literature of his race, has he justification? Is there glory in imitation? Is the painter greatest who copies old masters, or paints with their mannerisms?

Youth disdains barriers. He bursts from the wig and cloak of his grandfathers. He repels, perhaps, a little too brusquely, the models which conservatism points to with awe. He respects them as history, he repels them as models, for has he not life, abounding, fresh, contiguous life? Life that stings and smothers and overwhelms and exalts, like the salt, green, snow-tipped ocean surf; life, with its terrors and triumphs right here and now; its infinite drama, its allurement, its battle, and its victories. Life is the model, truth is the master, the heart of the man himself his motive power. The desire to create in the image of nature is the artist's unfailing reward.

To him who sees that difference is the vitalizing quality, not similarity, there is no sorrow at change. The future will take care of itself. In the space of that word "difference" lies all the infinite range of future art. Some elements are comparatively unchanging. The snow will fall, spring will come, men and women love, the stars will rise and set, and grass return again and again in vast rhythms of green, but society will not be the same. The physical conformation of our nation will change. It will lose its wildness, its austerity. It will become a garden where now the elk and the mountain lions are. The physical and mental life of men and women will be changed, the relation of man to man, and man to woman will change in detail, and the fiction of the future will express these changes.

To the veritist, therefore, the present is the vital theme. The past is dead and the future can be trusted to look after itself. The young men and maidens of that time will find the stars of their present brighter than the stars of '92; the people around them more absorbing than books, and their own outlook on life more reasonable than that of dead men. Their writing and painting, in proportion to its vitality and importance, will reflect this their natural attitude toward life and history.

The fiction of the future will be great in its mass of its minutiæ, humane in feeling, and hopeful in outlook. Above all else it will be sincere, this fiction of the future, and independent, but not disdainful of all past models. It will re-create, which is the office of all fiction. It will be self-cognizant, but not self-conscious, and it will be self-justifiable, as every really great literary expression has been and must ever be.

RUPERT HUGHES

"The Justification of Slum Stories"

While scarcely as famous as his nephew, the multimillionaire Howard Hughes, Rupert Hughes (1872–1956) deserves recognition as an author and literary critic. After attending Adelbert College (later Case Western

Reserve University), from which he graduated in 1892 and received his M.A. in 1894, Rupert Hughes became an editor of *Godey's Magazine* and contributed several articles under the pseudonym "Chelifer," including the following review essay *Godey's* 131 (October 1895): 430–32. In addition to *Maggie,* the essay considers Arthur Morrison's *Tales of Mean Streets* (1895), Henry W. Nevinson's *Neighbors of Ours: Slum Stories of London* (1895), J. W. Sullivan's *Tenement Tales of New York* (1895), and Edward W. Townsend's *Chimmie Fadden, Major Max, and Other Stories* (1895) and *Chimmie Fadden Explains* (1895). Reviewing *Active Service* and *The Monster and Other Stories* in 1900, Hughes traced his enthusiasm for Crane to the first edition of *Maggie:* "That Stephen Crane is a genius I have been convinced ever since I read that fatherless yellow-covered book: 'Maggie, a Tale of the Streets, by Johnston Smith.' To this belief I have clung in spite of many jolts and jars, for the retainer of this author must hang on like the watcher in the crow's nest of a ship, which plows splendidly forward, but with much yawing and buffeting and many a careen" (quoted in *Log 420*).

The highest office of the writer of fiction is the education of human sympathy. To widen, deepen, refine, mellow, generalize, particularize, stimulate, in one word, to educate the brotherly and sisterly feelings of mankind, is the sacred and Christly priesthood of the story-teller. The unpardonable literary sin of omission is the failure to entertain; the most heinous positive offence is the mismanagement of the reader's sympathy. Herein is the real responsibility of a writer of fiction, for there is no sillier heresy than the loud-mouthed anarchy that art has nothing to do with morals. If a novelist defends thievery, he is a criminal; if he stirs up sympathy for a certain thief, that is quite a different matter; if he glorifies licentiousness, he is Sir Pandarus;[1] if he becomes the apologist of a sore-tried backslider, he is following the hallowed footsteps of Him who said, "Neither do I condemn thee." The artist need not shackle himself with a Blue-book moral code. I can't see why every novel should be written with one eye on "the young person," whose chief innocence is ignorance. But the artist must have some

[1] *Sir Pandarus:* In both Chaucer's *Troilus and Criseyde* (ca. 1385) and Shakespeare's *Troilus and Cressida* (1609), Pandarus serves a go-between for the two lovers.

code of morals, and must feel responsible to it. Literary ethics is a principle necessary to valuable fiction.

One of the greatest fallacies among critics of high degree is the denial to tales of low life of all right to existence. They whine, "We have sorrow enough and to spare. Why show us the dark side of those things that we could never put an end to if we tried? Art is only meant to amuse." They are not consistent enough to stick to Joe Miller and the "Pickwick Papers,"[2] but they languish back to old romances or grasp at latter-day pseudohistorical novels and claim to be exactly satisfied. Then if fair maidens wail their souls out, or if heroic knights gasp in deep dungeons, or if fate goes hard with anyone in doublet and hose, they shed a melodious tear and purr as they weep. Yet other enemies of low life stick to society stories and revel in the pitiful existence of high-born dames mismated. No woe is too dire, so long as the characters are correct in attire, but the moment the scene changes and slouchy laborers are the Sirs Launcelot, and Queen Guenevere is a factory-girl, all is vulgar, plebeian, too heart-rending to be read. Even the comic side of low life is not bright enough, and humanity is no longer humanity, but a race divided into two species, society and simianity. Of course no one can object to stories of high life, but the whole loaf is better than half the bread.

Literature is the greatest of all democratizing forces. The liberation of man from despotries and serfdom has found its most tireless and skilful architect in literature. It will not stop now with the foundation of a few republics and constitutional monarchies. The freedom of man is not yet consummated. Nor will it be till the boundary-lines between the castes of to-day, social if not electoral, are blended into indistinguishable gradations of congeniality and luck in accumulating money. When Mrs. Crœsus realizes in the depth of her soul that the only difference between her daughter and her maid is the individual equation and certain accidental effects of environment; when Crœsus, Jr., acknowledges to himself that the young street-car driver is in all essentials his equal, and may twenty years hence occupy a higher social plane and a nobler place in his country — when these two consummations are

[2] *Joe Miller and the "Pickwick Papers": Joe Miller's Jests* had been a popular jest book for centuries, and the phrase "Joe Miller" had become a proverbial synonym for a jokester. Charles Dickens's *The Posthumous Papers of the Pickwick Club* first appeared serially in 1836 and 1837.

reached — mote I be there to see! — then actual democracy will assume the earth. It is inevitable that deep students of humanity should feel a keen sympathy for the denizens of "low" life, and that they should send novels, like tracts, out a-proselyting. A critic is only trying to dam Niagara with a goose-feather when he resists the tendency.

It is usually taken as one of the premises of devotion to beautifully purposeless art, that all stories of low life must be harrowing, pessimistic, dank with despair. Critics should keep off these premises. The sun shines for all. *Cantabit vacuus coram latrone viator.*[3] Down in the tenements laughter is not an unheard thing. Sometimes it is punctuated with hiccups, sometimes it is more boisterously crude than the wont of Fift' Avnoo; but it is just as enjoyable and more hearty. In these recent studies of low life this is especially noticeable. If you want comedy, genial, soul-warming humor, and rib-prodding wit, read Arthur Morrison's "That Brute Simmons," "The Red Cow Group," and "Squire Napper." Read the cockney repartee sparkling all through Henry W. Nevinson's "Neighbors of Ours." The polished, strictly correct steel of Henry James, the social chatter of Anthony Hope,[4] are no sharper or more enlivening wit than these homely duels. A dagger's a dagger whether it has diamonds in its hilt or an unchased blade.

A geranium in a tenement window is far more effectively beautiful than the same flower in a conservatory. So a bit of poetical feeling in a story of slum life is a rich sensation to jaded palates. Like gold smiling in the jagged glitter of quartz, they are the fairer for their contrast. So it would be hard to find in our literature a more rapturous and appealing prose poem than Nevinson's "The 'St. George' of Rochester," in which a rough old tar gasps out on his dying-bed the one romance of his life, a wild, sweet love-making with a betrayed girl of the upper classes who fled to his boat to hide herself from the world. And I can't imagine anything finer than his attempt to describe her:

"Oh, my soul! she was a reg'lar beauty, was that woman. Some'ow from first to last she always put me in mind of my spinnaker. P'r'aps it was through the spinnaker bein' so light and kind o' dainty, it bein' made to catch any breath of draught as might be, and left clean and

[3] *Cantabit vacuus coram latrone viator:* The traveler with empty pockets will sing in the robber's face.

[4] *Anthony Hope:* pseudonym of Sir Anthony Hope Hopkins (1863–1933), British author.

white, not smeared over with ochre and oil same as the other sails, as is all red and brown and 'eavy. And then she'd bend and curve this way and that, for all the world like the spinnaker when the wind's 'avin' a bit of a game with it, for all 'er bein' as tough as a steel-wire stay. And mind you, it's always the spinnaker as snaps the topmast through bracin' it for'ard like a whip. And I've never see the man, little or big, but that woman could 'a' done just whatever she 'ad a mind to with 'im."

To run upon such crystal beauty of thought in a slum story is like the rapture of a dive into the cold sea on a parching day. I think Browning calls it "the cool, silver shock of a plunge in a pool's living water."[5] Nevinson's book is full of the tenderer emotions of these people he welcomes as "neighbors of ours." He has strung ten stories on a slender chain of continuity. They are all told by one self-effacing cockney, and though the book is by no means one story, it contains occasional reappearances of characters. Mr. Nevinson displays very little dramatic feeling for climaxes and suspended "situations," but he gives the not unsatisfactory substitute of a genial, gossipy style, alight with good humor and hopeful sympathy even in the saddest moments. There is a constant flower-decking of poetical moods, too, that brightens the hardness of the life he describes. The best thing about these bits of color is the naturalness with which they bloom in the cockney jargon.

In some respects Mr. Morrison's "Tales of Mean Streets" is a greater work, though it lacks the warmth of color and geniality of Mr. Nevinson's writings. The stories are told with such a bold, free vigor, however, and the wit and humor are so grim and stolid and merciless, that the book is *sui generis*.[6] There is no lack of tenderness and sympathy with these people, among whom Mr. Morrison spent many years as secretary of a charity trust (witness the infinitely pathetic "Without Visible Means" and the kindliness of "All that Messuage"); but a certain sternness, even in humor, is most characteristic. "On the Stairs," for instance, is a story in which an old crone, eager to give her sick boy a fine funeral, adds to her savings the money given by a desperate young doctor to buy him necessary medicine. Of its grisly sort, this

[5] *Browning . . . water"*: Robert Browning (1812–1889), "Saul," lines 71–72: "The strong rending of boughs from the fir-tree, the cool silver shock / Of the plunge in a pool's living water . . ."

[6] *sui generis:* A unique type in itself, usually applied to literature, most of which can by classified by genres.

story is a horrible masterpiece. "All that Messuage," a more gentle tragedy, details the hardships of a laborer who thinks the perfection of ease is attained when his long savings enable him to buy and rent out a house. It shows the same fanatic hunger for a decent burial among wretches whose lives have known nothing approaching the ostentation of comfort. The somewhat longer story, "Lizerunt," is both comic and tragic, and each to a notable degree. I have already spoken of "That Brute Simmons," which is one of the funniest stories in the language. All the other tales in the book are of exceptional merit, and this one small volume must place Morrison immediately among the sparse group of the best English writers. His style has a curious mingling of archaisms and modern slang, and the close relation between the two is thus illustrated most vividly.

I regret to see that America cannot show the equal of these two books in its own studies of low life. But the cosmopolitan population of the country should offer the best field in the world, and doubtless competent tillers will not long be wanting.

Mr. J. W. Sullivan has written a volume of "Tenement Tales of New York," which contains much well-selected, well-observed material. The style in which they are told, however, is so baldly crude, there is so little verisimilitude in the talk of the characters and such complete lack of color and chiaroscuro in the narration, that they cannot be granted high value. Occasional bits of philosophizing are the best things in the book.

No better proof of the comedy possible in works on low life could be asked for than the writings of Mr. Edward W. Townsend, whose "Chimmie Fadden" has become a household familiar, and whose vivid language has infected the nation. Of course "Chimmie" figures principally in a scenery of high life, but the character-drawing is perfect and there are not a few visits to that seventh paradise, the "Bow'ry." Any writer who realizes the picturesqueness and charm of such a character deserves all he can realize on it.

But probably the strongest piece of slum writing we have is "Maggie," by Mr. Stephen Crane, which was published some years ago with a pen-name for the writer and no name at all for the publishers. But merit will out, and the unclaimed foundling attracted no little attention, though by no means as much as it deserves. The keenness of the wit, the minuteness of the observation, and the bitterness of the cynicism resemble Morrison's work. The foredoomed fall of a well-meaning girl reared in an environment of drunkenness and grime is told with great humanity and fearless art, and there is a fine use of contrast in the conclusion of the work, where the brutal mother in drunken sentimentality is persuaded with difficulty to "forgive" the dead girl whom she compelled to a harsh fate by the barren cruelty of home-life.

The subjects chosen by all these writers compel an occasional plainness of speech which may give a shock to spasmic prudishness, but there is nothing to harm a healthy mind, and they all should have the effect of creating a better understanding and a wiser, more active sympathy for the unfortunates who must fill the cellar of the tenement we call life. To do this is far better even than to be artistic.

WILLIAM DEAN HOWELLS

From "New York Low Life in Fiction"

William Dean Howells (1837–1920) deserves his reputation as the grand old man of American letters. Born the year Nathaniel Hawthorne published *Twice-Told Tales* and dying the year F. Scott Fitzgerald published *This Side of Paradise*, the life of Howells, the most vocal proponent of literary realism, spanned the period from romanticism to modernism. Though he witnessed the birth of modernism, he did not appreciate it fully. His attitude toward literature, that it be a faithful rendering of real life with a firm moral basis, changed little throughout his professional career as magazine editor, literary critic, and novelist. Howells is especially important for his role as mentor to a generation of younger writers, including Hamlin Garland and Stephen Crane. He was so well respected among contemporaries that his approval or disapproval could make or break a young writer. After reading the 1893 edition of *Maggie*, he immediately became a supporter of Crane's, promoting the book in his *Harper's Weekly* column. When *The Red Badge of Courage* appeared in 1895, Howells read it yet preferred *Maggie*. He used the publication of the 1896 Appleton edition of *Maggie* and the publication of *George's Mother* as the basis for an essay that appeared in the *New York World*, July 26, 1896: 18, from which the following has been excerpted. In modified form, the essay appeared as "An Appreciation" in the first British edition of *Maggie* (1896). British reviewers found Howells's enthusiasm for Crane's book a little overbearing. The *St. James's Gazette* (October 7, 1896), for ex-

ample, said that Howells's appreciation "is held like a pistol to the head of the timid British reviewer" (5).

It is a long time since I have seen the once famous and popular play "A Glance at New York,"[1] but I distinctly recall through the misty substance of some forty-five very faded years the heroic figures of the volunteer fireman and his friends, who were the chief persons of the piece. I do not remember the others at all, but I remember Mose, and Sikesy, and Lize. Good and once precious fragments of the literature linger in my memory, as: " 'Mose,' says he, 'git off o' dem hose, or I'll swat you over der head wid der trumpet.' And I didn't get off o' der hose, and he did swat me over der head wid der trumpet." Other things have gone, things of Shakespeare, of Alfieri,[2] of Cervantes, but these golden words of a forgotten dramatist poet remain with me.

I.

It is interesting to note that the first successful attempt to represent the life of our streets was in dramatic form. Some actor saw and heard things spoken with the peculiar swagger and whopperjaw utterance of the b'hoy of those dreadful old days, when the blood-tubs and the plug-uglies reigned over us, and Tammany[3] was still almost purely American, and he put them on the stage and spread the poison of them all over the land, so that there was hardly anywhere a little blackguard boy who did not wish to act and talk like Mose.

The whole piece was painted with the large brush and the vivid pigments of romanticism, and yet the features were real. So it was many long years later when Mr. Harrigan came to the study of our low life in

[1] *"A Glance at New York":* Benjamin A. Baker's *A Glance at New York* (1848) introduced the character of Mose the Bowery b'hoy. The figure quickly entered the oral culture, and Mose became America's first urban folk hero. Though the folktype had fallen from popularity in Stephen Crane's day, the character of Pete echoes the legendary Bowery b'hoys.

[2] *Alfieri:* Vittorio Alfieri (1749–1803), Italian tragic dramatist and poet.

[3] *Tammany:* Tammany Hall was a political organization founded in 1788 in New York City. Through the nineteenth century it grew in power and became increasingly dominated by Irish immigrants. Under the leadership of John Kelly during the 1870s and 1880s, it became a disciplined political machine that influenced New York City politics at all levels.

his delightful series of plays. He studied it in the heyday of Irish supremacy, when Tammany had become almost purely Celtic, and he naturally made his heroes and heroines Irish. The old American b'hoy lingered among them in the accent and twist of an occasional barkeeper, but the brogue prevailed, and the highshouldered, sidelong carriage of the Americanized bouncer of Hibernian blood.

The treatment, however, was still romanticistic, though Mr. Harrigan is too much of a humorist not to return suddenly to nature, at times from the most exalted regions of "imagination." He loves laughing and making laugh, and that always saved him when he was in danger of becoming too grand, or fine, or heroic. He had moments when he was exactly true, but he allowed himself a good many friendly freedoms with the fact, and the effect was not always that of reality.

It seemed to me that so far as I could get the drift of a local drama in German which flourished at one of the East Side theatres a winter ago, that the author kept no more faithfully to life than Mr. Harrigan, and had not his sublime moments of absolute fidelity. In fact, the stage is almost as slow as criticism to perceive that there is no other standard for the arts but life, and it keeps on with the conventional in motive even when the matter is honest, apparently in the hope that by doing the stale falsehood often enough it will finally affect the witness like a fresh verity. It is to the honor of the stage, however, that it was first to recognize the value of our New York low life as material; and I shall always say that Mr. Harrigan, when he was not overpowered by a tradition or a theory, was exquisitely artistic in his treatment of it. He was then true, and, as Tolstoi has lately told us, to be true is to be moral.

II.

The fiction meant to be read, as distinguishable from the fiction meant to be represented, has been much later in dealing with the same material, and it is only just beginning to deal with it in the spirit of the great modern masters. I cannot find that such clever and amusing writers as Mr. Townsend, or Mr. Ralph, or Mr. Ford[4] have had it on their

[4] *Mr. Townsend, or Mr. Ralph, or Mr. Ford:* Edward Waterman Townsend (1855–1942), *Chimmie Fadden, Major Max, and Other Stories* (1895); Julian Ralph (1853–1903), *People We Pass: Stories of Life among the Masses of New York* (1896); and James L. Ford (1854–1928), *Dolly Dillenbeck: A Portrayal of Certain Phases of Metropolitan Life and Character* (1895).

consciences to report in the regions of the imagination the very effect of the life which they all seem at times to have seen so clearly. There is apparently nothing but the will that is wanting in either of them, but perhaps the want of the will is the want of an essential factor, though I should like very much to have them try for a constant reality in their studies; and I am far from wishing to count them out in an estimate of what has been done in that direction. It is only just to Mr. Stephen Crane, however, to say that he was first in the field where they made themselves known earlier. His story of "Maggie, a Girl of the Streets," which has been recently published by the Appletons, was in the hands of a few in an edition which the author could not even give away three years ago; and I think it is two years, now, since I saw "George's Mother," which Edward Arnold has brought out, in the manuscript.

Their present publication is imaginably due to the success of "The Red Badge of Courage," but I do not think that they will owe their critical acceptance to the obstreperous favor which that has won. As pieces of art they are altogether superior to it, and as representations of life their greater fidelity cannot be questioned. In "The Red Badge of Courage" there is a good deal of floundering, it seems to me. The narration repeats itself: the effort to imagine, to divine, and then to express ends often in a huddled and confused effect; there is no repose, such as agony itself assumes in the finest art, and there is no forward movement. But in these other books the advance is relentless; the atmosphere is transparent; the texture is a continuous web where all the facts are wrought with the unerring mastery of absolute knowledge. I should say that "The Red Badge of Courage" owed its excellence to the training the author had given himself in setting forth the life he knew in these earlier books of later publication. He learned to imagine vividly from seeing clearly.

There is a curious unity in the spirit of the arts; and I think that what strikes me most in the story of "Maggie" is that quality of fatal necessity which dominates Greek tragedy. From the conditions it all had to be, and there were the conditions. I felt this in Mr. Hardy's "Jude,"[5] where the principle seems to become conscious in the writer; but there is apparently no consciousness of any such motive in the author of "Maggie." Another effect is that of an ideal of artistic beauty which is as present in the working out of this poor girl's squalid

[5] *Mr. Hardy's "Jude":* Thomas Hardy (1840–1928), *Jude the Obscure* (1895).

romance as in any classic fable. This will be foolishness, I know, to the foolish people who cannot discriminate between the material and the treatment in art, and who think that beauty is inseparable from daintiness and prettiness, but I do not speak to them. I appeal rather to such as feel themselves akin with every kind of human creature, and find neither high nor low when it is a question of inevitable suffering, or of a soul struggling vainly with an inexorable fate.

My rhetoric scarcely suggests the simple terms the author uses to produce the effect which I am trying to report again. They are simple, but always most graphic, especially when it comes to the personalities of the story: the girl herself, with her bewildered wish to be right and good; with her distorted perspective; her clinging and generous affections; her hopeless environments; the horrible old drunken mother, a cyclone of violence and volcano of vulgarity; the mean and selfish lover, a dandy tough, with his gross ideals and ambitions; her brother, an Ishmaelite[6] from the cradle, who, with his warlike instincts beaten back into cunning, is what the b'hoy of former times has become in our more strenuously policed days. He is indeed a wonderful figure in a group which betrays no faltering in the artist's hand. He, with his dull hates, his warped good-will, his cowed ferocity, is almost as fine artistically as Maggie, but he could not have been so hard to do, for all the pathos of her fate is rendered without one maudlin touch.

So is that of the simple-minded and devoted and tedious old woman who is George's mother in the book of that name. This is scarcely a study at all, while Maggie is really and fully so. It is the study of a situation merely: a poor, inadequate woman, of a commonplace religiosity, whose son goes to the bad. The wonder of it is the courage which deals with persons so absolutely average, and the art that graces them with the beauty of the author's compassion for everything that errs and suffers. Without this feeling the effects of his mastery would be impossible, and if it went further or put itself into the pitying phrases it would annul the effects. But it never does this; it is notable how in all respects the author keeps himself well in hand. He is quite honest with his reader. He never shows his characters or his situations in any sort of sentimental glamour; if you will be moved by the sadness of common fates you will feel his intention, but he does not flatter his portraits of people or conditions to take your fancy.

[6] *Ishmaelite:* Outcast.

In George and his mother he has to do with folk of country origin as the city affects them, and the son's decadence is admirably studied; he scarcely struggles against temptation, and his mother's only art is to cry and to scold. Yet he loves her, in a way; and she is devotedly proud of him. These simple country folk are contrasted with simple city folk of varying degrees of badness. Mr. Crane has the skill to show how evil is greatly the effect of ignorance and imperfect civilization. The club of friends, older men than George, whom he is asked to join, is portrayed with extraordinary insight, and the group of young toughs whom he finally consorts with is done with even greater mastery. The bulldog motive of one of them, who is willing to fight to the death, is most impressively rendered.

The Fallen Woman and Slum Fiction

ALBERT W. AIKEN

From The Two Detectives; or, The Fortunes of a Bowery Girl

Albert W. Aiken (ca. 1846–1894) was born in Boston. By his early twenties, he was writing plays and performing on stage. His play *The Witches of New York* was produced in 1869 and again in 1872, with Aiken performing in the second production. He published the story of the play in the *Saturday Journal* as "Orphan Nell, the Orange Girl" under the pseudonym "Agile Penne." He later expanded it as "Royal Keene, the California Detective; or, The Witches of New York" in 1872. Aiken temporarily retired from the stage in 1881 to devote himself to writing nickel and dime novels, but by 1885, he was back on stage performing in New York. He also ran a theater in Brooklyn called Aiken's Museum. Filled with dialogue and loaded with action, his numerous works reflect his dramatic experience. The following selection is from *The Two Detectives; or, The Fortunes of a Bowery Girl* (New York: Beadle & Adams, 1884), a work that originally appeared as *The Detective's Ward* (1871).

The Girl of the Streets

"Don't you dare to strike me!"

A girlish voice high in anger and fierce in determination.

The scene, an underground drinking-saloon known as "The Dive," situated on the Bowery, not a dozen blocks from Canal street, in the great city of New York; the time, night; the hour, twelve, and the actors in the scene — we will describe them.

In the center of the saloon, which was but a common basement fitted up with a bar and half a dozen small tables, stood a girl about sixteen years of age. She was slight in figure, with a pale face, lit up by great black eyes, that now were flashing bright with angry fires. Great masses of silken hair, black as the diamonds of the Pennsylvania mines and soft as the fleece of the merino, gathered in a simple knot at the back of her well-shaped head.

The face of the girl was white with passion; her bosom was heaving tumultuously, and the warm breath came quickly through the dilated nostrils. The full red lips, almost perfect in their beauty, were firmly shut together.

One passion alone swayed all her nature — anger.

Within a few feet of the girl stood the person to whom she had addressed her passionate warning. It was a man — an Italian, as one gifted in reading nationalities in the face would have guessed at once. The olive complexion, full black eyes, and crisp, curly hair of inky hue, told his race.

The Italian was a man of forty-five, dressed roughly, and an evil look lurked in the lines of his dark face.

Now his swarthy features were convulsed with anger, and his hand was raised, as if to strike the girl to his feet.

Two persons alone, besides the girl and the Italian, were in the saloon. One a woman, an Italian like the man, who stood behind the bar, leaning her elbows on the counter, and gazing upon the angry pair in the center of the room, with an expression of careless unconcern upon her olive-tinged features. The other was a man of that peculiar class, common to the great metropolis, and whom the world places under the generic head of "rough."

This man was sitting on a corner of one of the tables, swinging his legs carelessly, and smoking a cigar. He was dressed in a flashy light suit; a heavy chain — looking remarkably like gold, although it wasn't — dangled over his vest. In his ruffled shirt-bosom gleamed a huge pin; had it been a diamond, a "cool" thousand dollars would not have bought it; but as it was only an imitation, a ten-dollar bill had paid for it.

The rough had a coarse, brutal face; bulldog — that expresses it. A thick nose, broken evidently by some heavy blow; sinister-looking eyes, an ugly gray in color; coarse black hair, cropped tight to his head; a gigantic mustache, rusty black in hue, half-concealing the thick-lipped, sensual mouth — and you have the pen picture of Mr. Richard Hill, better known to his intimate friends — and the police — as Rocky Hill; a bully and a blackguard of the first water — a bright and shining light among the shoulder-hitters of Gotham.

"Why I no strike you, eh?" angrily demanded the Italian, who was called Giacomo, and was the proprietor of the little den known as "The Dive." By long custom, the Bowery boys had abbreviated the name of the saloon-keeper into "Jocky."

"Because if you do, it will be the worst blow you ever struck in all your life, *you* bet!" replied the girl defiantly.

"Better look out, Jocky — she's red hot!" cried the rough, who was enjoying the display of temper, as he would have enjoyed a dog fight or anything else brutal.

"You one cussed beggar!" exclaimed the Italian, gesticulating wildly. "You no do vat I wish? — *diavolo!* I will kill you dead!"

"I'm no beggar, Jocky!" returned the girl, in a passion. "I work hard for every crust of bread that you give me, you old miser! I *won't* be struck any more. I don't care what I do; the cops can take me as soon as they like. I'll give 'em something to take me for, too, if you go for to strike me again. I'm all black and blue now. I'd just as lief be dead as stop here with you. Who gave you the right to beat me? You ain't my father; I never had no father, and I don't care much!"

"Oh, you imp of ze devil!" cried the Italian in wrath. "I picked you out of ze mud-gutter, bring you up like a lady; give you beautiful clothes, and you no do as I tole you!"

"I ain't a-goin' to steal for nobody!" cried the girl quickly. "You dress me like a lady! Beautiful clothes! — this *is* a gay dress, *this* is!" And the girl surveyed the ragged gown that she wore, in contempt. "See here, Jocky; I've always acted square with you — I never went back on you; I always give you fair, just what I sell. I never say that I lost some of my money, like the rest of the girls do. All I ask is decent treatment. I ain't a dog, to be banged about; I wish I was a dog, sometimes — then I'd run away."

"Why don't you run away now?" asked Rocky, with a leer upon his brutal face.

"Where could I run to?" cried the girl, desperately. "Wouldn't Jocky, here, run after me and bring me back? There's only one thing for me to do."

"What's that?" asked Rocky.

"Jump into the dock. I'd do it, too, if I wasn't such a coward. Maybe I will, soon, for I ain't a-going to stand such a life as this much longer," and the girl sighed heavily as she spoke.

"Oh, you are ze imp of ze devil!" cried the Italian. "Why you no do as I tell you, eh?"

"I won't be a thief for anybody!" cried the girl. "Ain't it bad enough for to make me tramp the streets all day and nearly all night selling your mean shoe-strings, and hair-pins and buttons, without trying to make me do something that'll send me to the Tombs and up to the Island?[1] Maybe it would be better for me to go there; I'd be out of your reach anyway. But first and last, I *won't* steal for you nor nobody else!"

"Oh, you're a sweet one, you are!" exclaimed Rocky, in a tone expressive of the highest contempt. "Why don't you preach us a sermon now? Why, we ought to go right down on our blessed knees and worship such an angel as you are. Oh, my! ain't you cutting it fat, or nothing! You're giving us altogether too much pork for a shilling. Just think, Jocky, she's a-cutting up all this rumpus, 'cos I told her for to just slyly slip a bundle out of a woman's basket as she was a-follerin' on behind. Nobody would have see'd her; but she's a virtuous kid, she is!"

"Did I no tell you, you must do as Rocky say, eh?" cried the Italian, approaching still nearer to the girl with upraised hand.

The girl did not shrink from him in the least.

"I told you that I wouldn't and I didn't!" she replied, defiantly, her face plainly showing the angry passions raging in her heart.

"That's so!" cried Rocky.

"You no mind me, beggar, eh?"

"No!"

Like angry tigers, the two glared at each other with flaming eyes — the muscular, swarthy-faced man of forty, and the slight, pale-faced girl of sixteen.

Rocky looked on in delight; the woman leaning on the counter — the wife of Jocky — with unconcern.

[1] *Tombs . . . the Island:* The Tombs was the common name for the men's prison in Manhattan, patterned after an Egyptian tomb. Another house of correction was located on Blackwell's Island.

"Hi, hi!" ejaculated the rough, "why this is as good as a the-a-ter; oncore, *oncore!*" and he clapped his hands together in huge delight.

"You mud-gutter imp! did I not look out for you since you was a little child, so high as my knee? and now you no do what I want?" cried the Italian, foaming at the mouth with rage, and the big veins on his forehead and throat purple with angry blood.

"Oh, you've done a great deal for me, you have, you bet!" exclaimed the girl, contemptuously. "Ever since I could walk, I've worked all I knew how for you. I've earned every bit of bread that I've put in my mouth, twice over. And what have I ever got from you, except just enough to keep life in me? a gay life it has been, too!" and the girl laughed bitterly. "But now, I'm tired of being beaten; I'm too old for that, and don't you dare to strike me again! I'll work for you; work as hard as I know how to; but I won't steal for you. I don't know much, but I do know that it ain't right, and I won't do it."

"You no do it, eh?"

"No, I won't, Jocky; it's played out!" cried the girl, firmly.

The child of the streets used the language of the class who had surrounded her from childhood. It was more forcible than elegant.

"*Diavolo!* I kill you, some," exclaimed the Italian, making a terrible blow at her, that, had it fallen on the girl, would surely have felled her senseless to the floor. But the street life of the orphan had made her as quick as a cat. Anticipating the blow, she dodged under the arm of the Italian, and as he was carried past her by the force of his blow, she turned quickly and struck him with all the force of her clinched fist.

The blow took the Italian just under the right ear, and sent him reeling across the room, despite his size and weight.

Nerved as she was to desperation, the girl's strength was doubled.

"Bully for you!" yelled Rocky, in delight. "Round first! the old 'un gets a hot 'un under the ear. Round two, come to scratch, Jocky; time!"

The Italian staggered across the room, impelled by the violence of the blow he had received from the determined arm of the girl, until he brought up against the wall; that alone prevented him from falling.

Half-stunned by the blow, for it had lighted on a tender spot, Jocky felt of his neck in wonder. He could hardly realize that the slight figure of the girl could command strength to deal such a stroke.

"*Diavolo!*" the Italian cried, in a rage.

"Time!" yelled Rocky, in glee; "come up smiling, Jocky, or I'll throw up the sponge!"

Then the Italian seemed suddenly to understand what had happened. He drew a long, glittering knife from his bosom, and darted toward the girl.

In "The Dive"

With the keen-edged knife glittering in his hand, the Italian, mad with rage, rushed toward the girl.

The rough, seated on the table, and the woman, leaning on the bar, looked on calmly, without stirring a finger to protect the girl, or to save her from the death that seemed so near.

But as the Italian struck at her, with the quickness of a cat, she jumped to one side, thus evading the murderous stroke. And as the Italian turned, as if to repeat the rush, she caught up a chair, which stood near at hand. With a strength which one would not have suspected to have dwelt in her slight frame, she whirled the chair over her head and brought it down with terrible force upon the Italian.

Jocky threw up his arms to guard his head. The force of the blow hurled him headlong to the floor; but little injured, though, for his arms had saved his skull.

The girl then retreated a few steps, still grasping the chair in her hands — still prepared for another attack.

"Set 'em up ag'in," cried Rocky, in huge delight. "Round two — the old 'un goes to grass. Round three — *time!*"

But the Italian slowly rose to his feet, and showed but little inclination to again renew the attack. The desperation of the girl astonished him.

"*Diavolo!* you have broken my head!" he cried, in anger.

"Why don't you let me alone now?" exclaimed the maid, still keeping herself prepared for another assault — her face deadly white, and the full, red lips shut tightly together.

"Time, Jocky!" shouted the rough; "you ain't a-goin' to 'give it up so, Mr. Brown,' are you?"

"You better let me alone!" the girl cried, her eyes flashing, and her whole manner showing the desperation born of despair.

"Go for her, Jocky!" Rocky exclaimed. "Are you going for to let a girl back you down? Pretty sort of a rooster, *you* are! I wouldn't bet my stamps on you, nohow." The rough was disgusted, and expressed his feelings in his tone.

"Put down ze chair, you imp of ze devil!" the Italian cried, cautiously advancing toward the still defiant girl.

"I won't," she answered. "I give you fair warning that, if you attempt to strike me, I'll hit you with it again." In the eye of the girl the Italian read that she would keep her word, or, at least, attempt to do so.

"Come, Jocky, the audience is a-gettin' impatient!" exclaimed the rough. "If you ain't a-goin' to put up your bunch of fives, you'd better throw up the sponge and quit to onc't. I'd be ashamed for to have a girl back *me* down, I would!"

"If you no put down ze chair, *diavolo!* I will kill you!" cried the Italian, fiercely.

"You tried it on onc't, and you didn't do it!" the girl replied, still defiant. "You better not try it again. You've got my temper up, an' I had just as lief die now as not. This ends you and me. I don't stay here no more!"

The girl made a movement toward the door, but the Italian quickly anticipated her motion, and placed himself before it.

"You no go!" he cried, in rage.

"Well, now, this is interesting," said the rough, complacently. "Now hit him over the head with the cheer, 'cos you can't get out till you do."

"What! you tell her to hit me over ze head viz ze chair?" cried the Italian, in astonishment.

"In course; what a feller you are, for to want to spile the fun!" said the rough, in an aggrieved tone.

"Rocky, I gives you one dollar to take ze chair away from this devil's imp!" exclaimed the Italian, glaring upon the girl.

"A dollar! Now you're talkin'. I'm your man!" and Rocky got off the table.

"Don't you dare to come near me!" the maid cried, fiercely, retreating, and placing her back against the wall as she spoke.

"You jes' teach your grandmother to milk ducks," said the rough, with a grin. "You jes' put down that cheer, or I'll walk into you, lively, now. You can't skeer me!" And as he spoke, he slowly approached the girl.

"Keep away!" she cried, every muscle in her body trembling with excitement.

"Take ze chair, den I kill her, some!" exclaimed the Italian.

"Oh, jes' look at me now," said Rocky, shaking his head with a knowing air. "See me astonish her weak nerves."

Then the rough made a sudden dart forward, as if intending to seize the girl. With desperate force she brought the chair down, intending to fell the rough as she had made the Italian seek the floor; but the wily Rocky knew a trick worth two of that, for, as the chair descended, he suddenly darted back and avoided the blow. Then, before she could again raise the clumsy weapon that stern necessity had forced upon her, he seized it, wrested it from her hands, and sent it spinning across the floor.

The Italian uttered a shrill cry of triumph. The child was helpless in his hands.

"All done by the turn of the wrist!" exclaimed Rocky. "Old man, I'll trouble you for to fork over that dollar."

With white face, flaming eyes, and lips quivering with passion, the girl stood; her little fists tightly clinched, as though, even now, she was prepared to do battle with her enemies.

The door against which Jocky was standing was opened suddenly. As it turned into the room, the force of the concussion pitched the Italian forward into the saloon.

The girl started with joy. In the appearance of the strangers she saw a chance of escape.

HARRIET BEECHER STOWE

From We and Our Neighbors

While Harriet Beecher Stowe (1811–1896) established her reputation with *Uncle Tom's Cabin* (1852), she wrote many subsequent works, including *We and Our Neighbors: or, Records of an Unfashionable Street* (New York: Fords, Howard, & Hulbert, 1875), which went through numerous editions. A secondary character in the book, a young woman named Maggie, leaves her position as a domestic and finds employment in a retail store, where she meets a man who seduces her and then places her in a house of prostitution. The book's heroine, Eva Henderson, takes Maggie into her home as a servant, though Maggie must tolerate her family's abuse, especially that of Mrs. Maria Wouvermans or "Aunt Maria," who is known as the family dictator. Maggie leaves to save her benefactor

from such abuse. The following selection, taken from the first edition (317–27) comes after Maggie has left Eva, but it encapsulates, in flashback, Maggie's earlier experiences.

It was the week before Christmas, and all New York was stirring and rustling with a note of preparation. Every shop and store was being garnished and furbished to look its best. Christmas-trees for sale lay at the doors of groceries; wreaths of ground-pine, and sprigs and branches of holly, were on sale, and selling briskly. Garlands and anchors and crosses of green began to adorn the windows of houses, and were a merchantable article in the stores. The toy-shops were flaming and flaunting with a delirious variety of attractions, and mammas and papas with puzzled faces were crowding and jostling each other, and turning anxiously from side to side in the suffocating throng that crowded to the counters, while the shopmen were too flustered to answer questions, and so busy that it seemed a miracle when anybody got any attention. The country-folk were pouring into New York to do Christmas shopping, and every imaginable kind of shop had in its window some label or advertisement or suggestion of something that might answer for a Christmas gift. Even the grim, heavy hardware trade blossomed out into festal suggestions. Tempting rows of knives and scissors glittered in the windows; little chests of tools for little masters, with cards and labels to call the attention of papa to the usefulness of the present. The confectioners' windows were a glittering mass of sugar frostwork of every fanciful device, gay boxes of bonbons, marvelous fabrications of chocolate, and sugar rainbows in candy of every possible device; and bewildered crowds of well-dressed purchasers came and saw and bought faster than the two hands of the shopmen could tie up and present the parcels. The grocery stores hung out every possible suggestion of festal cheer. Long strings of turkeys and chickens, green bunches of celery, red masses of cranberries, boxes of raisins and drums of figs, artistically arranged, and garnished with Christmas greens, addressed themselves eloquently to the appetite, and suggested that the season of festivity was at hand.

The weather was stinging cold — cold enough to nip one's toes and fingers, as one pressed round, doing Christmas shopping, and to give cheeks and nose alike a tinge of red. But nobody seemed to mind the cold. "Cold as Christmas" has become a cheery proverb; and for prosperous, well-living people, with cellars full of coal, with bright fires and roaring furnaces and well-tended ranges, a cold Christmas is merely one of the luxuries. Cold is the condiment of the season; the stinging, smarting sensation is an appetizing reminder of how warm and prosperous and comfortable are all within doors.

But did any one ever walk the streets of New York, the week before Christmas, and try to imagine himself moving in all this crowd of gaiety, outcast, forsaken and penniless? How dismal a thing is a crowd in which you look in vain for one face that you know! how depressing the sense that all this hilarity and abundance and plenty is not for you! Shakespeare has said, "How miserable it is to look into happiness through another man's eyes — to see that which you might enjoy and may not, to move in a world of gaiety and prosperity where there is nothing for you!"[1]

Such were Maggie's thoughts, the day she went out from the kindly roof that had sheltered her, and cast herself once more upon the world. Poor hot-hearted, imprudent child, why did she run from her only friends? Well, to answer that question, we must think a little. It is a sad truth, that when people have taken a certain number of steps in wrong-doing, even the good that is in them seems to turn against them and become their enemy. It was in fact a residuum of honor and generosity, united with wounded pride, that drove Maggie into the street, that morning. She had overheard the conversation between Aunt Maria and Eva; and certain parts of it brought back to her mind the severe reproaches which had fallen upon her from her Uncle Mike. He had told her she was a disgrace to any honest house, and she had overheard Aunt Maria telling the same thing to Eva, — that the having and keeping such as she in her home was a disreputable, disgraceful thing, and one that would expose her to very unpleasant comments and observations. Then she listened to Aunt Maria's argument, to show Eva that she had better send her mother away and take another woman in her place, because she was encumbered with such a daughter.

"Well," she said to herself, "I'll go then. I'm in everybody's way, and I get everybody into trouble that's good to me. I'll just take myself off. So there!" and Maggie put on her things and plunged into the street and walked very fast in a tumult of feeling.

[1] *Shakespeare . . . you!": As You Like It,* act 5, scene 2, lines 43–45: "But O, how bitter a thing it is to look into happiness through another man's eyes!"

She had a few dollars in her purse that her mother had given her to buy winter clothing; enough, she thought vaguely, to get her a few days' lodging somewhere, and she would find something honest to do.

Maggie knew there were places where she would be welcomed with an evil welcome, where she would have praise and flattery instead of chiding and rebuke; but she did not intend to go to them just yet.

The gentle words that Eva had spoken to her, the hope and confidence she had expressed that she might retrieve her future, were a secret cord that held her back from going to the utterly bad.

The idea that somebody thought well of her, that somebody believed in her, and that a lady pretty, graceful, and admired in the world, seemed really to care to have her do well, was a redeeming thought. She would go and get some place, and do something for herself, and when she had shown that she could do something, she would once more make herself known to her friends. Maggie had a good gift at millinery, and, at certain odd times, had worked in a little shop on Sixteenth Street, where the mistress had thought well of her, and made her advantageous offers. Thither she went first, and asked to see Miss Pinhurst. The moment, however, that she found herself in that lady's presence, she was sorry she had come. Evidently, her story had preceded her. Miss Pinhurst had heard all the particulars of her ill conduct, and was ready to the best of her ability to act the part of the flaming sword that turned every way to keep the fallen Eve out of paradise.[2]

"I am astonished, Maggie, that you should even think of such a thing as getting a place *here,* after all's come and gone that you know of; I am astonished that you could for one moment think of it. None but young ladies of good character can be received into our workrooms. If I should let such as you come in, my respectable girls would feel insulted. I don't know but they would leave in a body. I think *I* should leave, under the same circumstances. No, I wish you well, Maggie, and hope that you may be brought to repentance; but, as to the shop, it isn't to be thought of."

Now, Miss Pinhurst was not a hard-hearted woman; not, in any sense, a cruel woman; she was only on that picket duty by which the respectable and well-behaved part of society keeps off the ill-behaving. Society has its instincts of self-protection and self-preservation, and seems to order the separation of the sheep and the goats, even before

[2] *flaming sword . . . paradise:* See Genesis 3.24.

the time of final judgment. For, as a general thing, it would not be safe and proper to admit fallen women back into the ranks of those unfallen, without some certificate of purgation. Somebody must be responsible for them, that they will not return again to bad ways, and draw with them the innocent and inexperienced. Miss Pinhurst was right in requiring an unblemished record of moral character among her shop-girls. It was her mission to run a shop and run it well; it was not her call to conduct a Magdalen Asylum: hence, though we pity poor Maggie, coming out into the cold with the bitter tears of rejection freezing her cheek, we can hardly blame Miss Pinhurst. She had on her hands already all that she could manage.

Besides, how could she know that Maggie was really repentant? Such creatures were so artful; and, for aught she knew, she might be coming for nothing else than to lure away some of her girls, and get them into mischief. She spoke the honest truth, when she said she wished well to Maggie. She did wish her well. She would have been sincerely glad to know that she had gotten into better ways, but she did not feel that it was her business to undertake her case. She had neither time nor skill for the delicate and difficult business of reformation. Her helpers must come to her ready-made, in good order, and able to keep step and time: she could not be expected to make them over.

"How hard they all make it to do right!" thought Maggie. But she was too proud to plead or entreat. "They all act as if I had the plague, and should give it to them; and yet I don't want to be bad. I'd a great deal rather be good if they'd let me, but I don't see any way. Nobody will have me, or let me stay," and Maggie felt a sobbing pity for herself. Why should she be treated as if she were the very off-scouring of the earth, when the man who had led her into all this sin and sorrow was moving in the best society, caressed, admired, flattered, married to a good, pious, lovely woman, and carrying all the honors of life?

Why was it such a sin for *her*, and no sin for him? Why could he repent and be forgiven, and why must she never be forgiven? There wasn't any justice in it, Maggie hotly said to herself — and there wasn't; and then, as she walked those cold streets, pictures without words were rising in her mind, of days when everybody flattered and praised her, and he most of all. There is no possession which brings such gratifying homage as personal beauty; for it is homage more exclusively belonging to the individual self than any other. The tribute rendered to wealth, or talent, or genius, is far less personal. A child or woman gifted with beauty has a constant talisman that turns all things to gold — though, alas! the gold too often turns out like fairy gifts; it is gold only in seeming, and becomes dirt and slate-stone on their hands.

Beauty is a dazzling and dizzying gift. It dazzles first its possessor and inclines him to foolish action; and it dazzles outsiders, and makes them say and do foolish things.

From the time that Maggie was a little chit, running in the street, people had stopped her, to admire her hair and eyes, and talk all kinds of nonsense to her, for the purpose of making her sparkle and flush and dimple, just as one plays with a stick in the sparkling of a brook. Her father, an idle, willful, careless creature, made a show plaything of her, and spent his earnings for her gratification and adornment. The mother was only too proud and fond; and it was no wonder that when Maggie grew up to girlhood her head was a giddy one, that she was self-willed, self-confident, obstinate. Maggie loved ease and luxury. Who doesn't? If she had been born on Fifth Avenue, of one of the magnates of New York, it would have been all right, of course, for her to love ribbons and laces and flowers and fine clothes, to be imperious and self-willed, and to set her pretty foot on the neck of the world. Many a young American princess, gifted with youth and beauty and with an indulgent papa and mamma, is no wiser than Maggie was; but nobody thinks the worse of her. People laugh at her little saucy airs and graces, and predict that she will come all right by and by. But then, for her, beauty means an advantageous marriage, a home of luxury and a continuance through life of the petting and indulgence which every one loves, whether wisely or not.

But Maggie was the daughter of a poor working-woman — an Irishwoman at that — and what marriage leading to wealth and luxury was in store for her?

To tell the truth, at seventeen, when her father died and her mother was left penniless, Maggie was as unfit to encounter the world as you, Miss Mary, or you, Miss Alice, and she was a girl of precisely the same flesh and blood as yourself. Maggie cordially hated everything hard, unpleasant or disagreeable, just as you do. She was as unused to crosses and self-denials as you are. She longed for fine things and pretty things, for fine sight-seeing and lively times, just as you do, and felt just as you do that it was hard fate to be deprived of them. But, when worse came to worst, she went to work with Mrs. Maria Wouvermans. Maggie was parlor-girl and waitress, and a good one too. She was ingenious, neat-handed, quick and bright; and her beauty drew favorable attention. But Mrs. Wouvermans never commended, but only found fault. If Maggie carefully dusted every one of the five hundred knick-knacks of the drawing-room five hundred times, there was nothing said; but if, on the five hundred and first time, a moulding or a crevice was found with dust in it, Mrs. Wouvermans would summon Maggie to her presence with the air of a judge, point out the criminal fact, and inveigh, in terms of general severity, against her carelessness, as if carelessness were the rule rather than the exception.

Mrs. Wouvermans took special umbrage at Maggie's dress — her hat, her feathers, her flowers — not because they were ugly, but because they were pretty, a great deal too pretty and dressy for her station. Mrs. Wouvermans's ideal of a maid was a trim creature, content with two gowns of coarse stuff and a bonnet devoid of adornment; a creature who, having eyes, saw not anything in the way of ornament or luxury; whose whole soul was absorbed in work, for work's sake; content with mean lodgings, mean furniture, poor food, and scanty clothing; and devoting her whole powers of body and soul to securing to others elegancies, comforts and luxuries to which she never aspired. This self-denied sister of charity, who stood as the ideal servant, Mrs. Wouvermans's maid did not in the least resemble. Quite another thing was the gay, dressy young lady who, on Sunday mornings, stepped forth from the back gate of her house with so much the air of a Murray Hill demoiselle that people sometimes said to Mrs. Wouvermans, "Who is that pretty young lady that you have staying with you?" — a question that never failed to arouse a smothered sense of indignation in that lady's mind, and added bitterness to her reproofs and sarcasms, when she found a picture-frame undusted, or pounced opportunely on a cobweb in some neglected corner.

Maggie felt certain that Mrs. Wouvermans was on the watch to find fault with her — that she wanted to condemn her, for she had gone to service with the best of resolutions. Her mother was poor and she meant to help her; she meant to be a good girl, and, in her own mind, she thought she was a very good girl to do so much work, and remember so many different things in so many different places, and forget so few things.

Maggie praised herself to herself, just as you do, my young lady, when you have an energetic turn in household matters, and arrange and beautify, and dust, and adorn mamma's parlors, and then call on mamma and papa and all the family to witness and applaud your notability. At sixteen or seventeen, household virtue is much helped in its development by praise. Praise is sunshine; it warms, it inspires, it promotes growth: blame and rebuke are rain and hail; they beat down and bedraggle, even though they may at times be necessary. There was a time in Maggie's life when a kind, judicious, thoughtful, Christian woman might have kept her from falling, might have won her confidence, become her guide and teacher, and piloted her through the dangerous shoals and quicksands which beset a bright, attractive, handsome young girl, left to make her own way alone and unprotected.

But it was not given to Aunt Maria to see this opportunity; and, under her system of management, it was not long before Maggie's temper grew fractious, and she used to such purpose the democratic liberty of free speech, which is the birthright of American servants, that Mrs. Wouvermans never forgave her.

Maggie told her, in fact, that she was a hard-hearted, mean, selfish woman, who wanted to get all she could out of her servants, and to give the least she could in return; and this came a little too near the truth ever to be forgotten or forgiven. Maggie was summarily warned out of the house, and went home to her mother, who took her part with all her heart and soul, and declared that Maggie shouldn't live out any longer — she should be nobody's servant.

This, to be sure, was silly enough in Mary, since service is the law of society, and we are all more or less servants to somebody; but uneducated people never philosophize or generalize, and so cannot help themselves to wise conclusions.

All Mary knew was that Maggie had been scolded and chafed by Mrs. Wouvermans; her handsome darling had been abused, and she should get into some higher place in the world; and so she put her as workwoman into the fashionable store of S. S. & Co.

There Maggie was seen and coveted by the man who made her his prey. Maggie was seventeen, pretty, silly, hating work and trouble, longing for pleasure, leisure, ease and luxury; and he promised them all. He told her that she was too pretty to work, that if she would trust herself to him she need have no more care; and Maggie looked forward to a rich marriage and a home of her own. To do her justice, she loved the man that promised this with all the warmth of her Irish heart. To her, he was the splendid prince in the fairy tale, come to take her from poverty and set her among princes; and she felt she could not do too much for him. She would be such a good wife, she would be so devoted, she would improve herself and learn so that she might never discredit him.

Alas! in just such an enchanted garden of love, and hope, and joy, how often has the ground caved in and let the victim down into dungeons of despair that never open!

Maggie thinks all this over as she pursues her cheerless, aimless way through the cold cutting wind, and looks into face after face that has no pity for her. Scarcely knowing why she did it, she took a car and rode up to the Park, got out, and wandered drearily up and down among the leafless paths from which all trace of summer greenness had passed.

Suddenly, a carriage whirred past her. She looked up. There he sat, driving, and by his side so sweet a lady, and between them a flaxen-haired little beauty, clasping a doll in her chubby arms!

The sweet-faced woman looks pitifully at the haggard, weary face, and says something to call the attention of her husband. An angry flush rises to his face. He frowns, and whips up the horse, and is gone. A sort of rage and bitterness possess Maggie's soul. What is the use of trying to do better? Nobody pities her. Nobody helps her. The world is all against her. Why not go to the bad?

EDGAR FAWCETT

From The Evil That Men Do

While Edgar Fawcett established his contemporary reputation as a writer of novels that indicted New York's high society, his finest book, *The Evil That Men Do* (New York: Belford, 1889), takes the tenement districts for its setting. This novel tells the story of Cora Strang, a young woman from the country who comes to New York, where she works a series of low-paying jobs. Unemployed, with no place to live, she calls on her friend Em Cratchett. The following selection (86–96) tells the story of Em's battle to support her family and live an upright life. Later in the book, Em dies. Cora obtains a position as personal maid in a well-to-do household where she is discovered by Caspar Drummond, who seduces Cora with the promise of marriage and then abandons her to marry a society woman. Cora goes to a disorderly house and turns prostitute, takes to drink, becomes a streetwalker, and ultimately meets her death.

Em Cratchett was a sewing-girl who supported a bed-ridden mother, an idiot brother seven years old, and two sisters, aged about nine and eleven. This family had once occupied two rooms in the Prince Street tenement-house, and there Cora had got to know Em even better than she knew Effie and Ann Flynn. Em had been forced to find cheaper quarters, and a tenement-house close to Grand Street had supplied them. It was a den of filth, but its two yet smaller rooms were a dollar and a half lower per month, and to Em that meant a great sum. She had more than once told Cora that she believed if it wasn't for the strong tea, starvation would have killed her long ago. She took it as black as ink, and no doubt it buoyed her up among the fearful sights and smells on every side. Not a cent, with Em, but counted; and when they raised the electric light within a few yards of her windows it brought her one more chance to save. For a monstrous iron structure had been built, of late, just over the way, and its bulk had darkened the sunshine, so that morning seemed like afternoon and three o'clock in the day was like dusk. But when night came the electric light flooded Em's front room with its keen, pale splendor, as though it had been the marvellous moonlight of another planet. One evening a sudden thought seized the girl, and she tried those acute white rays to sew by. Always afterward she did her work at night, and took what rest she could get between morning and the hours that followed. Soon others in the house imitated her — such as were not too slothful and drunken among the womenkind. Those cold and colorless beams poured in upon bent shapes and wan faces, night after night. The late feasts of luxury and dissipation in other parts of the town were copied here with tints of frightful parody and irony. These were revellers with cups of gall for their wine, and spectres of want to serve as footmen. Sin rioted in the reeking house, whose very stairs had rotten creaks when you trod them, as though fatigued by the steps of sots and trulls. To enter some of the rooms was to smell infection and to face beastliness. Fever lived in the sinks and closets along the halls, where festered refuse more rancid and stenchful than stale swill, and so vile that to name it would be to deal with words which are the dung of lexicons. Those halls had nooks of gloom whence miasma might have fled in fright before the human grossness that spawned there. Little children dipped their chastity in poison between the scurfy-grained wainscots of every corridor, and twisted their soft lips into the shaping of oaths that would scare brothels. Now and then, in the lull of midnight, while the sewing-machines clattered from rooms like Em's and her toilful sisters, high yells would ring out as the beaten wife cowered and shivered, murder had been done here. There was a room with the ghost in it of a hanged desperado, which had so lowered its rent by its uncanny pranks that an Italian couple with six little ones had got it cheap after quitting the steamer. Malaria forever kept busy her minions of disease, and the just historian of this noxious house must have collected his annals ill if he forgot to tell how often the pine-wood coffins of the Potter's Field undertaker had been hustled over its noisome floors.

Em Cratchett gave a great start as Cora's hand touched her on the shoulder. She was working away at her machine, and had heard no one enter because of its clamors. "Well, I do declare!" she exclaimed, and looked up at her visitor as though she had been a wraith. "For the Lord's sake, Cora, where did *you* drop from?"

Cora made it all plain in a few words. "You ain't a bit to blame," decided Em, after she had mused a little. "You didn't do a thing you hadn't ought to a done — not a *thing!*"

"Oh, yes, Em, I left Owen —"

"Stuff! Couldn't he a follered you if he'd a mind, an' took you from them other parties? He always *was* a big, sassy bully, with a cheek onto him like the hull of out-doors — 'cept when he was drunk, an' then he was a heap sight worse. If I was as pretty as you I'd turn up my nose at a beau like that. I don't have any of 'em look at me nowadays, but when I did they never tried to bulldoze me, *you* can bet, without they got consid'rable left doin' it!"

This reference to a time when Em bore the dower of comeliness was fraught with a terrible pathos and satire. She had once been rosy as a sea-shell and straight as a reed, but you would not guess it to-day if you noted her pallor and her stoop. Not much past twenty, she was a mere haggard wreck of womanhood. She used to say, with a laugh bitterer than any sob could sound, that she had kept her character and now she was being rewarded. Lots of the girls whom she had grown up with had lost theirs. "But after all," she would dismally announce, "I can't see just where the difference lays. Such as us, unless we can marry some good man and quit the awful grind and fret of our lives, must die young anyhow. I guess them that go crooked has a good deal the best time. I don't see why so many *does* go if 'taint that way. They get nice clothes and food, for a few years at least. If most of 'em didn't drink along with the other badness they wouldn't end half so quick. But there's very few women that coins dollars out o' their self-respect without cravin' to deaden the self-disgust it brings."

Em, for all that she had made a sad failure of her life, was no cynic philosopher. She had striven with that sublimity which is not embalmed in epic tragedies, but which has obscurity for its theatre, and for foot-lights the uncouth old flickering lamps of side-streets and alleys. Yet defeat had answered her struggles, and this defeat had in a way been typical of her entire class; for she was one of hundreds who learn that they cannot support themselves in decency by being honest and biding separate from the tens of thousands who reap the lucre of profligacy. She had not yielded, and now she began to doubt if for soldiers like herself the battle held any palm of victory. A mind conscious of its own virtue? Ah, yes, there was perhaps guerdon[1] in that! But how did it serve when both spirit and flesh were enfeebled by fatigue, and the shadow of untimely death had already spread its veil across the sun? And then if there were only those who cared whether a girl went astray or not! But how many ever did care? How many even knew whether if she had sunk or no into the black gulf, or still was busy keeping the wash and lap of its waves from that slim raft she clung to?

"Of course you can stay here, Cora," said Em, a few minutes later. "But you ain't used to our kind o' livin', though you have roughed it, sure enough. I swear to God," the girl went on, lifting one bony hand in the elfin glare, "that I ain't got but forty cents to my name this night, though I ain't missed a single workin'-hour for six weeks. We've had to keep the stove red-hot just as you see it, or we'd half froze up here; and then there's the victuals, of course, and milk for Stevie, 'cause the child spits out everything else you give him, and medicines for mother —"

"Oh, I thought I'd hear something o' that kind. My sakes, I wish I was dead an' buried in the same grave with your poor father, an' not layin' here a nuisance to my own flesh an' blood!"

Worriment had swept over Em's hollow-cheeked face as these words were whined forth from a bed in one of the corners. At once she left her chair and went toward the bed, saying while she did so:

"Now, mother, you just know that's all fiddlestick! If you *was* to go I'd give right up, an' be after you in no time."

Here she tucked the bed-clothes deftly and nimbly round a skinny, white throat, not very much larger than the stem of a stout grape-vine, and peered down, smiling, for a moment, at eyes that made one think of two big brilliants aglow in a skull. For years Mrs. Cratchett had been as much the derision of death as the victim of disease. Her trouble was a spinal paralysis that left her wholly helpless. But no resignation

[1] *guerdon:* A reward or compensation.

had come with the affliction. Her tongue was waspish in its discontents and sarcasms; there could not have been a more embittered invalid. She loved to moan forth melancholy remarks about graves or shrouds or the decomposition of the dead, and such tendency would reveal itself more strongly while the room where she and Em slept was illumined by that eerie radiance from the street. Its glamorous and cheerless power seemed to provoke her morbid meditations and make her deliver them aloud with a sinister gusto. Em had implored her, at first, when the plan of night-sewing had first been formed, to have a bed put up in the next room beside the children's, and thus escape the worst disturbances of the machine. But no; she appeared to see a new chance for obstinate revolt, and declared that the machine "soothed" her, and that she didn't want to "pop off" while out of her daughter's sight. She was always prophesying a moment when she would "pop off;" she had been expecting it and brooding over it for years, and if she had said nothing at all on the subject Em would have missed the doleful utterance as though it had been a lack of the familiar pungency in her own too frequent cups of tea. Long ago the unhappy woman could have gone to a hospital for incurables, but she had rebelled against that idea with characteristic mulish hardihood. Doctors of eminence had seen her, and wondered what latent vigor kept her alive. It might be said that she had been alive only mentally for a longer time than common credence would have cared to accept. Mortification hovered over her like a vulture above its carrion. But somehow the thrust-in beak had never left its plunderous mark. A subtler disintegration, however, had bitten with invisible fang the tissues of her brain, and she lay, the object of a malignity more severe than that which any physical dissolution could have wreaked.

"Excuse me for not seeing you when I came in, Mrs. Cratchett," promptly said Cora. "But perhaps you were asleep, an didn't see *me*."

"Oh, no, I wasn't," said the sick woman. "I heard everything you told Em. I don't blame a spry, pink-faced gal like you for likin' goin' to balls. I used to love 'em. But Lord! what's human happiness, even for such as lives onto Fifth Avenue an' hangs their winders with lace? Why, we're all nothin' but skeletons with a little flesh put round the bones, that's all we are. An' somewheres, either in the lumber-yard or the tree itself, is waitin' our —"

"Let me boost your head up a little higher, mother," Em here broke in, with the natural wish to curtail speeches of so pessimistic a dolor. And while she proceeded to rearrange the flabby and flimsy pillow, a laugh sounded in a near doorway, quaint and chuckling, as if some gnome out of a fairy-tale had made it in a dream.

"Oh, come here, Stevie," said Cora, and she held out one hand to a tiny shape with straw-colored hair and huge blue eyes made empty by idiocy.

But the boy would not obey her, and was running back into his bed-chamber, when his two sisters, Maggie and Katy, emerged thence and pushed him forward between them. The girls were both in their night-gowns, made from some coarse gray flannel, whose thick woof could not hide the wasted outlines of their bodies. They both knew Cora well and were fond of her, and rebelled when their elder sister ordered their return to bed. They had little, pinched, bluish-white faces, and the younger was disfigured by a deep scar on one cheek, which had come to her a year or two before in a fight with an Italian boy of about thirteen years old — a juvenile Don Juan of the slums — who was now in a reformatory for this and similar assaults. Maggie went to the stove, declaring it was so cold in her room that she could not sleep there, while Katy put one pipe-stem of an arm round Cora's neck, and whispered that she was "awful hungry."

"I don't think much about the cold," rang Mrs. Cratchett's nasal tones from the bed. "What's the use, when we're all bound to be a good deal colder 'fore we git through with this queer job they call life?"

"Hee, hee!" laughed Stevie, in his eldritch way. He had climbed upon his mother's bed and sat beside her, with his legs crossed and his arms folded, like a droll little pixie, while the searching and crystalline light gave his queer, long face an unearthly vacancy. He always laughed like this whenever his mother spoke one of her funereal sentiments; he of course had no conception of what they meant, but something in the tone of her voice he would always appear to recognize and to make a signal for his impish mirth. His mother's illness had begun the day he was born, and it had been alleged that the "dear dead husband" whom Mrs. Cratchett was so fond of talking about had kicked her in a drunken rage. Stevie was born idiotic, and his mother became the lingering and piteous invalid we have seen her.

"We ain't got bed-clothes enough in there," said Maggie, as she crouched in front of the stove. She had her mother's querulous temperament, and gave Em twice as much trouble as Katie. "Besides, 'tain't fair to have one room all het up an' the other like an ice-house. Now Cora's come," she added, "p'raps we may have a little beer."

"Beer! beer!" cried Stevie, who knew the word; and he clapped his little waxy hands together as if in applause at Maggie's courageous hint.

"Hold your tongue, Mag," reproved Em. "The less beer you git the better for yourself and them that's got charge of ye."

"She was drunk on beer, night 'fore last, Maggie was," whispered Katie to Cora, but in a tone loud enough for all the rest to hear what she said.

Maggie scowled, with the round red-hot stove glaring at her side as though it were the head of an evil spirit.

"You hold your nasty little tattlin' tongue!" she cried to Katie; and then, forgetting the "company manners" which Cora's presence should both have evoked and fostered, she poured out a sudden torrent of profanity, each oath of which seemed to fly up and hammer the low ceiling with peril to its cracked and swollen plaster.

Em shrieked her name in expostulation, and then rushed toward her with an evident design of taking her by either shoulder. But the project failed, and miserably; for Em stood still in a sudden dazed way, and putting one hand over her heart slipped limply as a dropped garment to the floor. Cora was kneeling beside her in an instant and scanning the worn young face that this wild, fairy sort of light made deathly.

"Em!" she called in alarm, "Em! What *is* it?" and almost at once the fallen lids were lifted again, and a laugh rippled through the chalky lips. "Oh, it's nothin' — just one o' my spells. I take 'em sometimes, but they pass off as quick as they come."

"Yes," said Mrs. Cratchett from the bed, "an' some day, Cora, *she'll* pass off *in* one of 'em. It *might* be that she'll go even 'fore *I* do, an' if she does, that *would* be a joke!"

"Hee! hee!" cachinnated little Stevie, who seemed to think it would indeed be a fine joke, though he really thought nothing at all about it except that his mother's voice had sounded in the old sardonic tones.

"Em," Cora murmured, after the collapse had given place to a half-recumbent languor, "if you've a bit of brandy it would be a good thing just to take a swallow, and —"

"Oh, no," hurried Em, whispering the words, with a glance at Maggie, who crouched sullenly by the stove; "I never dare keep *anything* like that here, because of *her*. She's only a child, but that curse is on her, and she'd drink anything she could get. . . Oh, *Cora!*" and here Em burst into tears, laying her head on the shoulder that was so close to it. "There's been changes even with such as us since you was away from us. I didn't want to tell you, for I know your lot ain't any too easy; but I've just got to stay chained to my work day in an' day out an' see that little thing go right down to — to the very dogs. . . Well," she broke off, with a great gulp that showed she was heroically swallowing her tears, "don't mind my blubberin' a bit. There, now; it's over. Lemme get up. . . Oh, yes, I can; don't be afraid." And she rose with even a certain nimbleness.

"Some day she won't get up. She'll lay there," said Mrs. Cratchett. And Stevie did not laugh, this time, for some reason that his own topsy-turvy little brain best knew about.

"Em," now said Cora, in a low and very earnest voice, remembering Katie's recent words to her, "I wish you'd let me get you a bite of something. I —"

"A bite? Why, Cora, I've had my supper. We all did. It's my breakfast, you know. All I need, now, is a cup o' tea; an' the tea-pot's there on the stove. We'll both have one."

"I hate tea," grumbled Maggie; "it's such wishy-washy stuff."

"I like coffee, though," said Katie wistfully. "You can get a big cup to a place a little ways from here, with a hunk o' sugar into it an' lots o' milk, for three cents. An' if ye've got five you can get a piece o' the yaller mushy pie — punkin, I guess it is — throwed in."

"Em," Cora pursued, after this epicurean outburst was ended, "I do believe the tea hurts you a good deal more than it heals. Now I'll tell you what it is. I'd like a cup of hot coffee myself, and if you've got a dish of any kind I'll just try to bring in some little relish from that place Katie spoke of.

The place was not hard to find. It was a narrow little eating-house with a tawny dish of baked beans in the window, flanked by a few sanguinary beefsteaks. At first the proprietor would not give Cora a dish of his corned-beef hash to take away, though this was the viand she selected after a swift survey of the bill. Soon, however, he relented, and it may have been that her face, with its curves and tints, down in such a street as this, where ugliness had marked nearly everybody for its cadaverous, bloated or pimpled own, proved a cogent promoter of his clemency. She got him to fill a small tin can that she had brought, too, with the coffee of which Katie was so fond.

'The whole thing has only cost me forty cents,' Cora thought, as she sped back again through the still midnight street, 'and I dare say it'll give lots of comfort to *them*.' She paused suddenly and asked herself what change had come over her spirits, why she felt this glow as of good-cheer at her heart. Then, soon, the answer made itself plain.

She was confronting a worse misery than her own. It was just as though Em and she stood on the same dank, earthly stairway that plunged down into opaque darkness. There was yet a little light where she stood, but Em seemed all swathed in shadow. And moreover she could fancy that she heard Em's voice calling to her from those pitchy depths, "Don't come down any further — don't."

Oh, but how could she help it? And no matter into what gloom of destitution and penury the staircase led, so that it staid undarkened by sin!

J. W. SULLIVAN

"Minnie Kelsey's Wedding"

J. W. Sullivan (1848–1938) was born in Carlisle, Pennsylvania, where he learned the printer's trade. He settled in New York City around 1882 and became foreman of the proof room of the *New York Times.* In New York he was active in reform movements, advocating the importance of initiatives and referendums with *Direct Legislation by the Citizenship through Initiative and Referendum* (1892). He also joined the labor movement, befriending Samuel Gompers (the president of the American Federation of Labor), and rising to the presidency of the Central Federated Union of New York City. Writing fiction was a sideline, yet his writings, like his political activism, sought to elevate the common man. He published two collections of short stories, *So the World Goes* (1898) and *Tenement Tales of New York* (New York: Holt, 1895), the source of the present story (45–65).

In the tenement districts of the West Side of New York city the streets are as noisy, the sidewalk children as countless, the dwellings as forbidding, the stairways as steep and dark, the apartments as narrow, the poor as comfortless, as on the East Side. The West Side tenement population, transferred to an open country site, would form an important city of the world.

On a certain corner in a West Side tenement neighborhood stands a cheaply constructed five-story brick, poor-man's hive. Of the common pattern of corner tenements, its ground floor is a liquor-saloon; its side-street apartments have windows to half the rooms; each floor is divided among four families; the big rooms of each apartment are about twelve feet square; the bedrooms are wide enough for a large trunk and a small bed side by side, while the general living-room is the kitchen.

On the afternoon of a sunshiny winter's Sunday, in a fifth-story chamber of this house, a young girl sat at the one window. No fire in the room, she had wrapped herself warmly in an old shawl, and had thrown high the shades that the sun might drive out the chill. She was preoccupied. Eye or ear apparently brought her at the moment no message. Had she looked out, she could have seen down the side street some little way, and, by pressing her face against a window-pane, a patch of river and the Jersey City heights afar, forming in the clear air a horizon of toy houses on snow-clad hills. In the street beneath, Sunday-dressed working-people sauntered by; and on the corner opposite, a group of half-grown boys wearily killed the holiday, skylarking when the policeman was out of sight, and dispersing when he drew near. Within the house echoed the domestic noises of twenty families — doors slamming, children wailing, women scolding, boys hallooing, all mingling with the endless clatter of kitchen labors. As the apothecary grows insensible to his shop smells, the denizen of a tenement is dulled to its sounds and odors and scenes.

The girl was engrossed, not in the view before her eyes, but in mental pictures. She was contemplating, as might a bird aloft in air, a pleasant little country town, and reviewing incidents of her own life which had happened there. Memory and affection were weaving in her their network of witchery. In that dear old place it had always been summer, the grass was always green, the fruits were ever ripening, the flowers always in bloom. How she now again recognized every friendly tree, every pleasant prospect of the by-lanes, even the sticks and stones about her old home. Mere acquaintances were now friends, her playmates were in her very presence, and again she heard their sweet voices and gazed into their kindly eyes. Musing over these heart pictures, she was thrilled with a melancholy pleasure. In this mood, events but sped past her mind while scenes remained, intermingling. In flashes she recalled her father's death, her widowed mother's struggles, her mother's death and funeral, the sale of the few household effects, the letter that came from her only relative, offering her a home in New York, and then her journey to the great city, her bewilderment in its noisy streets, her disappointment on seeing into its tenement-house life. And thus, at length, was she brought back to her actual existence.

Her relative, an aunt, she had found a pale, slender little woman with three young children. Her aunt's husband was a rough working-man, spending his week's wages in drink on Sunday, and sitting about sulky and reticent week-day evenings — a man who lived in his working-clothes, and never shaved his stubby whiskers but to prepare for his week's spree.

The girl had come expecting, as her aunt had written, to become a sales-woman in a store and to help a little with the children in the evenings. In a week after her arrival she was hard at work in a factory, and had learned that half a night's drudgery in her new home often awaited her after shop hours.

A year had sped by since her coming to the city, bringing her to the age of eighteen. When one works long hours every day but Sunday, time flies. The hours, the days, may seem wearisome, but the year, seen backward, is a brief span. Factory life is humdrum, one day like every other, with little to break the monotonous chain. A factory girl, engaged month in and month out at a single process in subdivided labor, may envy the prisoner in a treadmill. He at least digests his food. She sits, or, worse, stands hour after hour, her eyes fixed on her work, her hands following motions which time makes automatic. After a while she forgets all other processes than her own in the work. To her there is no raw material, no completed product. Nothing but at her left hand a pile of work in the nineteenth stage, which she is to pass to her right, manipulated by her into the twentieth. Even of this she at length loses sight. Her tedious motions are no more part of her than is the steel machinery. They leave behind them only the unreal, the unseen — warp and woof of her own time and vitality. Into the interminable web of her life these are worked, and that web she sees stretch behind her, colorless, figureless, unprofitable, intangible. For the hour, its making has brought her bread and a covering. In it she has lost roses from her cheeks, sprightliness from her girl's nature, buoyancy from her youth, force from her mentality. She, her own chameleon, has taken on from it morbidness, bloodlessness, colorlessness, dullness. Yet some of the horrors in her place of toil never lose their dreadful reality. Penned up as she is in a hot, stifling room ten hours a day, the vitiated air dries up the blood in her veins. The heavy, discordant buzz of the machinery grinds into her very brain. The nauseating odor of decomposing machine oil sickens her. Dust thickens the atmosphere and chokes her.

An hour after beginning work, a low fever seizes her and tortures her until, long after lying down to her night's rest, she falls into a disturbed sleep.

So was time going with this young girl. She had taken to asking herself desperate questions: Was this to be her life? Only this, and then death? Or would a new life-era come with a marriage, to bring her a man perhaps like her uncle? Was she to have no girl's enjoyments? Was her routine, day by day, to be work, only broken by the tramp to the infernal factory and back to the detested tenement house? Was she never to love a good man, never to be the loved mistress of a home such as she had seen in the country, where homes are?

She turned away from the window and looked about the cheerless little room. In it stood, end to end, two narrow beds. One she shared at night with the youngest child; the other was for the two elder children. To-day, for once, she was alone: the mother gone with her brood to call on friends, the father off drinking. The girl looked at the wall covered with cheap colored prints and pictures from the newspapers; at the little make-believe mantel-shelf covered with a bright red cloth, set with a few photographs; at her own poor attempts to make her poverty bright and her contracted quarters spacious. As her gaze wandered about the mean room, she sighed in weariness; as it rested on her mother's picture, she sighed in grief.

Presently she pulled down the window-shade to the lowest half-pane. She made sure the door was bolted. She took down her hair, combed it slowly, braided it, and looked at it in a hand-glass; it was fair, wavy, and fine. She tossed off her shawl, tightened her dress about her shoulders and waist; and again looked in the glass; she was straight, her waist was round, her shoulders comely. She trimmed her finger-nails, washed her hands and face, and once more looked in the glass: her features were not rough, her expression was pleasing, her hands were not yet misshapen from work. She brought out from her trunk a pair of new shoes, examined them, put them on, smoothing them and trying her feet in them, and then, drawing aside her skirt, looked down as she walked a step or two back and forth; her feet were small and shapely. She stood with her back against the wall and straightened herself, breathed deeply, and was thoughtful. She again looked in the glass and showed her teeth; they were her best feature. She walked the length of the room, down and up, jauntily, and hummed a little tune.

Drawing the shawl about her again, she opened her trunk, knelt before it, reached down among the clothing, and brought up some pictures. They were photographs of actresses. She looked at them long and closely, noting the dresses of the women, and their figures, pose, and expression. She rummaged still further down in the trunk and brought out another photograph. It was that of a handsome, serious-looking man of thirty. While she was inspecting it intently — with inquiry rather than affection — someone loudly rapped at the door of her room.

"Minnie! Minnie Kelsey!" from feminine voices.

The young girl plunged the photographs to the depths of the trunk. Running to the door, she opened it slightly to peep at the callers, and then wide to let them in.

Two young factory girls, neighbors. They were in high feather. To-morrow night the ball of the factory employees was to come off. Had Minnie decided to go? No, she admitted she had not.

"Minnie, if you'll go," urged one, "I'll lend you my cousin's blue dress. She's away now, you know. How it would become you! It wouldn't do for you to stay at home! You're the prettiest girl in the factory, and you're kind o' different too, coming from the country. Oh, go! You'll have a splendid time at the ball."

The other girl chimed in. Minnie's objections were met with counter-persuasions. The cousin's dress was at her friend's disposal because the cousin was now wearing clothing belonging to the friend. The friend herself had a new hat which she would so much like Minnie to wear. Her black hat suited her own dress. Minnie must not put on airs. Minnie had often done her favors in the shop; now she could repay them. Poor people must do for one another.

Minnie was eighteen. She had never been to a ball. She had danced a little with the girls in the big factory hallway, during the dinner half-hour, at times to the music of a street organ or of a strolling band, and again to their own singing. To dance in a ballroom would be enchanting. She hesitated in promising, but would try to go. Assent enough for the visiting girls; they chatted about what they should wear at the ball. Next came the gossip of the factory. Then it was time to go.

When Minnie Kelsey returned from the factory next evening, her aunt told her that a bundle which had been left for her was on her bed. Minnie found the bundle in her room to be a large and pretty paper box, its contents a stylish hat and a new dress of much richer material than any she had ever worn. After admiring the clothes, she put them back in the box, and, going to the window, looked pensively down on the lamp-lit streets. She tried to plan what to do. She had been carried along by events until she could hardly decide not to go to the ball. She had not dared to mention the matter to her churlish uncle. Her desire to go would be sufficient reason for him to forbid her going. She had said nothing of it to her aunt, who feared her husband. Now, here all was ready. The girls were probably at the home of one of them near by, waiting for her. To put aside the fine clothes, go to bed with the children, lie awake half the night thinking of the ball, and in the morning at the factory be the sole one unable to join in the cheery talk over the event of the year — why must she suffer that?

Supper over, and her evening's drudgery done, she put the children to sleep, bade her aunt good-night, and looked the fine clothes over again. Not at all tired, she tried the pretty blue dress on. It fitted her marvelously. In the box were bows and ribbons to match, and these she put on, to see how they would become her. The little hand-glass told her that never had she looked so well. She tried on the pretty hat. The glass whispered more flattery than before.

Tired and sleepy? Not a bit, though she had quite given up all thoughts of going to the ball. But it would be a pity if the girls were yet waiting for her. She would run over and tell them to wait no longer. And, since she had the new clothes on, she would let the girls see how, by pure accident, they fitted her so well.

She tied on a veil, turned out her room light, moved quickly into the hallway and to the head of the stairs, and in a moment more had slipped unobserved down into the street.

At her friend's, her greetings were little screams of delight from the girls, and compliments from the young men. In a moment she found herself half coaxed, half carried along with the party, and, before she could make up her mind not to go, was in a street-car on the way to the ball.

Her escort was the handsome man of thirty whose picture was in her trunk. He had been polite to Minnie when meeting her at her friend's, had given her his picture, and had been paired off with her in the girls' talk. Always well dressed, he gave the girls presents, which they accepted, knowing they cost him little effort; for, whatever he did for a living, it was plain it was not hard work.

Very grand indeed appeared the glittering ballrooms to Minnie. Before a great mirror in the ladies' room she rearranged her bows and ribbons and hair, and when she passed into the dancing-hall the girls smilingly told her she had never looked so well. They were all having a splendid time at the ball.

Though dancing little himself, her escort took care during the evening that she should dance often with good partners. Between dances he walked with her about the hall, or sat with her chatting, good-humoredly telling who many of the young men were, how such balls are made to pay, and making the time pass agreeably. He seemed to know everybody and to have the run of everything perfectly, but was more interested in describing things to her than in enjoying them himself. His manners were easy and graceful, and he spoke to her amiably in a pleasant voice.

The hall was large, and twenty quadrille sets were on the floor at a time. The young people enjoyed themselves heartily — "Hands-all-around" coming off with shouts of delight. "Swing partners" sent couples whirling as long as the music let them. Some youths cut up antics, a favorite caper being for one to shake the sole of an uplifted foot at another. The girls enjoyed the dancing undisguisedly, laughing aloud in their delight. Minnie danced merrily. She was having a splendid time at the ball.

A young man who had engaged her to dance in a quadrille not making his appearance when it began, Minnie seated herself in a corner of the hall. Presently, through an open doorway, she heard a party of young men in an adjoining room talking. Said one:

"I see Tom King has a new girl tonight. One every season."

"Yes, a pretty girl, of course. He's got her dressed out in style, too."

"King gets his money at the races. Come easy, go easy. She looks like an innocent young thing, but she's got on clothes she can't buy herself, that's sure."

Minnie felt herself growing faint. She went to the ladies' room to collect her thoughts. In a little while one of her party came in to tell her that all were going out to a restaurant to supper. This was at least, she thought, an opportunity to get away.

In a few minutes the party were seated in the restaurant. All were gay, except Minnie, who could hardly be forced to speak. Tom King ordered wine. Minnie alone declined to take any. Before long one of the girls grew talkative.

"Minnie," she said, "I know why you are so quiet. You're in love."

An outburst of laughter.

"Minnie," continued the girl, "let me congratulate you on your fine clothes. That new dress of yours fits you beautifully, and becomes you more than anything else you ever wore. I wish I could have a dress like that."

Minnie felt the tears coming to her eyes.

"New dress, new hat," laughed another girl; "dear knows what next. Oh, what luck some girls have!"

Tom King rose to pay the bill. The rest of the party passed out. When King and Minnie reached the sidewalk, the others were out of sight. King said:

"Minnie, I think you had better go home."

"I am going home," she replied.

They went the whole way in silence, walking a good distance, and then riding in a street-car. At Minnie's corner King said:

"Minnie, someone said something to you about me this evening."

"They did, sir, or rather it was said in my hearing. They said you had a new girl every season, and that you bought her fine clothes. I know now you bought this dress and hat, and got another girl to deceive me into wearing them. I have been made the victim of my own vanity. I have suffered in my reputation through you. I shall never again be caught in such a trap."

"Minnie, you are a good girl."

"And I mean to be — I'm home. Good-night, sir."

She hurried in through the front door of the tenement-house and went up the long flights of stairs. As she reached the top floor she saw in alarm that her uncle's apartments were lighted up and open. She heard the man's hoarse voice. He was drawling out in a drunken fit of indignation:

"Ed Brady told us all about it down in Gilligan's liquor-store. She was there with that gambler, Tom King, dressed out the most expensive in the room. We know she never got them fineries honest. She's the first connection o' mine ever did the likes o' that. When I lay my hands on her she'll be sick of ever disgracing a decent family like mine."

Minnie's first impulse was to rush into the room and explain all, but the man's brutal looks frightened her. She turned and lightly ran down the stairs, but tripped and fell. Her uncle came out on the landing, and his harsh voice called out her name, but she made her way quietly down the dark stairway. He heard her footsteps, however, and followed, awakening the inmates of the tenement, bawling:

"You stop there! You stop there, girl! You can't git away from me. I'll break every bone of your body. You'll disgrace an honest family, will you? To think of it! A respectable working-man to have such cattle come right in with his own children."

Gaining the street, Minnie ran around the block. By chance her uncle ran around the same block the other way, and thus they met midway around. He was swearing to himself. He stared at her; she was about to scream; but he passed on. When he was a few steps beyond her she began running. In a moment he was after her.

"To think I didn't know her in her guilty finery!" he roared.

She sped on down the street. She heard his heavy footfalls as he ran after her. As she panted on, half-distracted, she saw the river only a little further ahead. The wild thought came to her that there was an end to all. She would rather die than live. Friendless, homeless, a castaway — why prize life?

Still she heard the man's boots clattering after her on the sidewalk, and now and again his oaths and threats. She quickened her speed. She reached the river street. But there a police officer moved out from the house shadows and stood in her way. She stopped, but said nothing. The uncle, running up, aimed a blow at her. An arm interposed, that of a man beside the officer, and Tom King's voice said:

"Not so fast! Get away from here!"

The reply was a shower of oaths and another attempt to strike the girl. King suddenly knocked him down.

"Shut up!" he said, "or I'll choke you."

The uncle's manner changed. He whined. His antagonist told him to get up. As he rose he received a kick which started him away. The patrolman made no attempt to interfere. King said:

"You know me, officer. The lady is now in my charge."

Poor Minnie stood still, a victim of the tide of affairs. Again King spoke:

"Minnie, this is what I wanted to say sooner this evening; I want a wife. Will you marry me?"

Minnie was silent.

"You can't have a worse life anywhere than with your folks. Your uncle plundered you of your wages. I've plenty of money. I'll share it with you — dress you well, give you a nice place to live."

Minnie made no answer, but when Tom King gently took her hand, put her arm under his, and walked away, she went along with him. But she took the lead to the house of a clergyman she knew, where Tom explained his knocking up the good man at so unseemly an hour by the fact that Minnie had no proper home to go to, and he proposed to take her to his.

Selected Bibliography

This bibliography is divided into two parts, "Works Cited" and "Suggestions for Further Reading." The first part contains all primary and secondary works quoted or discussed in the general or chapter introductions. The second part is a selective list of materials that will be useful to students who want to know more about Stephen Crane's life and culture or are interested in reading some of the major critical studies of his work. With one or two exceptions, a book or article that appears in "Works Cited" is not recorded again under "Suggestions for Further Reading." Thus, both lists should be consulted.

WORKS CITED

"An American Tragedy of the Streets." *St. James Gazette* October 7, 1896: 5.
Beer, Thomas. *Stephen Crane: A Study in American Letters.* New York: Knopf, 1923.
Bennett, Arnold. "Book Chat." *Woman* September 30, 1896: 7.
Crane, Jonathan Townley. *The Arts of Intoxication: The Aim and the Results.* New York: Carlton, 1871.
———. *Popular Amusements.* New York: Carlton, 1869.

Crane, Stephen. *The Correspondence of Stephen Crane.* Eds. Stanley Wertheim and Paul Sorrentino. 2 vols. New York: Columbia UP, 1988. Cited as C.

———. *The Works of Stephen Crane.* Ed. Fredson Bowers. 10 vols. Charlottesville: UP of Virginia, 1969–76. Cited as *Works.*

Dos Passos, John. *Manhattan Transfer.* New York: Grossett, 1925.

Franklin, Benjamin. "Information to Those Who Would Remove to America." In *Writings.* Ed. J. A. Leo Lemay. New York: Library of America, 1987.

Garland, Hamlin. *Crumbling Idols: Twelve Essays on Art and Literature.* Ed. Robert E. Spiller. Gainesville: Scholars' Facsimiles and Reprints, 1952.

———. *Main-Travelled Roads.* New York: Harper, 1891.

Gullason, Thomas A. "The First Known Review of Stephen Crane's 1893 *Maggie.*" *English Language Notes* 5 (June 1968): 300–02.

———. "The 'Lost' Newspaper Writings of Stephen Crane." *Syracuse University Library Associates Courier* 21 (Spring 1986): 57–87.

Hogarth, William. *Engravings by Hogarth.* Ed. Sean Shegreen. New York: Dover, 1973.

Holloway, Jean. *Hamlin Garland: A Biography.* Austin: U of Texas P, 1960.

Hughes, Rupert. "The Rise of Stephen Crane." *Godey's Magazine* 133 (September 1896): 317–19.

Jackson, Kenneth T., ed. *The Encyclopedia of New York City.* New Haven: Yale UP, 1995.

Johannsen, Albert. *The House of Beadle and Adams and Its Dime and Nickel Novels: The Story of a Vanished Literature.* 2 vols. Norman: U of Oklahoma P, 1950.

Kibler, James E., Jr. "The Library of Stephen and Cora Crane." *Proof* 1 (1971): 199–246.

Lanza, Clara. "Women Clerks in New York." *Cosmopolitan* 10 (February 1891): 487–92.

Lawrence, Frederick M. *The Real Stephen Crane.* Ed. Joseph Katz. Newark: Newark Public Library, 1980.

Manning, William Henry. *Bowery Bob, Detective.* New York: Beadle, 1891.

Melville, Herman. *Correspondence.* Ed. Lynn Horth. Evanston: Northwestern UP, 1993.

———. *Journals.* Eds. Howard C. Horsford and Lynn Horth. Evanston: Northwestern UP, 1989.

Meyerowitz, Joanne J. *"Women Adrift": Independent Wage Earners in Chicago, 1880–1930.* Chicago: U of Chicago P, 1988.

Nagel, James. *Stephen Crane and Literary Impressionism.* University Park: Pennsylvania State UP, 1980.

O'Neill, Eugene. *The Hairy Ape, "Anna Christie," The First Man.* New York: Boni, 1922.

Parker, Hershel, and Brian Higgins. "The Virginia Edition of Stephen Crane's *Maggie:* A Mirror for Textual Scholars." *Bibliographical Society of Australia and New Zealand Bulletin* 19 (1995): 131–66.

Perkins, Bradford. *The Great Rapprochement: England and the United States, 1895–1914.* New York: Atheneum, 1968.

Phillips, David Graham. *Susan Lenox: Her Fall and Rise.* New York: Appleton, 1917.

Riis, Jacob A. *The Children of the Poor.* New York: Scribner's, 1892.

———. *How the Other Half Lives: Studies among the Tenements of New York.* 1890. Ed. David Leviatin. Boston: Bedford, 1996.

Rollins, Alice Wellington. *Uncle Tom's Tenement: A Novel.* Boston: Smythe, 1888.

Sedgwick, A. G. Rev. of *The Red Badge of Courage, Maggie,* and *George's Mother,* by Stephen Crane. *Nation* 63 (July 2, 1896): 15.

Spencer, Herbert. *The Data of Ethics.* New York: A. L. Burt, 1879.

———. *The Principles of Ethics.* 2 vols. London: Williams, 1892–93.

Stanley, Henry M. *In Darkest Africa: or, The Quest, Rescue and Retreat of Emin, Governor of Equatoria.* New York: Scribner's, 1890.

Stokes, I. N. P. *The Iconography of Manhattan Island, 1498–1909.* 6 vols. New York: Dodd, 1915–1928.

Traill, Henry D. "The New Realism." *Fortnightly Review* 67 (January 1897): 63–73.

"Two Books by Stephen Crane." Rev. of *Maggie: A Girl of the Streets* and *George's Mother. Critic* 25 (June 13, 1896): 421.

United States, Dept. of Labor. *Working Women in Large Cities.* Washington: GPO, 1889.

Walsh, John. *Poe the Detective: The Curious Circumstances Behind the Mystery of Marie Roget.* New Brunswick: Rutgers UP, 1968.

Walters, Ronald G. *American Reformers, 1815–1860.* Rev. ed. New York: Hill, 1997.

Wells, H. G. "Another View of 'Maggie.'" *Saturday Review* 82 (December 19, 1896): 655.

Wertheim, Stanley, and Paul Sorrentino. *The Crane Log: A Documentary Life of Stephen Crane, 1871–1900.* New York: Hall, 1994. Cited as *Log.*

———. "Thomas Beer: The Clay Feet of Stephen Crane Biography." *American Literary Realism* 22.3 (1990): 2–16.

Wyman, Margaret. "The Rise of the Fallen Woman." *American Quarterly* 3 (1951): 167–77.

Zola, Émile. *L'Assommoir.* Trans. John Stirling [Mary Sherwood]. Philadelphia: Peterson, 1879.

———. *Pot-bouille.* Trans. John Stirling [Mary Sherwood]. Philadelphia: Peterson, 1882.

SUGGESTIONS FOR FURTHER READING

Bibliography

Dooley, Patrick K. *Stephen Crane: An Annotated Bibliography of Secondary Scholarship.* New York: Hall, 1992.

Stallman, R. W. *Stephen Crane: A Critical Bibliography.* Ames: Iowa State UP, 1972.

Biography and Letters

Crane, Stephen. *The Correspondence of Stephen Crane.* Eds. Stanley Wertheim and Paul Sorrentino. 2 vols. New York: Columbia UP, 1988.

Wertheim, Stanley, and Paul Sorrentino. *The Crane Log: A Documentary Life of Stephen Crane, 1871–1900.* New York: Hall, 1994.

Critical Studies

Brown, Bill. *The Material Unconscious: American Amusement, Stephen Crane and the Economies of Play.* Cambridge: Harvard UP, 1996.

Cunliffe, Marcus. "Stephen Crane and the American Background of *Maggie.*" *American Quarterly* 7 (1955): 31–44.

Fine, David M. "Abraham Cahan, Stephen Crane and the Romantic Tenement Tale of the Nineties." *American Studies* 14.1 (1973): 95–107.

Gandal, Keith. *The Virtues of the Vicious: Jacob Riis, Stephen Crane, and the Spectacle of the Slum.* New York: Oxford UP, 1997.

Gibson, Donald B. *The Fiction of Stephen Crane.* Carbondale: Southern Illinois UP, 1968.

Gullason, Thomas Arthur. "The Sources of Stephen Crane's *Maggie.*" *Philological Quarterly* 38 (1959): 497–502.

Halliburton, David. *The Color of the Sky: A Study of Stephen Crane.* Cambridge: Cambridge UP, 1989.

Hapke, Laura. *Girls Who Went Wrong: Prostitutes in American Fiction, 1885–1917.* Bowling Green: Bowling Green State U Popular P, 1989.

———. "Maggie's Sisters: Nineteenth-Century Literary Images of the American Streetwalker." *Journal of American Culture* 5.2 (1982): 29–35.

Hussman, Lawrence E., Jr. "The Fate of the Fallen Woman in *Maggie* and *Sister Carrie.*" *The Image of the Prostitute in Modern Literature.* Eds. Pierre L. Horn and Mary Beth Pringle. New York: Ungar, 1984. 91–100.

Katz, Joseph. "The *Maggie* Nobody Knows." *Modern Fiction Studies* 12 (1966): 200–12.

LaFrance, Marston. *A Reading of Stephen Crane.* New York: Oxford UP, 1971.

Nagel, James. *Stephen Crane and Literary Impressionism.* University Park: Pennsylvania State UP, 1980.

Parker, Hershel, and Brian Higgins. "The Virginia Edition of Stephen Crane's *Maggie:* A Mirror for Textual Scholars." *Bibliographical Society of Australia and New Zealand Bulletin* 19 (1995): 131–66.

Robertson, Michael. *Stephen Crane, Journalism, and the Making of Modern American Literature.* New York: Columbia UP, 1997.

Slotkin, Alan Robert. *The Language of Stephen Crane's Bowery Tales: Developing Mastery of Character Diction.* New York: Garland, 1993.

Solomon, Eric. *Stephen Crane: From Parody to Realism.* Cambridge: Harvard UP, 1966.

Wertheim, Stanley. *The Merrill Studies in Maggie and George's Mother.* Columbus: Merrill, 1970.

———. *A Stephen Crane Encyclopedia.* Westport: Greenwood, 1997.

Ziff, Larzer. *The American 1890s: Life and Times of a Lost Generation.* New York: Viking, 1966.

Printed in the United States
by Baker & Taylor Publisher Services